Structural Dynamics @ 2000:
current status and future directions

Structural Dynamics @ 2000: current status and future directions

Edited by

D. J. Ewins

Professor and Director
Centre of Vibration Engineering
Imperial College of Science, Technology and Medicine
University of London, UK

D. J. Inman

G. R. Goodson Professor and Director
Center for Intelligent Material Systems and Structures
Virginia Polytechnic Institute and State University, USA

RESEARCH STUDIES PRESS LTD.
Baldock, Hertfordshire, England

RESEARCH STUDIES PRESS LTD.
16 Coach House Cloisters, 10 Hitchin Street, Baldock, Hertfordshire, England, SG7 6AE

and

325 Chestnut Street, Philadelphia, PA 19106, USA

Marketing:

UK, EUROPE & REST OF THE WORLD
Research Studies Press Ltd.
16 Coach House Cloisters, 10 Hitchin Street, Baldock, Hertfordshire, England, SG7 6AE

Distribution:

NORTH AMERICA
Taylor & Francis Inc.
International Thompson Publishing, Discovery Distribution Center, Receiving Dept.,
2360 Progress Drive Hebron, Ky. 41048

ASIA-PACIFIC
Hemisphere Publication Services
Golden Wheel Building # 04-03, 41 Kallang Pudding Road, Singapore 349316

UK & EUROPE
ATP Ltd.
27/29 Knowl Piece, Wilbury Way, Hitchin, Hertfordshire, England, SG4 0SX

Library of Congress Cataloging-in-Publication Data

Structural dynamics @ 2000 : current status and future directions / edited by D.J. Ewins,
D.J. Inman
      p. cm. -- (Mechanical engineering research studies. Engineering dynamics series ; 11)
   Includes bibliographical references and index.
  ISBN 0-86380-251-6 (alk. paper)
   1. Structural dynamics. I. Ewins, D.J. II. Inman, D.J. III. Series
 TA654 .S763 2000
 624.1'7--dc21

                                              99-088666

British Library Cataloguing in Publication Data
A catalogue record for this book is available from the British Library.

ISBN 0 86380 251 6

Printed in Great Britain by SRP Ltd., Exeter

Preface

Structural Dynamics is a subject that touches almost every engineering discipline and yet, curiously, it is one that does not have a single focus for its practitioners. Almost every engineering institution has a vibration or noise or shock group or division, and yet there are very few organisations which seek to draw all these disparate applications together. Those who work in this field know very well that the subject shares many phenomena across aeronautical, civil, mechanical and electrical engineering, as well as chemical and information technologies, and yet specialist group meetings tend to be established within just one of these areas at a time.

It was at one such conference that the need for a more general and more fundamental conference for structural dynamicists was recognised and the idea for a workshop to discuss the future directions of Structural Dynamics was born. The specific need identified was for a Forum to discuss the whole field of Structural Dynamics, to assess its current capabilities and to decide where it was and where it should be directed in the future. The meeting needed to focus on long-term strategic issues rather than the more conventional short- to medium-term incremental developments that are the basis of most specialist conferences.

In April of 1999, a group of some 40 like-minded structural dynamicists assembled at Los Alamos where they spent a whole week locked in animated debate on a wide range of issues in the historic surroundings of the original Los Alamos school building, playing out the result of 5 years of planning and preparation of an event which became known as "SD2000". The objectives of the SD2000 Forum were to conduct a rigorous review of the subject and to make a critical self-assessment of where 'we' are, where we need to be, and how we should proceed in order to get there. The specific tasks included a requirement to encapsulate the week's deliberations in a book that could serve as a summary of all the above issues and also provide a guide to current and future practitioners of the subject and its many application areas.

Extensive debate on the issue of the future directions for developing the subject of Structural Dynamics beyond 2000 resulted in their classification into three groups: (i) Developing Technologies; (ii) Emerging Technologies and (iii) Culture and Education. In the first category, those of the current technologies which are considered to have significant further potential by evolutionary development are identified. Specific strategies for these areas have been summarised and the direction for more detailed planning is outlined and ready for subsequent detailed planning. In the second category, new and especially interdisciplinary technologies are reviewed and explored. Here, ideas and suggestions for speculative research are presented. The third category presents some interesting perspectives on the whole discipline and its (and our) inter-relationships with other disciplines and communities. In particular, a proposal is developed to the effect that Structural Dynamicists could and should take more of a leading role in many areas where Structural Dynamics is one of a number of considerations. Two quotes from the discussions serve well to illustrate this theme:

"Structural Dynamics is the glue which integrates many multi-disciplinary phenomena",
and
"Structural Dynamicists should take the lead in multidisciplinary activities since the blame for most dynamics-related structural failures is placed at the door of the structural dynamicist"

The final output of the SD2000 Forum is not confined to this text. A series of initiatives, including the Grand Challenges described here, have been launched to take the ideas and spirit of the meeting on into the next century. A Virtual Forum – SD2K - has been created, based around the SD2000 meeting and its activities, to provide a means for the continuation of the work started at Los Alamos in 1999 and to develop the subject of Structural Dynamics to become a major contributor to the development of better, safer, quieter, more reliable structures and - therefore - world in which we all live. For further insight, interested readers are invited to visit the SD2K website at: www.sd2k.org

D.J. Ewins, D.J. Inman. December 2000

Acknowledgements

As is always the case with projects such as this one, many more people have contributed to the outcome than are identified by their authorship of sections of the text contained in this volume. Accordingly, it is fitting to record the thanks of the initiators of SD2000 to several people and organisations whose support along the way - in passing or as fellow travellers - provided part of what was required to reach the destination.

In the very early days, Frank Fahy, Isaac Elishakoff and Sam Ibrahim were all participants in the brainstorming sessions (usually accompanied by some beverage, and often in pleasant if not exotic surroundings) that gave free rein to even more innovative ideas than those finally adopted for the actual meeting. Once the ideas began to turn into more concrete ideas, and the likely cost of the venture became apparent, Gene Remmers was instrumental in making the necessary connections with ONR, the outcome of which was an eventual contribution towards the documenting of the Forum's output.

However, by far the most significant contribution came from the Los Alamos National Laboratory who not only provided the venue for our meeting but also the entire funding complete with technical, administrative and moral support. Although already listed as authors in the text itself, additional recognition must be made of the unfailing support for SD2000 from the first moment they learned of it to the present day - and beyond! - given by Chuck Farrar and Scott Doebling. They simply made it happen. Recognition must also be given to a number of their managers - the two with whom we had most contact being John Rimner and Steve Girrens, who not only listened to all manner of requests for support for the project but acceded to them, apparently willingly, and even enthusiastically. We trust that they now feel that their decisions were vindicated by what has emerged.

Not for the first time, it is fitting to conclude this list of thanks with a record of our personal appreciation for the many, many tasks - often requested at very short notice, not all of them reasonable and often unnecessarily complicated by the self-constructed chaos surrounding their bosses - undertaken by Liz Hearn (now Liz Savage), Margaret Reeves, Kathie Womack, and Beth Howell. Their contributions have been just as critical as those of the participants.

Frontispiece

The Los Alamos School House c.1930

A poem which reflects the spirit of SD2000

Sometimes

Sometimes things don't go, after all,
from bad to worse. Some years, muscadel
faces down frost; green thrives; the crops don't fail,
sometimes a man aims high, and all goes well.

A people sometimes will step back from war;
elect an honest man; decide they care
enough, that they can't leave some stranger poor.
Some men become what they were born for.

Sometimes our best efforts do not go
amiss; sometimes we do as we meant to.
The sun will sometimes melt a field of sorrow
that seemed hard frozen: may it happen for you.

Sheenagh Pugh (b. 1950)

Taken from Sheenagh Pugh: Selected Poems (Seren, 1990)

CONTENTS

PART I   INTRODUCTION

PART IV QUESTIONS, DISCUSSIONS, CHALLENGES BEYOND 2000

APPENDICES

Part I

Introduction to SD2000:
Origins, Concepts, Plans

D.J. Ewins
Professor of Vibration Engineering,
Imperial College of Science, Technology & Medicine, London, UK
D.J. Inman
Centre for Intelligent Materials and Structures
Department of Mechanical Engineering, MC 0261
Virginia Polytechnic Institute and State University
Blacksburg, VA 24061, USA

ORIGINS

The SD2000 Forum was a one-off event that took place at the turn of the millennium in an attempt to review, and - if necessary – to redraw, the directions of the subject of Structural Dynamics. The origins of SD2000 can be traced back to a small but luxurious clearing in the Brazilian rainforest which was, in fact, the venue of another conference, one of the many such meetings that are held every year to chart the incremental progress of the subject. However, that was one with a difference in that it created the germ of the idea which led, 6 years later, to the gathering which became known as SD2000.

Like most technical subjects, Structural Dynamics depends significantly on the people who are its practitioners, on their skill and care, on their innovation and imagination, and also on their vision. The group who assembled at Los Alamos in April of 1999 - a group who had in large measure selected themselves by a process of surviving a number of obstacles - were confronted with the task of projecting their thoughts and experience into the future of their subject and of divining the directions in which they should encourage the rest of the community to follow. This book is an attempt to capture their collective efforts in that unique activity.

BASIC CONCEPTS

It was one of the basic premises that the format of SD2000 would need to be quite different from that used for the usual type of technical conference, which is primarily concerned with the reporting and discussion of recent incremental developments in the subject, and rather less – except for the occasional keynote paper – with the longer-term strategic issues of the future directions in which the subject should be steered. In order to realise the overall idea of SD2000, the meeting was to be small; it would be totally self-contained, lasting for a period of about a week; it would have a schedule that permitted and even encouraged the

2

repeated re-visiting of issues that are not immediately resolved or decided and it would be driven towards delivering a statement at its conclusion.

Thus there emerged the specifications for a meeting which would have the following characteristics

- No more than 40 participants (in order to facilitate discussion of the whole group);
- Participants who are prepared to stay together as a group for the duration of the Forum;
- Participants who are prepared to debate and argue their case with the prospect of being forced to re-think their own position;
- Participants selected so as to span as wide a spectrum of Structural Dynamics as possible;
- A remote venue (to inhibit temptations for partial attendance);
- Structured schedule which includes a wide range of activities, from plenary sessions, to small group discussions and individual one-to-one debates.

It was also realised that it would be important to define the objectives of the Forum very carefully so that the meeting would be able to direct its efforts towards a specific set of goals, rather than simply comprising a set of uncoordinated debates of the issues without achieving a clear outcome.

OBJECTIVES

Accordingly, the first result of considerable pre-meeting debate was to define a set of goals and objectives, one of which was to be a very clear and visible outcome from the Forum, in a format which could be readily disseminated and shared by the rest of the structural dynamics community: namely, this volume – Structural Dynamics @ 2000.

The aims and objectives set at the outset of the meeting were as follows.General Goals:

- To review the state-of-the-art of the major branches of Structural Dynamics.
- To reflect on the extent to which these technologies meet the needs of the user.
- To ensure that critical issues are on 'our' agenda.
- To discuss ideas for future work and collaborations.

More Specific Objectives for the Forum were set as:

- To give all participants the opportunity to learn more about aspects of Structural Dynamics other than their own specialisation.
- To give all participants the opportunity to discover links with other areas or other techniques, or into new applications.
- To give all participants the opportunity to devote more time and space to reflect on their subject than is normally available.
- To explore ways in which the Structural Dynamics community can have more positive influence in engineering design.

- To define the challenges which face the Structural Dynamics community in the coming years.

However, the major output was to be an encapsulation of the essence and detail of the meeting and was to take the form of a book, comprising a mixture of the formal presentations that were made and a digested summary of the various discussions and debates that continued throughout the week. Thus we had as the primary objective:

- To produce a Document which contains a summary of the State-of-the-Art of the subject of Structural Dynamics and a Plan for its development in the coming years.

It was considered important that the output from the Forum should serve the community as a whole and so some thought was given to the format that the primary deliverable – the book – should take in order to achieve this goal. It was decided that the book should include not only records of both current techniques and the debates which took place at the Forum, but also that it should present as comprehensive an analysis and interpretation of the subject as possible. It was hoped that this latter section would provide researchers and practitioners with an up-to-date perspective of the most important and promising directions for future research in the subject, a feature that would be invaluable both to researchers seeking sponsorship and to sponsoring agencies who have to fight for an appropriate slice of the pie for Structural Dynamics.

PRACTICALITIES

Once the idea of SD2000 had taken shape, attention was turned to the practicalities of making it happen. Considerable attention was given to the task of designing a meeting so as to maximise the chances of success, and it is considered appropriate to relate these issues in some detail for the benefit of those who might seek to pursue similar goals in the future. The issues which arose in this phase included:

- Venue
- Funding
- Organisation
- Participants
- Format and schedule

Venue

As mentioned, a venue was sought which would not only provide the right location to ensure the required lack of distraction, coupled with sufficient remoteness to inhibit partial attendance, but would also provide an atmosphere that had the potential to inspire the participants to make SD2000 a very special event. Attendance of one of the organisers at a workshop at Los Alamos in 1997 revealed the potential of that location as a venue for SD2000 and, a surprisingly short time later, the connections were made and the venue was set. In the event, the actual

4

site used for the meeting was the historic building that had been the School House in the early part of the 20th century, a building which was familiar to Dr Robert Oppenheimer, who often vacationed in the area in the 1930s, and which provided him with the location for another, yet more historic, gathering that became the Manhatten Project in the 1940s. In the event, the original Los Alamos School House (Frontispiece) provided us with an exceptional venue for our deliberations.

Funding and Organisation
Funding for the Forum also posed a problem for the organisers, who had as original aspirations the hope that sponsorship for the entire project might be found so that no potential participant would be prevented from attendance on financial grounds. Early approaches to US Defense Agencies received encouraging responses. One such agency agreed in 1996 to provide half the required funding if another sponsor could be found to provide the other half. This proved more difficult, but once the connection with the Los Alamos National Laboratory was made (see above comments regarding the venue), the 'other' sponsor was identified early in 1998. Unfortunately, the initial offer from the Defense Agency was withdrawn at this stage but the LANL group were able to find sufficient funds for all but the travelling expenses for the participants. Thus, the Forum was viable, and fully funded by LANL for everyone from the moment of landing at Albuquerque to departing 7 days later. Due recognition of the response from LANL must be made here, as well as elsewhere, because without this support, the Forum, its output and this volume would simply not have been realised.

The organisation for the Forum was managed by the LANL team together with a very small Organising Group of four that was enlarged as the Forum approached and involved on a daily basis throughout the week's deliberations in order to optimise its effectiveness. From the formal announcement of the project, and initial contact with potential participants, communications between the organisers and the members of the workshop were very effectively conducted through the SD2000 website, set up and run through the LANL.

Participants
It was accepted from the outset of the planning that the success of the Forum would depend considerably on the participants, and the appropriate selection of these key players exercised the organisers during the early planning stages. There were several criteria: participants needed to be selected, first and foremost, to cover as wide a spectrum of the subject as possible; they needed to be individuals who would participate fully throughout all the planned activities of the meeting and who would be prepared not only to defend their own specialist positions but also to be open to new ideas and, possibly, to reassessment of their own ideas of the future directions that research in structural dynamics should be encouraged to pursue. It was anticipated that most participants would be from academe, but a crucial element were to be recruited from industry/industrial research organisations in order to temper any tendency of the academics to become carried away by what might be considered to be too fanciful ideas! Most people who were approached from the lists drawn up by all those involved in the planning responded positively

and enthusiastically, and some new names were added following suggestions from those contacted directly. A full complement of prospective participants was quickly drawn up. Perhaps inevitably, there were a few last minute withdrawals with the result that certain areas were less well represented than had been planned, but in the event, everyone who participated clearly attended to the interests of the subject as a whole and the nature of their input overrode any concerns about subject coverage. A target of 35-40 primary participants was set at the outset: a number which deviated only very slightly throughout the entire process of planning and running the Forum.

Format and Schedule

Alongside the practical issues of venue, funding, organisation etc., there was an ongoing debate concerning the format and schedule that should be followed for the actual meeting itself. The structure of the Forum had to be designed so as to meet the objectives: just having a talking shop with no specific direction would be to waste a valuable opportunity presented by the gathering. The ultimate objective – to prepare a Guide for the future directions that the subject of Structural Dynamics should be encouraged to follow – provided a suitable goal, and one that could be brought out whenever discussions were in danger of becoming bogged down in detail, and of losing sight of the longer term aims.

Thus the general format for the technical activities was planned to include the following elements:

- Presentations of philosophical and/or state-of-the-art reviews of various aspects of the subject to prepare the background for a detailed debate of individual areas;
- Formulation of a series of key questions that the various participants proposed as being those that needed to be addressed in the coming years;
- Digestion and review of these questions and issues into an agreed agenda of topics or themes that should form the basis of future directions in structural dynamics research; and
- Proposals for specific actions that individuals or agencies might take up in the immediate future in order to provoke the non-trivial tasks of winning support for the major proposals in the previous item.

It was envisaged that all these objectives would be contained in a single volume which would be published after the Forum.

In addition to the 'work' aspects of the meeting, it was also considered important to have matching recreational opportunities as well, not only to leaven the load of the intense intellectual activity that such a workshop can develop, but also to give participants time to reflect on the formal proceedings, and perhaps to discuss them in a less structured way than in the main part of the meeting. Some of this 'relaxation' time was used for talks by local Los Alamos experts on a range of topics from the new large-scale computing initiatives to the local history and archaeology. It was considered important to keep the group together and so, wherever possible, there was only one activity gong on at a time: all meals were taken together and outings were to be available to all. The daily schedule was set as having a structured morning session, followed by an afternoon outside – picnic lunch, visit, tennis, etc. etc. – and then a second, early-evening, Forum session leading on to dinner at a different venue each day.

DETAILED PLANNING

It was decided that the participants should be grouped into 4 categories, with corresponding roles to play in the workshop. The first and second groups comprised 19 of the 'elders' of the tribe and would be labelled 'philosophers' or 'specialists'. The third group of 14 would be the 'youngsters', typically 10 years younger than members of the first groups and likely to be their successors. The fourth group comprised 4 senior people from industry or industry/government research laboratories who have the benefit of access to both sides of the research issue – that of its prosecution and also of its need for justification on the grounds of its usefulness, be that long-term or short-term.

The roles to be played by these various groups were planned to be as follows:

- First group: to prepare, in advance of the meeting, essays on a chosen theme, which then would be expounded in a 30-minute presentation during the Forum;

- Second group – the specialists: to prepare a state-of-the-art review of their particular speciality, again in advance of the meeting, and to make a presentation during the Forum with the remit that at both stages the reviews were to be aimed at specialists in *other* disciplines;

- Third group: to prepare in advance a series of 3 or 4 key issues or "questions" which they proposed should be included in the lists of topics to be debated during the Forum;

- Fourth group: to lead the debates of the various questions and to help to extract from these discussions the essential features of the subject which constitute its major future prospects and contributions. These would constitute the formal output from the Forum and provide the material for subsequent follow-on activities that would turn these into more specific, detailed, argued proposals for new research to be undertaken.

In terms of the structure of the agenda used to carry out these various activities, the plan was to have the formal set-piece 30-minute presentations by the first two groups in the morning sessions of the first three days. These would all be plenary sessions. In the first and second afternoons, there would be (probably 3) breakout small group discussion sessions that would report back to the assembled gathering at the end of each day. On the fourth and fifth days, there would be further discussion sessions, both in small groups and in plenary sessions, as the task of formulating the contents of the Document and future actions became more pressing.

An essential requirement for this style of activity was the need to appoint some 'scribes' to take notes of the discussion sessions (a task that could not be expected of those directing each such session) and three of the younger participants volunteered for this task, a role which turned out to be absolutely critical to the demand to capture and incorporate all the ideas raised throughout the week's discussions.

Throughout this design process, and the associated detailed planning, as well as throughout the Forum itself, it was necessary to keep always in view the end objective of producing a document that would both encapsulate the essence of the week's debates and provide the community as a whole with a cogent, argued, prioritised, convincing set of outline plans for the future directions of the subject. To this end, it was found to be necessary to ensure that at each stage of the Forum – for each set of presentations or each discussion session - the purpose of that activity was made clear, and that it had a well-defined objective and, often, deliverable. This was to be a strong underlying theme of the Forum: the need to keep everyone's eyes on the ball!

IMPLEMENTATION – A FEW COMMENTS

Most of the output from the Forum will be described in the following Chapters of this book, and they constitute delivery of the primary objective of the whole project. However, it is perhaps appropriate to end this description of the conception and gestation of the SD2000 Forum with a few comments on how the birth went.

In summary, the meeting ran very closely to its planned form. Certainly, all the major tasks were undertaken and completed, and the declared deliverable of a single comprehensive document being produced was self-evidently achieved.

As to the running of the meeting: the program sketched out for the first half of the week was followed quite closely, while the detailed structure of the debates on the fourth and fifth days were changed as they unfolded. This flexibility was necessary in order to ensure that all the detailed ideas that were generated in the many hours of very effective small-group discussions were considered and reconsidered as the final version of the Forum Report was being drafted. It is believed that, in the end, a very comprehensive, and cogent, synthesis of wide-ranging and sometimes conflicting views was forged out of this highly interactive process. It is also hoped that the reader of this volume will feel that the combined product of some 1 man-year's intensive deliberations has resulted in a work of some interest and value to all members of the Structural Dynamics community.

Part II

Uncertainty Modeling in Dynamical Systems: A Perspective

L. A. Bergman
Professor of Aeronautical and Astronautical Engineering
University of Illinois at Urbana-Champaign

1.   INTRODUCTION

I suspect that all of us here have, at one time or another, been vexed by performance that quantitatively failed to match our analytical predictions. Despite our best modeling efforts and careful manufacture and assembly, we find ourselves cast into the position of reconciling our analyses in an attempt to bring the predicted behavior of our systems into line with observations. This apparently occurs with sufficient regularity to warrant giving the activity a unique identity in the literature: model updating, or some variation thereof.

Having found myself in precisely that position more times than I really care to admit to over the course of a career that spans more than thirty years, I've had plenty of time to think about these problems and have decided to use this opportunity to offer some insights into how we, as a community, might improve our capability to recognize and accommodate uncertainty in vibrating systems. The modeling of uncertainty and its effects upon the response of dynamical systems of engineering interest has been of concern for many years. Professor Stephen Crandall coined the term "random vibration" more than 40 years ago, though the analysis of noise-driven systems was certainly   underway decades earlier (see, for example, the work of S. O. Rice, 1944-45), particularly in the field of what was then called "radio engineering".

In structural dynamics, we hold the view that uncertainty enters the typical system in at least one of three ways: through the input to the system, the coefficients of the differential equations governing the evolution of the system, and the initial and, possibly, boundary conditions. However uncertainties arise, their effect on system response and reliability can be profound.

One can envision a number ways in which uncertainty can be included in the analysis and design of a dynamical system, the "worst-case" and probabilistic approaches being two that have gained acceptance within the community. In the former, one examines some response of the system to a particular class of inputs or over a range of parameters and identifies the extreme case; while in the latter, some response of the system to a class of inputs represented by a stochastic process of known probability distribution is itself a stochastic process defined in terms of a probability distribution, from which statistics can be determined. As the notion of uncertainty is most familiar in the context of control, perhaps it will be instructive to couch the following discussion in that language, in particular as applied to structural control.

2. A PROBABILISTIC VIEW OF ROBUSTNESS

Model uncertainty, if ignored, can seriously degrade the performance of an otherwise well-designed control system. If the level of uncertainty is extreme, the system may even be driven to instability. In the context of structural control, performance degradation and instability imply excessive vibration and structural failure. Robust control has typically been applied to the issue of model uncertainty through worst-case analyses. These traditional methods include the use of the structured singular value, as applied to the small gain condition, to provide estimates of controller robustness (Doyle *et al.*, 1991). However, this emphasis on the worst-case scenario is without a probabilistic basis, from which a much more intuitive insight into controller robustness can be gained. A number of methods have been devised through which the robustness problem can be viewed in the context of probability, including Monte Carlo simulation, first and second order reliability methods adapted directly from system reliability resulting in the determination of a "stability index", and others based upon the use of μ-synthesis procedures. A brief description of the problem follows.

The distribution of poles in uncertain dynamical systems and its relationship to the robustness issue have been topics of some interest in recent years. Stengel and Ray (1991) were among the first to use large-scale Monte Carlo simulation to estimate the robustness of uncertain controlled structural systems. Using this approach, one constructs a distribution of root loci simulating the stochastic behavior of the closed-loop pole locations. Because this is a graphical method, one gains an intuitive understanding of system robustness. The results reported were quite promising, but the large number of realizations required to attain a high degree of accuracy in the distribution of the tails of the closed-loop poles makes this approach computationally unattractive.

Spencer *et al.* (1992,1994) introduced a systematic approach for determining the probability that instability will result from the uncertainties inherently present in a controlled structure. This probability measure is a direct indication of the robustness of the closed-loop system. As described by Spencer, this investigation into the probability of failure of controlled structures led to a method for characterizing the stability of the system based upon an eigenvalue criterion, namely the probability that the real part of every eigenvalue will be contained strictly in the left-half plane. First and second-order reliability methods (FORM/ SORM), known to be accurate for series-type system reliability problems, were used for estimating the probability of system instability, and a series of numerical examples were constructed in which the stability of a controlled single degree-of-freedom system with four uncertain parameters was analyzed.

Traditional methods used to assess controller robustness, involving the use of the singular value of some mapping as applied to a small gain condition, are often conservative in nature. Therefore, most recent efforts involved comparing robustness estimates using the FORM methodology with those estimates obtained using the more traditional techniques. Field *et al.* (1996a,b) introduced an approach to adapt the robustness measure μ gained from the structured singular value analysis into a probabilistic framework.

We thus have three methods that can be compared: Monte Carlo simulation, FORM, and probabilistic μ ; and we can evaluate their comparative performance by application to the active control of a structure subjected to, for example, a seismic ground motion. In order to provide more realism, we can also consider the effects of controller time delay, which can be a significant contributor to system instability. Utilizing the three methods of assessment, we can also characterize the robustness qualities of several control law designs and conclude which will be most effective for this particular class of problems.

Figure 1 illustrates the SDOF model, as first reported by Chung *et al.* (1988).

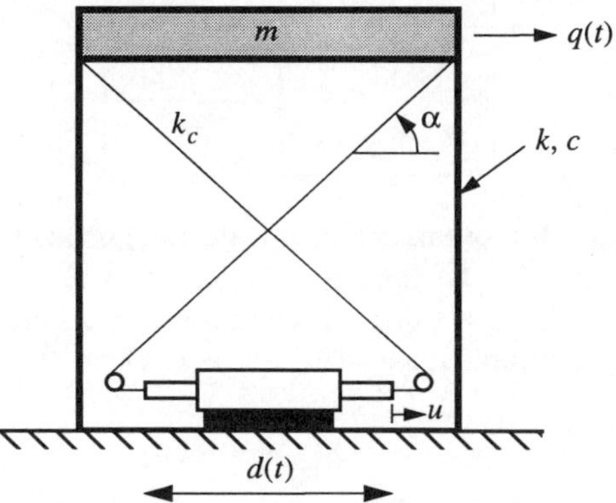

Fig. 1 Single degree-of-freedom structure with active tendon

The control is applied to the structure using prestressed active tendons connected to a servocontrolled hydraulic actuator. The equation of motion of this system is given by

$$m\ddot{q}(t) + c\dot{q}(t) + kq(t) = -(4k_c\cos\alpha)u(t-\tau) + md(t) , \qquad (1)$$

where m, c, and k are the mass, damping, and stiffness values, respectively, of the SDOF structure and $\ddot{q}$, $\dot{q}$, and q are the acceleration, velocity, and position, respectively, associated with the rigid floor of the structure. Additionally, k_c and α are deterministic parameters associated with the structure of the controller, u is the position of the controlling actuator, and $d(t)$ represents the acceleration of the ground. In addition, when utilizing output feedback, the measurement signal is corrupted by noise

$$y(t) \;=\; \ddot{q}(t) + n(t)\,. \tag{2}$$

The parameter means are presented in Table 1.

Table 1 Statistical means of the parameters of the structure

Parameter	Mean, m
m (lb-s^2/in)	16.69
c (lb-s/in)	9.02
k (lb/in)	7,934
τ (ms)	20
k_c (lb/in)	2,124
α (degrees)	36

The optimal controller for this problem is the Linear Quadratic Regulator; *i.e.*, a state feedback gain matrix, $\mathbf{K}$, that is obtained from the solution of a standard algebraic Riccati equation. Note that this controller is the same one that is obtained for the more usual (deterministic) system, where $d(t) = 0$ and the objective is to minimize

$$J \;=\; \int_0^\infty [x^T(\tau)\mathbf{Q}x(\tau) + u^T(\tau)\mathbf{R}u(\tau)]\,d\tau\,, \tag{3}$$

for some initial defined state, *i.e.*, $x(0) = x_0$. We'll call this design the *nominal LQR* control. In the presence of time delay, a simple modification of the optimal LQR gains can be introduced to account for any phase additions introduced by the delay. The corresponding solution is simply a revised gain set attempting to correct for phase additions due to controller delay and is termed the *phase-corrected LQR* control (LQRPC).

Usually in structural systems, state information is unavailable for direct measurement. In addition, measurement noise signals are present and tend to degrade the performance of the closed-loop. Modeling these added terms provides a much more realistic control problem. Here we'll consider two optimal output feedback design methods for the nominal system. The notion of optimality should be related to the particular assumptions on the system inputs as well as the cost objectives.

The first is the H_2 (or LQG) optimal design, where it is assumed that both $d(t)$ and $n(t)$ are unit intensity, uncorrelated white noise processes. The objective is to minimize the steady-state variance of the output

$$\lim_{T_f \to \infty} E\left[\frac{1}{T_f} \int_0^{T_f} z^T(\tau)z(\tau)\,d\tau\right]^{\frac{1}{2}} = \|w \to z\|_2. \tag{4}$$

The second is the H_∞ optimal design, where it is assumed that both $d(t)$ and $n(t)$ are energy bounded signals. The design goal is to minimize the "worst case" amplification of the energy of the output

$$\sup_{w \neq 0} \frac{\left[\int_0^\infty z^T(\tau)z(\tau)\,d\tau\right]^{\frac{1}{2}}}{\left[\int_0^\infty w^T(\tau)w(\tau)\,d\tau\right]^{\frac{1}{2}}} = \|w \to z\|_\infty. \tag{5}$$

Here, $d(t)$ is an exogenous disturbance (*e.g.*, a seismic excitation), $n(t)$ models sensor noise, $y(t)$ is the measured output vector, and $u\ (t - \tau)$ is the input, subject to controller time delay. Let the input vector, $w(t)$, and the regulated output vector, $z(t)$, be given by

$$w(t) = \left\{ \begin{array}{c} d(t) \\ n(t) \end{array} \right\} \text{ and } z(t) = \left\{ \begin{array}{c} Q^{\frac{1}{2}}x(t) \\ R^{\frac{1}{2}}u(t) \end{array} \right\}, \tag{6}$$

where Q and R are positive semidefinite and positive definite matrices that weight the state of the structure and control input, respectively. There are two important things to note at this time. First, in order to apply the usual finite-dimensional H_2 and H_∞ design methods, the time delay must be approximated by a rational transfer function. Second, the input/output pair given by (6) is identical for all of the control designs considered. This allows the best comparison of the robustness characteristics of each individual design since, in each case, an identical closed-loop transfer function is used.

Let's first consider the robustness of the system subject to a deterministic time delay, $\tau = 20$ ms. MCS, FORM and μ-analysis robustness estimates of the four controllers are presented in Table 2, where m, c, and k are modeled as independent uniform random variables with a specified coefficient of variation (COV) of 25%.

Table 2 Robustness estimates (τ = 20 ms) by FORM and μ, $\mathrm{COV}_{m,c,k}$ = 25%.

Controller	p_f^{μ}	p_f^{FORM}	p_f^{MCS}
LQR	0.3450	0.0	0.0
LQRPC	0.0	0.0	0.0
H_2	0.8586	0.2686	0.2683
H_∞	0.9987	0.8792	0.8788

The conservative nature of the μ-analysis method is readily apparent from the data since, in all cases, $p_f^{FORM} \leq p_f^{\mu}$. However, while the two assessment methods provide different estimates of robustness, they predict a similar pattern. The data clearly illustrate the improved level of accuracy obtained when using the FORM method.

With the addition of the random time delay, the robustness estimates using the μ-analysis method become *even more conservative* as indicated in Table 3. Again, although both methods predict similar trends, the robustness estimates obtained using the FORM method are clearly better than those utilizing the μ-analysis method. This leads to the general observation that, when considering systems of this type and in this framework, the well-established reliability methods are more useful.

With regard to the robustness of the various control designs, it is clear from the data that the LQR and LQRPC control are the most robust in the presence of uncertainty. This can be attributed to the inherently good robustness properties of LQ regulators (*e.g.*, guaranteed phase and gain margins, *etc.*). None of the nominal output feedback designs can match the robustness of the LQ designs. Hence, the need for output feedback techniques that can directly incorporate uncertainty into the design process is apparent.

Table 3 Robustness estimates by FORM and μ, random τ, $\mathrm{COV}_{m,c,k}$ = 25%.

Controller	p_f^{μ}	p_f^{FORM}	p_f^{MCS}
LQR	0.8575	4.69e-3	0.0
LQRPC	0.6395	0.0	0.0
H_2	0.9987	0.8342	0.8339
H_∞	1.0	0.9971	0.9968

3.   SUMMARY

Traditional methods used to assess controller robustness may have precluded a probabilistic understanding of robust control. In this presentation, some ideas and results related to assessing controller robustness as a probability measure have been presented. With the use of these methods, one can classify the robustness characteristics of various control designs for a specific problem.

The first-order reliability method (FORM), used here to estimate robustness, is an approximation that relies upon two assumptions. First, the failure surface defining the onset of instability must be fairly linear. If it is not, the first-order curve fit to the failure surface may become inadequate. Second, the most probable failure condition must be sufficiently governed by a single mode. Highly uncorrelated failure modes may lead to inaccuracy of the FORM approximation, and more sophisticated FORM/SORM methods may be needed. However, for the problem considered herein, robustness estimates using the FORM method are adequate. This may be attributed to the high level of correlation between failure modes, typical of a structural system.

The use of the structured singular value, μ, to assess controller robustness proved to be conservative when applied in a probabilistic framework. The main premise of this method operates only upon the maximum structured singular value and, hence, evaluates the "worst-case" scenario only. In addition, the fundamental robust stability criterion used is valid for uncertainty that may contain much more than the class of uncertainty considered here. The examples considered assumed only real parametric uncertainty, a small subclass. As a result, the analysis gave conservative, although qualitatively accurate, results. That is, μ-analysis can correctly predict which controller is more robust probabilistically, yet the predicted probability of instability can be quite conservative. Therefore, it is possible to conclude that the FORM method provides superior robustness estimates when applied to structural control problems.

Implicit in all this is the assumption that the mean and variance of the parameters m, c, k, τ are well known and available and, in the case of a real structure, this is rarely the case. Effective utilization of the probabilistic framework will require a significant effort to characterize the material and physical properties of structural components as well as of the inputs to the system in a stochastic sense, far more than has been accomplished to date. Whether the benefits to be derived would justify the needed effort is an open question.

4.   ISSUES PROPOSED FOR DISCUSSION

A.   Is the structural dynamicist equipped to view the his/her world stochastically?

B.   Are current analytical methods sufficiently capable?

C.   .What would an adequate base of probabilistic data consist of?

5.   REFERENCES

Chung, L., A. Reinhorn and T. Soong (1988). "Experiments on Active Control of Seismic Structures," *Journal of Engineering Mechanics*, Vol. 114, No. 2, pp. 241-256.

Doyle, J., A. Packard and K. Zhou (1991). "Review of LFT's, LMI's, and μ," *Proceedings of the 30th IEEE Conference on Decision and Control*, England, pp. 1227-1232.

Field, R.V. Jr., P.G. Voulgaris and L.A. Bergman (1996a). "Methods to Compute Probabilistic Measures of Robustness for Structural Systems," *Journal of Vibration and Control*, Vol. 2, No. 4, pp. 447-464.

Field, R.V. Jr., P.G. Voulgaris and L.A. Bergman (1996b). "Probabilistic Stability Robustness of Structural Systems," *ASCE Journal of Engineering Mechanics*, Vol. 122, No. 10, pp. 1012-1021.

Spencer, B., M. Sain, J. Kantor and C. Montemagno (1992). "Probabilistic Stability Measures for Controlled Structures Subject to Real Parameter Uncertainties," *Smart Materials and Structures*, Vol. 1, pp. 294-305.

Spencer, B., M. Sain and J. Kantor (1994). "Reliability-Based Measures of Stability for Actively Controlled Structures," *Structural Safety and Reliability*, Schueller, Shinozuka and Yao (eds.), Balkema, Rotterdam, pp. 1591-1598.

Stengel, R. and L. Ray (1991). "Stochastic Robustness of Linear Time-Invariant Control Systems," *IEEE Transactions on Automatic Control*, Vol. 36, No. 1, pp. 82-87.

Exciting Vibrations:
A philosophical approach to resolving structural dynamics problems

David J. Ewins
Professor of Vibration Engineering,
Imperial College of Science Technology and Medicine, London, UK

SUMMARY AND SUBJECT OF THE ESSAY
This essay discusses three underlying principles of resolving structural dynamics problems:
- on the need for a systematic approach to identifying and understanding the problem to be solved (**'exciting'** here refers to the satisfaction of tracing the source of the problem);
- on recognising that all that really matters is how large the vibration response will be (or is) and so a knowledge of the real source of **excitation** becomes a primary requirement; and
- on the inevitability of having to undertake tests to supplement predictions since, no matter how good our computational skills become, our demands of them will always outstrip their performance: as a result we have to **excite** our structures deliberately in order to understand exactly how they behave in order to be able to design them to possess acceptable characteristics.

1.   FIRST POINT - WHAT *IS* THE PROBLEM?

For our first point, we shall consider the critical phase of tackling any problem, which is the part where we actually define what the problem is. This is a very important part of the overall problem-solving process and one which is often sabotaged by a tendency to initiate an over-hasty application of an available or familiar solution procedure before the correct approach has been properly identified. Such precipitate action can lead to much frustration and delay in obtaining a satisfactory solution and in order to avoid the associated disappointment which this can bring, we advocate here a more measured and systematic approach.

1.1   Defining the questions

The first issue concerns the need to define the correct questions. It is often the case that the hardest part of solving a difficult problem lies in the correct identification of the specific questions to be addressed. The solution, once the questions are properly formulated, is frequently the relatively straightforward application of quite well developed tools. We need to know, for example, whether to 'solve' a vibration problem by adding damping, by removing the excitation or by changing the natural frequencies. The correct solution depends on the correct identification of the root problem.

How can we achieve this? By being methodical and precise in describing the various features and phenomena that we observe. So, for example, we might assemble a list of questions which seek to get to the bottom of the problem:

- Is the problem ... that the component breaks?
- Is the problem ... the fact that the machine stops (because the component breaks)?
- Is the problem ... the fact that the vibration level was too high?
- Is the problem ... the fact that the excitation frequency was too close to a critical natural frequency?
- Is the problem ... that the excitation forces were too high? ...or that the structure was too responsive? ...or that there was insufficient damping?

Clearly, we must understand what the problem really is so that we can set about fixing that and not trying to deal with a consequence of the actual problem. There is no point in simply replacing the broken component with new one if the cause of its failure is not removed: that will only lead to a second failure. More often than not, the problem comes down to the fact that the excitation is too large (and we shall discuss that in more detail in the second part of this chapter) but, of course, such a simple statement hides important issues about the definition of the complex phenomenon of 'excitation' and also of how we must determine what constitutes 'large'. So, we have established here the need to define the problem.

1.2   Adopt a structured approach

When formulating these questions, they should follow a logical sequence:

- What is the apparent problem?
- How to define what is causing it?
- If we need a model of the structure - how shall we obtain it? How to check that it is valid? How to use it to identify the problem?
- How to see whether we have correctly identified and corrected the problem?

We can often make use of 'road maps' to guide the analysis (dissection) of a problem and to formulate such a sequence of questions. This we can do with the aid of road maps and the like, perhaps themselves structured, or graded:

- route maps of major processes
- street maps of detailed procedures
- floor plans of algorithms.

It is interesting to note that it is often appropriate to categorise the various issues we are dealing with here and that these often fall naturally into groups of three. This is found in many situations and, indeed, there is reference to 'the power of three' in some works of a philosophical nature. [It is perhaps no accident, therefore, that this essay has chosen to focus on three points or issues, or that the name of the original project at the site of the Forum was 'Trinity'.] While reflecting on this point, it is perhaps an appropriate time to present the corresponding classification for the types of analysis that we might undertake as part of our studies of structural dynamics. Of course, there are three types, or levels:

- Level 1 - Preliminary analysis, where order-of-magnitude answers are required to ascertain whether or not there might be a problem that demands a more thorough analysis. This is followed, in those cases where the results of the preliminary analysis indicate that there is a case to answer, by a more extensive, standard analysis:
- Level 2 - Standard analysis, where a conventional assessment is conducted to predict or to measure the extent of the problem; to determine the level of vibration or noise, for example. This may then be followed up in special cases by
- Level 3 - Advanced analysis, based on state-of-the-art techniques, which are often difficult and/or expensive to apply but which are reserved for those applications which demand the closest inspection because they represent critical cases where structural or competitive failure is at stake.

1.3   Use precise and consistent terminology and notation

As an aid to the development of systematic and effective analysis of the problem, we should be sure to use precise definitions, and we should try to foster a set of terminology and notation which is more common across the various branches of the field. By way of examples, we can mention the frequent interchangeability of two or more words used for a specific item. Such a practice causes a loss of precision in the language and, eventually, demands longer expressions to define something precisely. Most of us have a number of 'pet hates' in this respect but there is a serious side to the issue. For example, we should be careful not to confuse **data** with **information**. The acquisition of a large quantity of data does not necessarily equate to the gaining of a corresponding amount of information and so the two words should be used distinctly. Indeed, it is part of the skill of the structural dynamicist (and other engineers) to maximise the information that can be extracted from the minimum of data.

Another distinction that should be clearly defined is the difference between **validation** and **verification**. The two words are widely used interchangeably and yet there are two distinct processes to which we need to refer and these two processes (and hence the words used to describe them) are not (and should not be) the same. One process is that of establishing that (a) the data supplied really has come from the transducer with which it is identified (and not from a different one by some error of labelling), or that (b) a computer routine correctly executes the algorithm which it is supposed to perform. To my mind, these are examples of **verification** and they indicate that things are working or recorded correctly. However, there is no associated suggestion that the data provided by the transducer, or that the results of the computer routine are useful, or valid. That depends on the appropriateness of the measurement process or the input data to the computer routine and is a different matter altogether. The second process, **validation,** refers to the question of whether the results which are obtained (such as from a measurement or a computation) are good enough for the purpose for which they were sought. An eigenvalue routine can be working perfectly correctly (i.e. has been verified as correct) but the usefulness of the natural frequencies which are produced using it depend on the correctness of the model and the parameters used as well as on the correctness of the computations. These natural frequencies will only be valid if the model is good enough, and that has nothing to do with the eigenvalue routine. So, verification and validation are clearly different issues. Similar comments can be made for **comparison** and **correlation** and, no doubt, there are many other examples.

One area where different words (which actually mean different things) are used almost without discrimination is the experimental world where we use **test**, **measure** and **experiment** with equal ease. It is interesting just to see how the Oxford dictionary differentiates these and there we find:

measure // n. & v.

..

v.

1 tr. ascertain the extent or quantity of (a thing) by comparison with a fixed unit or with an object of known size.

..

[Middle English via Old French mesure from Latin mensura, from metiri mens-
'measure']
experiment // n. & v. n.
1 a procedure undertaken to make a discovery, test a hypothesis, or demonstrate a
known fact.
..
v.intr. // (often foll. by on, with) make an experiment.
..
[Middle English from Old French experiment or Latin experimentum (as
experience)]
test // n. & v. n.
1 a critical examination or trial of a person's or thing's qualities.
2 the means of so examining; a standard for comparison or trial; circumstances
suitable for this ..
v.tr.
1 put to the test; make trial of (a person or thing or quality).
..
[Middle English via Old French from Latin testu(m) 'earthen pot', collateral form
of testa test2]

While this discussion may seem like pedantry, I submit that the reliability
and effectiveness of our communications can depend heavily on the precision of
the language we use, and that we have in our language the capacity for greater
precision and efficiency than we generally exploit.

Turning finally to more specific and technical terms, another worrying trend
is found in an almost-universal tendency not to differentiate between **natural
frequency** and **resonance frequency** (or should it be **resonant frequency**? - no!).
Clearly, these are different quantities - one referring to a system property and the
other to a response characteristic - and in many cases, this difference is important,
even if the numerical differences between resonance frequencies and natural
frequencies are often insignificant. We, of all people, should be more precise in
our use of the language we use in our subject.

So, to summarise this first part of the essay, it is my submission that in
order to tackle our various problems with greater effect, we need:

- a structured approach to identifying the real questions;
- a more uniform use of notation, terminology and definition;
- a precise use of language.

All of these are available: it is a matter of availing ourselves of them.

2.   SECOND POINT - THE 'BOTTOM LINE'

The second point to be discussed in this essay concerns the feature which is the
focus of our interest in most structural dynamics problems. It is suggested here
that we need to have a clear picture of what **is** the feature of ultimate interest and
that this is generally the level of **response** suffered by the structure which is the

subject of our studies - be that a building or vehicle whose integrity is at stake, a machine whose reliability is of concern, or a human being who is the end user (sufferer) of many dynamics-related phenomena.

2.1   Response as the ultimate measure

These, and many other examples, indicate that the 'bottom line' on which we generally need to base all our decisions is a knowledge of the response levels and not simply of the system's modal properties of natural frequencies and mode shapes. Such modal properties are all very useful but as a means to an end rather than an end in themselves. It could be argued that FRFs are much more useful than modal properties since they demonstrate the real behaviour of the structure in a much more explicit and practically-useful form. Indeed, the widespread and extensive development of modal analysis has served to bridge the gap between modal and response properties very effectively. Thus, it is suggested that the maximum emphasis possible should be placed on the response function descriptions which are used in modal analysis, and that these data should not be discarded once the modal analysis has revealed the underlying modal properties: it must always be remembered that the modal properties are a subset of the information contained within the response functions.

If we accept the pre-eminent significance of response characteristics, this means that the excitation becomes as important a parameter as the mass and stiffness descriptions with which we are so preoccupied in our analysis of system properties. Accordingly, we should make sure that a balanced effort is made in the attempts to model the dynamic behaviour of real structures. It is possible that new developments on smart structures mean that we do not need to anticipate these excitation forces in quite the same way - perhaps, they can simply be identified by the structure and counteracted automatically at source as a means of reducing response levels to an acceptable degree. Be that as it may, it still remains the case that the response is the ultimate measure of acceptability, or not, and must feature prominently in our studies.

By the same argument, if we are primarily concerned with the response levels, then it follows that the damping properties of the structure are also very important. Many vibration problems are related in some way to resonance and it is clear that the response suffered in this area is directly affected by damping. Hence, we need to be as thorough in our estimation of damping levels as we are in our analysis of the excitation forces. Both of these requirements place a considerable additional burden on the structural dynamicist: in one case (damping) because the necessary modelling techniques are very much more demanding than are those required for the inertia and flexibility behaviour, and in the other case (excitation) because this usually requires an excursion into a different discipline, such as fluid dynamics or electromagnetics.

2.2   Questions concerning response levels

If we follow the recommendations provided in the first part of the essay, we shall set about addressing the problem of response prediction by posing a series of questions. Perhaps the relevant sequence of such questions, assuming that a model of the structure already exists, is as follows:

- What will be source of the excitation?
- How do I determine what the excitation is/was by measurement?
- Why will the excitation be at that level?
- What controls the excitation level?
- What level of excitation is acceptable?
- What do I do to change the excitation level?

It can be seen that these questions add up to a requirement for a comprehensive understanding of the nature of the excitation forces, and of their analysis, leading to the possibility of modifying them as a means of reducing the response they inflict on the structure.

2.3   Determining the excitation as a key to response

Predicting the excitation may require access to a technology that we do not readily have. For example, fluid dynamics is a major source of excitation in many machines and structures, including offshore platforms (waves) buildings (wind) aerospace structures and many machines, including turbomachinery, which have major and complex sources of excitation from the working fluid. In all of these examples, relatively advanced methods of analysis may be required to establish the excitation forces imposed on the structures of interest to us. Indeed, in one example - that of the excitation of blade vibration in turbomachines - the aerodynamics of relevance (unsteady) is more complex than that used for the performance analysis (steady flow aerodynamics). Thus, the implications of the emphasis which, it is proposed here, should be put on the dynamic response of structures may be quite significant in terms of the efforts that must be made to achieve the necessary balance.

In some cases, attempts may be made to establish the excitation in complex situations by measuring it, either directly or indirectly. Direct measurement may be as difficult as direct analysis in many cases, and even more expensive as it is usually extremely difficult to place transducers close to the source of the excitation. Instead, there is a widespread interest in using the [excitation/system properties/response] relationship as a means to deduce excitation forces by an identification process. The basic concept is simply stated: if we measure the response of a structure or machine under operating conditions, and if we have previously obtained a mathematical model of the structure (by measurement or by analysis), then it should be possible to determine the excitation forces which must be causing the observed response. As with many inverse problems, this process is inherently ill-conditioned and suffers from the weakness that it cannot yield meaningful results if the set of assumed (but initially unquantified) excitation forces is incomplete. If one force which exists in reality is omitted from the set of forces to be determined by this inverse identification approach, then there will be the strong risk of physically-meaningless results being obtained, even though they may well satisfy the narrow requirement that they are capable of reproducing the observed response. It must always be remembered that we are obliged to use incomplete data in these processes and the fact that we do not have a measure of the response at point X does not mean that this quantity is zero. It **does** mean that we cannot tell whether our identified forces reproduce the correct response there (at point X) as well as at the few points we have measured. The determination of realistic excitation forces in most typical structures and machines is a frequently-underestimated task, but a critical one nevertheless.

There are, however, subtleties to the excitation which we need to appreciate so that we can exploit them whenever possible. There is a tendency to regard the excitation forces as having simply a characteristic magnitude and frequency. The magnitude of the response is directly related to the magnitude of the excitation, and is strongly related to the relationship between the frequency (content) of the excitation and the characteristic frequencies of the structure. This is clear. However, it must also be noted that the excitation in most practical situations is not confined to a single or even a few points, but is often distributed widely across the exposed surface(s) of the structure. Thus, we must take due note of the 'shape' of the excitation forcing as well as its magnitude and frequency. This can be an extremely critical feature because the magnitude of the response, in addition to its dependence on the force amplitude and frequency, can be dramatically influenced by the spatial distribution of the forces. In the same way that response magnitude is very sensitive to the relationship between the excitation force frequency and the structure's natural frequencies, so also is that response strongly influenced by the relationship between the structure's mode shapes and the excitation's spatial distribution (which is, in effect, a 'mode shape' of forces rather than displacements).

This aspect can be illustrated by reference to certain the vibration properties of rotating machines. A typical turbomachine blade installed in its working habitat of a compressor or turbine finds itself a participant in a very large number of modes of vibration (those of the bladed assembly of which it is one component). These modes, when charted on the typical Campbell or interference diagrams used in this application, find themselves vulnerable to an alarming number of potential resonances as a result of excitation by non-uniformities in the gas flow - known as 'engine-order' forced response. If all those coincidences of a natural frequency (from a mode of the bladed assembly) and an excitation frequency (from an integer multiple of rotation speed) converted into a significant resonance, then there would be no suitable operating region within the machine's speed range. However, because each of the potential resonances involves a particular mode shape together with a particular spatial feature of the excitation (an excitation 'shape'), it is discovered that the vast majority of the potential resonances result in near-zero response levels because these two shapes are incompatible with each other (i.e. they are orthogonal). This phenomenon constitutes a major feature in the vibration response characteristics of turbine blades and is responsible for the vibration environment of these critical components being tolerable. Nor are such phenomena unique: there are many other examples - the fact that only certain modes of a rotating shaft can be excited by out-of-balance forces (forward whirl modes can be excited by out-of-balance; backward whirl modes cannot) is another example of orthogonality between a mode shape and an excitation shape overriding the potential danger of a resonance resulting from the coincidence between an excitation force frequency and a structure's natural frequency.

So, to summarise the second part of this essay, it is my submission that we need:

- To focus more on response levels than modal properties;
- To pay more attention to the analysis of excitation forces and damping; and
- To understand and thus to control and exploit the spatial aspects of excitation as well as the magnitude and frequency properties.

All of these are possible: it is a matter of affording them the necessary priority.

3. THIRD POINT - 'HAIR OF THE DOG'

The third point to be made here is, simply, the case for continued development of experimental as well as theoretical methods for dealing with structural dynamics problems. The thesis is that no matter how refined our theoretical prediction techniques become, they will not be able to satisfy the associated demands placed on them. As a result, it is inevitable that as long as we are concerned with the design of structures that are to be built and used, and therefore to exist, it will be necessary to conduct practical tests on them or on their components. While the nature and detail of these tests may well evolve to accommodate the ever-improving theoretical predictions, their future demand is considered to be inevitable. What must happen is that the concept and detail of experimental methods must be refined so as to maximise the benefit that can be derived from them, and that is the challenge for the experimentalist in the coming decade. So, in order to be able to design safe, reliable, noise-free structures we have to deliberately provoke them to vibrate (a version of the old Scottish belief that you can avoid evil consequences resulting from a dog bite by applying a few hairs from the offending dog to the wound).

3.1 The failure of predictions to provide all

The trend towards better computer techniques offers the promise of designs which rely only on theory-based predictions, and do not need validation by any kind of experimental measurement. This ideal state is unlikely to be reached in the foreseeable future for the simple reason that no matter how well we are able to predict these characteristics, it will not be good enough. Expectations will always lead capabilities and so we shall always be chasing that next improvement. If the requirement now is to predict vibration response levels to within, say, 10%, then as soon as that objective is realised it will be replaced by a requirement to be within 3%, and so it will continue. However, the only way in which the performance of these predictions can be assessed is by quantitative observation of the actual behaviour and that can only be done objectively by suitable test techniques.

There are a number of ways in which the theoretical predictions (as we shall call this type of result) fall short of their mission. The first is that their numerical values are inaccurate, and can differ from reality by a margin which is unacceptable from the engineering viewpoint. The second is that they fail to predict all the events or phenomena which occur in practice. It is not satisfactory if those events which are predicted have an acceptable accuracy if others, which may be just as important, are overlooked. Third, they can only be as accurate as the input data used and, in many cases, it is necessary to supply numerical values for parameters which cannot be predicted (material density, Poisson's ratio, etc). In these and other respects, theoretical predictions remain restricted in the extent of their capabilities.

3.2   Practical limitations mean that tests will 'always' be required

In reality, there will be limitations imposed on our ability to predict a structure's dynamics simply because of our inability to manufacture such structures to the precision required to ensure repeatable behaviour. One area in particular is in the characterisation of joints. We are already at the stage where, although we can predict the modal properties of most complex-shaped components quite well, we cannot predict their response anywhere nearly as accurately, and we cannot predict their modal behaviour when connected together to form the engineering structure in which we are ultimately interested without some additional information on the joints – information which can usually only be supplied by tests. It is believed that this situation currently poses a serious limitation to our ability to predict the dynamic response of many real engineering structures and in order to develop a mathematical model upon which to base the design of a critical component or structure, then it will continue to be necessary to build a hybrid model which relies on theoretical modelling for those areas which are amenable to this approach but which makes recourse to empirical data for other regions (such as joints), which are notoriously difficult to analyse by theory. Of course, one alternative approach to resolving this difficulty (because it is true that empirical data are expensive to obtain and they are generally not available until very late in the design cycle) is to change the design of these joints and other areas that are difficult to model so that they **can** be analysed using the same techniques as are used with success on other structural elements. That is sometimes opposed on the grounds that it is the 'tail wagging the dog' (it is not clear if it is the same dog whose hair is mentioned in the title) but it is an inevitable suggestion for a way to avoid this particular - and very real - problem.

3.3   Testing remains an essential technology

From the above comments, we see that tests remain an essential feature in the vibration engineer's armoury of tools to tackle the various problems he faces in designing, developing or curing the dynamic properties of a wide range of structures and machines. In coming to terms with this situation, we can see that there are (at least) three distinct types of test, which can be classified as follows:

- Tests to measure parameters which are, in effect, unpredictable: empirical data, material properties, etc. (these are 'measurements', pure and simple);
- Tests to validate predictions of dynamic behaviour, but made under controlled excitation conditions and designed to check specific features only, usually of the structural model, rather than of the excitation forces; ('experiments', according to the earlier definitions) and

- Tests to record what actually happens in service, generally under uncontrolled and unknown excitation conditions (these may include those tests carried out to measure excitation indirectly).

It is important to be clear which type of test is being carried out in each case and what is the specific objective sought from it. For example, full validation of the accuracy of the theoretical model of the structure's dynamics cannot be expected without knowledge of the **complete** excitation and response quantities.

It is another important feature of all measurements that we should ensure that all the available information is extracted from the measured data (see earlier comments regarding the difference between 'data' and 'information'). There is always the danger that we shall extract (or 'see') what we are looking for in test results, especially in a validation test, and fail to recognise (or 'see') other indications, especially if these risk contradicting the behaviour that we are seeking to validate. This failing is not necessarily the result of a wilful ignoring of contrary signs but it is a well-known characteristic that 'once we have seen what we came for, we stop looking any further'. That is a weakness in any investigation, and especially so in experimental studies of structural dynamics where there is always a real risk that our theoretical models may fail to predict all the phenomena which actually occur.

In the same vein as ignoring unexpected, unwanted or inexplicable results is the need to ensure that 'impossible' results are properly investigated and not simply dismissed as being due to 'noise' or 'non-linearity' - two common scapegoats for data which do not conform to our preconceptions. This comment is prompted by two recent experiences of the author relating to the apparently successful measurement of (a) frequency response functions of an unstable rotor system (why did the excitation used to measure the response functions not induce an unstable response in the structure?) and (b) frequency response functions of heavily non-linear structures in which the intermediate (unstable and unattainable?) section of the overhanging region of the curve was successfully measured. There are usually good explanations for such 'impossible' results and these should be sought out.

3.4   Need to make testing more efficient

The final point to make regarding the ongoing and crucial role to be played by experimental techniques in structural dynamics is the urgent need to make testing just as efficient as theoretical analysis. To this end, we need to optimise the testing procedures in just the same way that we have optimised the numerical algorithms used for computational studies. Of course, there have been a number of dramatic technique developments in the recent past - the FFT is a good example from 30 years ago - but this comment is less about the measurement processes themselves and more about the design of the test and the selection of the most appropriate data to be measured. Although there are always exceptions, there is a widespread philosophy where tests are involved 'to measure as much data as possible and then to see what information can be extracted after the test is complete'. The author can recall a series of tests which were made on an annual basis in which the preceding year's data had only partly been analysed (because of its volume) by the time it was required to plan the next year's tests. Clearly, this is sub-optimal in terms of the benefit being derived from the experiments which, as is often the case, were very expensive to conduct.

The philosophy of optimising tests can best be illustrated in the case of tests which are undertaken to validate theoretical models that have been constructed for design calculations of a particular structure. These are typically modal tests in which the modal properties of the test structure are derived from measurements of its response to a controlled excitation and are then compared with the corresponding data predicted by the theoretical model. The resulting comparisons are intended to provide some systematic feedback to the modeller so that the model can be refined (or updated) until the agreement between the test data and the predictions are within an acceptable tolerance. An obvious way to proceed in this application, in order to maximise the benefit which can be gained from the measured data and to minimise the amount of data to be acquired, is to use the theoretical model which is to be validated to 'rehearse' the test in a computer simulation of the test itself. In this way, the effectiveness of having certain items of data when trying to perform the validation and updating procedure can be explored much more cheaply than is possible with actual test data. In principle, unnecessary or ineffective data can be identified prior to the proposed tests being carried out so that time is not wasted measuring data which serve relatively little useful purpose.

It is becoming clear that the timing of these validation tests (in the overall design cycle) is an increasingly critical factor - test data frequently arrive too late to permit full benefit to be gained from them. Thus, we need to re-think the strategy of validation tests and this may tend to press towards earlier and earlier testing which means, in turn, that effort should be put on validating the individual components as they are manufactured, rather than waiting for the fully assembled structure which can only be done when the last item is available. Thus, validation testing needs to work through a series of philosophical issues in an attempt to improve the efficiency of the process by an order of magnitude. Not until this has been done, is it more likely that testing will be accepted as an integral part of the design process.

So to summarise the third part of this essay, it is my submission that we need:

- To accept that experimental techniques will continue to play a central role in the development of structures which are acceptable from the dynamics viewpoint;
- To be more objective and analytical in our design of tests;
- To optimise tests for validation by using the model to be validated in a rehearsal of the proposed tests; and
- To incorporate dynamic testing as an integral part of structural design, and not an afterthought.

All of these are possible: but may require holding back the tide.

Dynamics of Structures with Joint Connections

L. Gaul
Professor & Director
Institute A of Mechanics, University of Stuttgart, Pfaffenwaldring 9,
7 05 50 Stuttgart, Germany

R. Nitsche
Research Assistant
Institute A of Mechanics, University of Stuttgart, Allmandring 5 b,
7 05 50 Stuttgart, Germany

1.   Abstract

Mechanical systems are often assembled through connections at joints. The friction in the joints is a dominant factor if no other forms of damping is present. For improving the performance of systems, numerous studies have been done to predict, measure, and/or enhance the energy dissipation of friction. This chapter reviews the procedure for describing the nonlinear transfer behavior at bolted joint connections. It gives an overview of modeling issues and solution techniques. An application is outlined with regard to the use of joints as semi-active damping devices for vibration control of flexible structures.

2.   Introduction and Review of Literature

The dynamic behavior of fabricated structures is influenced by the load or moment transfer behavior of interfaces of bolted, riveted and compression joints. Studies in dynamics of frictionally constrained bolted joints have been carried out extensively over the past five decades. It is an important and challenging area of research, since friction is present in all mechanical systems incorporating parts with relative motion and the behavior of systems with friction is nonlinear. Friction phenomena in the joints are comple because they are caused by different physical mechanisms. On occasions friction is a desirable property, as it is in the case of brake squeal, but it is undesirable in joints used for purposes of servo control. Dry friction can also enhance system performance due to its damping and isolation properties as external influence. Internal friction can destabilize systems such as rotating machinery [1, 2]. Internal damping descriptions by material damping have been recently reviewed by Gaul [3].

The review paper [4] focuses on the beneficial uses of dry friction and surveys research which has sought of enhance those beneficial features for designing joints and friction contacts.

Three survey papers have adressed friction in engineering systems. A thorough review of friction modeling and on friction-induced vibration is given by Ibrahim [5, 6]. These studies also treat aspects of friction damping, which is a major topics in this chapter. The third survey paper [7] focusses the analysis tools and compensation methods for the control of dry-friction damped systems. The review articles [8] and [9] provide extensive bibliographies of recent papers in the friction damping area. The reader is referred to these papers for additional material on dry friction modeling and analysis.

Friction between rigid bodies is related with contact mechanics. A review on this topic is given in [10–13].

Friction phenomena have been treated by tribologists, material scientists, and mechanical engineers almost independently and with different approaches. It seems to be useful to bring the microscopic approach of tribology together with the more macroscopic phenomenological approach mechanical engineers. A valuable survey on contact mechanics is given in [14]. A broad overview on rolling contact of elastic bodies can be found in [15]. Dynamical contact problems with friction are treated in [16] and [17]. Numerical aspects of the dynamical contact problem with the help of the boundery element method can be found in [18].

An analytical treatment of a single-degree-of-freedom (SDOF) dry friction damped system is given by Den Hartog [19]. He exploited the piecewise linear characteristic of the Coulomb friction law to obtain the exact steady-state solutions. The exact solution technique was also used in [20] to receive so-called 'linear Coulomb damping' for a SDOF system, for a normal force growing linearly with slip displacement. The same system was investigated in

[21] where the influence of displacement-dependent component of the friction force on the sticking condition is studied.

Multi-degree-of-freedom systems with local hysteretic behavior for clamped systems and systems with constraints are treated in [16, 22–24].

A modal approach was used in [25–27] describing continuous systems with friction contact. In [28–31] the finite element method and the direct solution of the wave equation of substructures was used to describe systems with friction contact.

This chapter reviews the role of friction damping in joint connections. The next section outlines some of the basic models of friction for bolted joint connections. A brief overwiew of solution techniques is given in section 4. Section 5 describes the use of *semi-active* joints for vibration control, which is a rather new and promising concept, and section 6 contains some concluding remarks.

3.   Modeling Issues

A general rule concerning modeling holds for joints as well [32]:

The joint models must be as simple as possible and as accurate as necessary.

Therefore in many applications, the Coulomb friction model is very useful in spite of its simplicity. It can explain several phenomena associated with friction in joints and it is commonly used for friction compensation in motion control systems [33].

The Coulomb friction model is a *phenomenological* description of sliding friction. Phenomenological contact models are based on experimental observations and decribe the *global* relation between the tangential force and the relative displacement in the frictional interface. *Constitutive* models are based on interface physics in the contact area. They describe the friction phenomena in a *local* manner, e. g. the relation between traction and displacement. This helps to overcome deficiencies of the Coulomb law when applied to flexible members since it is strictly valid for the contact between rigid bodies.

3.1   Phenomenological Description

Before mathematical models are presented, a brief phenomenological description of sliding friction is given. A friction interface in a joint connection is considered. The friction force acts as internal force in the tangential direction of the contacting surfaces. Below a critical value it is a reaction force causing elastic deformations with no relative motion between the contacting surfaces. At a critical value, the force obeys a constitutive equation such as Coulomb's law and is acting against the relative velocity.

The force required to keep the joint moving depends on the relative velocity and many other factors. Such phenomena can be described by a model which regards friction as a static function of velocity. Therefore, they are called *static friction models* and are described next.

Static Friction Models
The simplest models describe the friction force as a function of the velocity difference v between the sliding surfaces. Such models are often called *static models*.

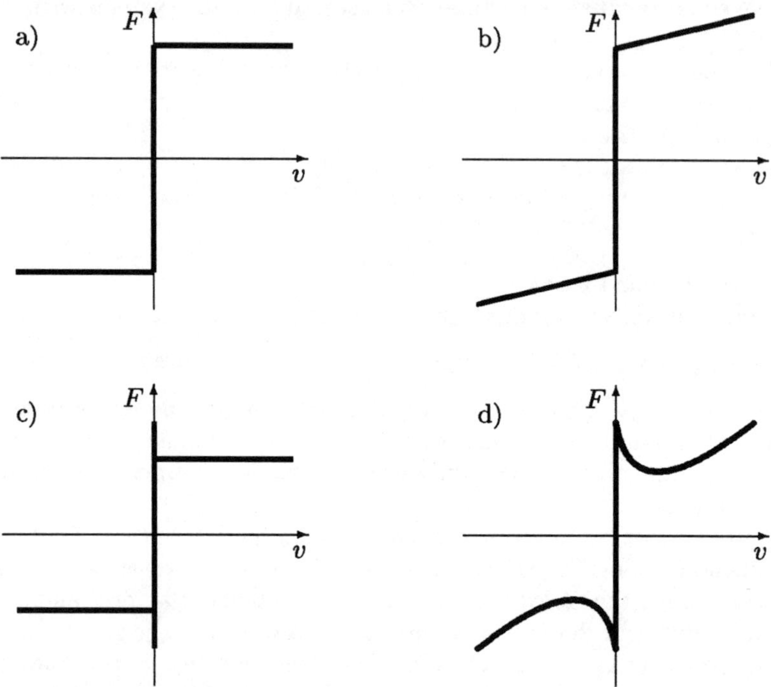

Fig. 1: Static friction models: (a) Coulomb model (1), (b) Coulomb friction model with viscous friction, (c) Stiction plus Coulomb friction, (d) Stribeck's friction model (3).

Signum-Friction Models: The most frequently used model is Coulomb's law, developed in the 18th century. According to Figure 1(a), the friction force is given by

$$F = F_C \operatorname{sgn}(v) = \mu F_N \operatorname{sgn}(v) \tag{1}$$

where F_C is the friction force, F_N the normal load and μ the friction coefficient. For vanishing velocity the contact force is a reactive force governed by the equation of motion. The Coulomb friction model can be improved by adding viscous friction as illustrated in Figure 1(b).

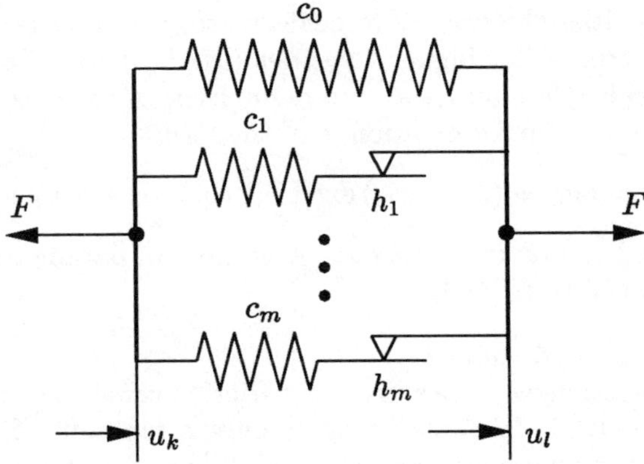

Fig. 2: Generalized elasto-slip model.

Experiments have shown that the parameters F_C and F_N often depend on velocity. It should be noted that the sgn-notation is only correct in case of sliding. Sticking must be treated separately.

Elasto Slip Model: *Stiction* describes the state with the contact force at rest. A drawback of the sgn-based friction laws is that they do not account for deformation that can occur prior to interfacial slip. One way to account for this phenomenon is to model the friction interface as an elastic spring c_i in series with an ideal Coulomb element with threshold forces h_i, which is a so-called *Jenkins-* or *Masing-Element* [34]. The transfer behavior of a joint can also be described arranging Jenkins-Elements in parallel (as can be seen in Figure 2). The contact force is given by

$$F(u, \dot{u}) = c_0 u + \sum_{i=1}^{m} \begin{cases} r_i(t) & |r_i(t)| < h_i \\ h_i \, \mathrm{sgn}\, \dot{u} \end{cases}$$

$$u = u_k - u_l \,, \ |r_i| = |c_i(u - u^+) - h_i \, \mathrm{sgn}\, \dot{u}^+|$$

(2)

where u^+ displacement prior to velocity reversal.

A hysteresis loop for the contact force (2) with periodic displacement of amplitude $\hat{u}$. Figure 7. As can be seen, the loop consists of piecewise straight lines according the Coulomb elements. Figure 1(c) describes a model such that the maximum reaction force in stiction can be higher than the friction force.

The classical models can be combined in different ways (see Figure 1), they include components of different velocity dependence. Stribeck [35] observed that for low velocities the friction force decreases continously with increasing velocities and increases for higher velocities (Figure 1(d)). This phenomenon is named *Stribeck effect* and the minimum friction force is called *Stribeck friction* [36]. A common formulation is of the form

$$F(v) = \left(F_C + (F_S - F_C)\exp\left(-(v/v_s)^{\delta_s}\right)\right)\operatorname{sgn}(v) + F_v v \qquad (3)$$

where v_s is called the *Stribeck velocity*. A number of parameterizations of (3) have been proposed (c.f. [37]).

Dynamic Friction Models

In the previous section we have shown that friction cannot be always described sufficiently by a frictional force which is only a function of velocity. Such models have both fundamental and practical drawbacks. These can be avoided by recognizing that friction is indeed a dynamic phenomenon which needs a modal applicable to a dynamical system. The strain-rate state models are examples of such behvior [38].

The LuGre Model: The LuGre Model introduced in [39] is a dynamic friction model. The friction interface is visualized of as contact between bristles. A detailed description of the model and its application can be found in [36]. The model was used to describe the nonlinear load and moment transfer behavior of joints [40] and in turbine plates [41].

The starting point of the model derivation is the force generated by solid-to-solid contact, as shown in Figure 3. At the microscopic level the surfaces are very irregular and make contact at a number of asperities which can be thought of as a contact between bristles. The bristles deflect like springs and give rise to the friction force, when a tangential force is applied. If the contact is sufficiently large some of the bristles deflect to the extent that there is loss of contact. Due to the irregular shape of the surfaces, contact phenomena are highly random [39, 42]. Therefore the model is based on the average behavior of the bristles that make up the contact. This model was designed to reproduce all observed friction phenomena over a wide range of operating conditions. The governing equations are

$$F_f = \underbrace{(\sigma_0\varphi + \sigma_1\dot{\varphi} + \sigma_2 v)}_{=:\,\mu(\varphi,\dot{\varphi},v)} \qquad (4a)$$

$$\dot{\varphi} = v - \sigma_0\,\frac{|v|}{g(v)}\,\varphi\,,\ \varphi(0) = \varphi_0 \qquad (4b)$$

$$g(v) = F_c + F_\Delta\,\exp\left(-(v/v_s)^2\right) \qquad (4c)$$

where the frictional force is F_f, and the relative sliding velocity at the friction interface is v. The internal friction state φ represents the average deflection of the bristles as can be seen in Figure 3. The parameter F_c is the Coulomb friction threshold whereas the sum $F_c + F_\Delta$ corresponds to the sticktion force. The Stribeck velocity v_s determines how $g(v)$ varies within its bounds $F_c < g(v) \leq F_c + F_\Delta$. The stiffness of the bristles is described by σ_0, while the two parameters σ_1 and σ_2 control the dynamic dependence of friction on velocity. The variable μ, which is defined in (4a), can be interpreted as a state dependent friction coefficient. With this model it is possible to describe presliding displacement, stick-slip motion including the Stribeck effect and other rate dependent friction phenomena [36, 39].

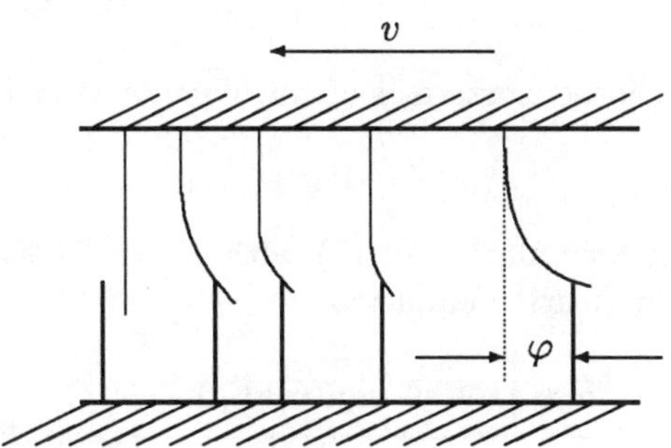

Fig. 3: Model of the friction interface [36]. For simplicity the bristles of the lower part are assumed to be rigid.

The Valanis Model: The nonlinear transfer behavior of an isolated joint connection has been experimentally investigated by Gaul and coworkers [43, 44]. To obtain the behavior in the local slip regime (micro-slip) and the global slip regime (macro-slip) by a single model and to simulate the response under cyclic loads as well as transient behavior, the *Valanis Model* known from endochronic plasticity [45] is investigated. The model is governed by the first order differential equation [44, 46]:

$$F'(z) + \lambda F(z) = E_0 q'(z) + \lambda E_t q(z) \tag{5}$$

where F is a generalized force, q is a generalized displacement, and E_0, E_t and λ are material parameters. The relation between the internal variable $z(t)$ and physical time t is given by

$$\dot{z}(t) = \left| \dot{q}(t) - \kappa \frac{F(t)}{E_0} \right| \tag{6}$$

where κ is another dimensionless parameter. By restricting the parameter κ:

$$0 \leq \kappa < 1, \tag{7}$$

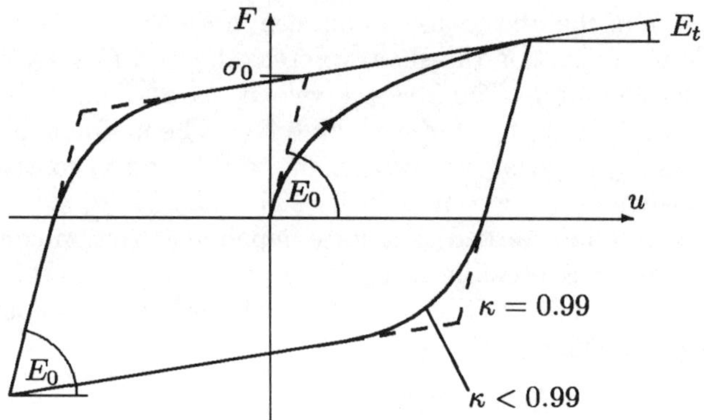

Fig. 4: Closed hysteresis of the Valanis Model of a joint.

an equivalent formulation of (5) with $\dot{F} = F'\dot{z}$ leads to the nonlinear first order constitutive equation

$$\dot{F} = \frac{E_0 \dot{u} \left[1 + \frac{\lambda}{E_0} \operatorname{sgn}(\dot{u})(E_t u - F)\right]}{1 + \kappa \frac{\lambda}{E_0} \operatorname{sgn}(\dot{u})(E_t u - F)} = \dot{F}(u, \dot{u}, F). \tag{8}$$

If we interpret $q = u$ as the relative displacement in the friction interface and F as the friction force, the result (8) can be used to describe the 'transfer' behavior of a joint connection [43, 44, 47]. The parameters in (8) have the following physical meanings: the elastic modulus of the sticking is E_0 while the tangent modulus E_t describes the slope of slip motion (Figure 4). The parameter κ controls the influence of microslip. High values of κ corresponds to small influence of microslip. The parameter σ_0 denotes the stick limit equivalent to yield stress as being defined in [48]:

$$\sigma_0 = \frac{E_0}{\lambda \left(1 - \kappa \frac{E_t}{E_0}\right)}. \tag{9}$$

The parameters in (8) can bepbtained from measured hysteresis as shown in Figure 4. The trivial case of linear elastic behavior ($\kappa = 1$) is excluded by definition (7).

The numerical efficiency and accuracy of the presented joint models (4), (8) is very high as we do not have to detect zero velocity. This differs from static friction models. Hence the models can easily be implemented into a dynamical simulation environment, such as multibody systems [47] or in a finite-element program [44].

3.2 Constitutive Description

The finite-element method is often used to describe the local behavior between surfaces which are in contact, i. e. a joint connection. As we regard the relation between stress fields and displacement fields, this is a constitutive approach. Integration over the contact area yields global force-deformation properties [42].

Joint Description by Contact Mechanics with Statistical Surface Roughness Description

Laws of normal and tangential contact including friction are developed on the basis of a statistical surface model in [42]. The surfaces in the contact area of the joint are rough and have contact only at their asperities. The true area of contact is much smaller than the apparent contact area resulting in very high local normal pressure. Therefore in early investigations plastic deformations of contact at the asperities were considered. This simplification was extended by introducing a plasticity index [49], a measure of the elastic part of the true contact area. However all these investigations assume initial contact. In case of repeated contact the surface will bear equal or lower normal loads elastically [50]. The same argument holds for tangential loads, initial junction growth comes to an end in case of oscillating loading.

In what follows, the following assumptions are made [42]:

- Only elastic contact of metallic surface is considered, i. e. plastic deformation of initial contact is not included.
- Isotropic surfaces containing asperities with spherical summits.
- Contact points do not interact, i. e. normal loads are below flow pressure of the solid material.
- Wear is not allowed, i. e. surface parameters do not change with time [51].
- No lubrication, i. e. dry friction.

The normal contact on the microscale can be described by Hertzian theory [14]. The tangential microscale load-displacement function is formulated by a combination of Mindlin's elastic shear theory and constant shear strength τ_{max} at the outer contact radius. For the implementation of the induced normal and tangential contact forces in a macroscopic model, the local relations have to be distributed over the apparent area of contact. This is done by using a statistical model [52] of a rough surface. This model uses finite difference approximations of the slope and curvature of surface asperity profile. The physical parameters of the surface model can be determined by measuring the height distribution of a real surface. A contact virtual work formulation given in [42] leads to a contact algorithm, was implemented in research and commercial finite-element programs.

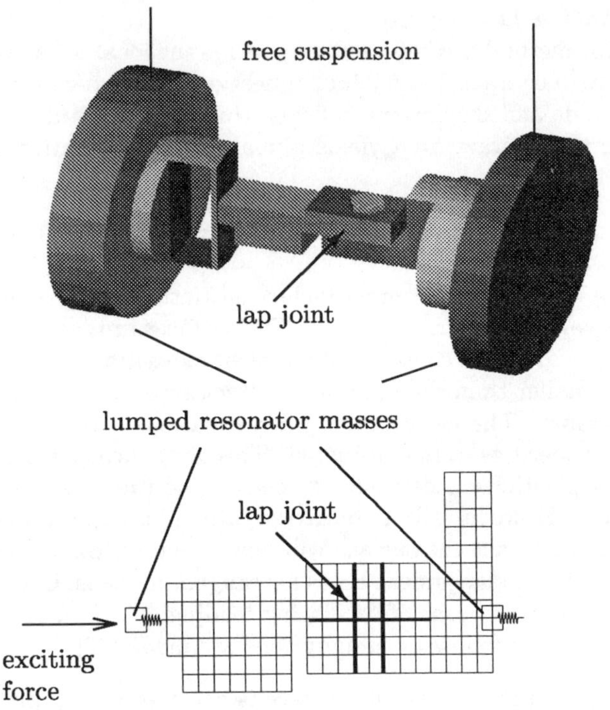

Fig. 5: Lap joint resonator and plane FE discretization [42].

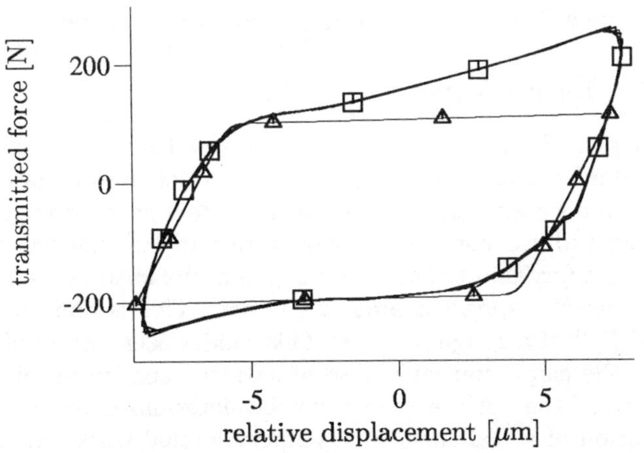

Fig. 6: Comparison of the measured □ and calculated △ hysteresis loop of lap joint resonator [42].

An example of a lap joint resonator is used in [42] to investigate the transfer behavior including energy dissipation by microslip effects (Figure 5). The resonator is excited by a harmonic force acting mass at left. The dissipated energy is determined from hysteresis loops derived from the transmitted force versus the relative displacement of the lap joint. Figure 6 shows the comparison of a typical measured hysteresis and a calculated one. The deviations in the macroslip range are due to a rather crude modeling of the bolt which bears a major part of the load after gross slip occurs.

4. Solution Techniques

This section briefly describes two common solution techniques applied to the analysis of structures with joint connections and systems with friction, an equivalent linearization by the harmonic balance procedure and the time integration technique.

One of the earliest solution techniques, namely the exact solution method uses the *piecewise* linear characteristic of the Coulomb friction law to obtain *exact* steady state solutions for harmonic excitation [19]. A number of authors have used and extended the exact solution method, e. g. [20,21,53–58], which is limited to a few degrees of freedom (DOF).

4.3 Computational Time Integration

One of the purposes of a joint model is to use it in simulations. It is necessary that these simulations can be performed efficiently. Models with disontinuities are difficult to simulate, since it is necessary to find the time where the discontinuous events occur [59]. Simulation software must have a structured way of finding zero crossings, otherwise the step size of the integration routine has to be decreased. For all static models the problem of detecting zero velocity is present. To alleviate this problem, most studies make use of the sticking condition which is checked at each time when the slip velocity in the frictional interface crosses zero. If the static friction force is sufficient to prevent slipping, time integration is resumed using a reduced set of dynamics pertaining to the sticking system [60]. The Karnopp model [61] avoids the problem of detecting zero velocity by defining an interval as zero velocity, $|v| < \alpha$. For velocities within this interval the internal state of the system, e. g. the velocity, may change and be non-zero but the output of the block is maintained to be zero by a dead-zone. By doing so it is not necessary to detect exactly when the interval is entered. A coarse estimate of the first time inside the interval suffices, which implies that a larger time step can be used without event handling. It should be noted that time integration can be very inefficient if one is only interested in the steady state response to periodic excitation. If the system is only lightly damped, one must integrate the systems of equation until the transient response has died out, which can take hundreds of forcing periods.

If we regard structures with several joint connections, it is very difficult to implement sticking criteria within time integration routines. A modal

40

condensation technique can be applied to reduce the system order to the nonlinear DOF's at the joints only [31]. The decision on whether a particular friction interface is sticking or slipping depends on the sum of the tangential forces in the contact plane. They have to overcome the static friction force in an interface momentarily at rest. However, the normal and tangential forces themselves depend on the friction forces at other frictional interfaces [4]. A number of authors have analyzed these types of systems, e. g. [62–64].

4.4   Harmonic Balance

A method to describe a linearized steady harmonic response of a joint connection in the frequency domain is presented. A measured closed hysteresis loop is assumed [31]. If we define $f(u)$ as loading and $g(u)$ as unloading function of the joint hysteresis, we obtain a tangential force in the friction interface of the form [65]:

$$F = \frac{1}{2}[f(u) + g(u)] + \frac{1}{2}[f(u) - g(u)]\operatorname{sgn}(\dot{u}). \tag{10}$$

Using the harmonic balance method [66] we obtain an equivalent stiffness and viscous damping coefficient:

$$c_{eq}(\omega,\hat{u}) = \frac{1}{\pi\hat{u}}\int_0^{2\pi} F(\hat{u}\cos\omega t, -\hat{u}\omega\sin\omega t)\cos\omega t\, d(\omega t), \tag{11}$$

$$d_{eq}(\omega,\hat{u}) = \frac{1}{\pi\hat{u}\omega}\int_0^{2\pi} F(\hat{u}\cos\omega t, -\hat{u}\omega\sin\omega t)\sin\omega t\, d(\omega t). \tag{12}$$

In Figure 7 the ellipse of the linearized tangential or friction force is shown, with the complex stiffness $c(\omega,\hat{u}) = c(\omega,\hat{u}) + i\omega d(\omega,\hat{u})$. Additionally the polygon of the nonlinear model is shown in Figure 7, which corresponds to Figure 2. A linearization of the generalized elasto-slip model, as shown in Figure 2, leads to the equivalent stiffness

$$c(\omega,\hat{u}) = c_0 + \sum_{i=1}^{m}\frac{c_i}{\pi}\left(\alpha - \frac{\sin 2\alpha_i}{2}\right) \tag{13}$$

and equivalent viscous damping coefficient

$$d(\omega,\hat{u}) = \sum_{i=1}^{m}\frac{c_i}{2\pi\omega}\left(1 - \cos 2\alpha_i\right) \tag{14}$$

with $\cos\alpha_i = 1 - 2\frac{h_i}{c_i\hat{u}}$. Effects of weak nonlinearities on modal analysis due to such linearized joint models are studied in [27].

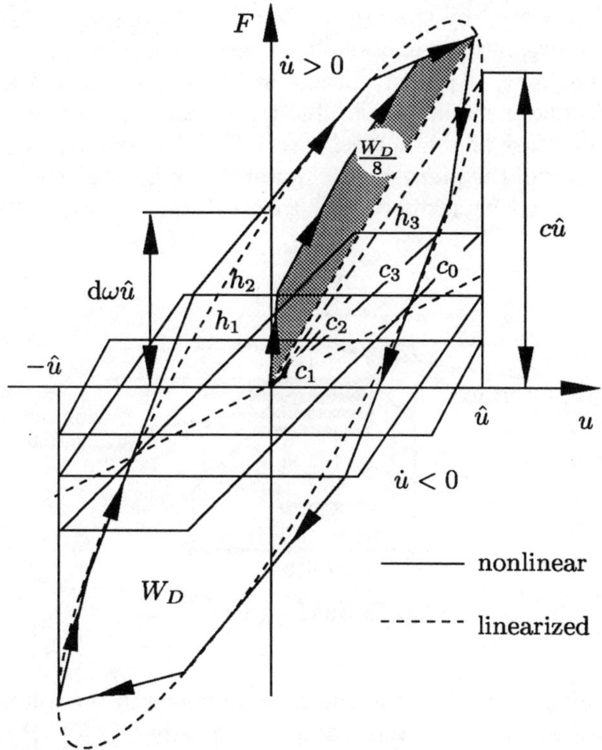

Fig. 7: Hysteresis loop of a lap joint model (2) with 7 parameters according to Figure 2.

5.   Semi-Active Joints

A common simplification used in studies of the transfer behavior of joints is that the normal force in the frictional interface is of constant magnitude. This section discusses the design of systems for which the normal force in the friction interface is not constant.

The normal force can be varied by an active device. The concept of actively controlling the normal force in real time based on feedback from sensory outputs of the connecting parts is termed *semi-active*. It was proposed in [67] to augment the damping properties of passive elements. In [68, 69] several means of realizing semi-active damping are suggested. It is interesting to note that semi-active damping cannot add energy to the system. Hence the closed-loop system is more stable than a fully active one and is insensitive with respect to the 'spillover problem' which is an important consideration in the control of flexible structures.

A possible realization of such a semi-active joint is shown in Figure 8, proposed and patented by the first author [70, 71]. Using a piezoelectric disc plate, used as a 'washer ' it is possible to vary the normal force in the friction interface of the joint. Applied voltage at the stack disc induces thickening of the piezoelectric material which, due to the constraining effect of the bolt, results in an increase of the normal force. For this application the nonlinear dynamic behavior of the piezoelectric material can be neglected, so the normal force is assumed to be proportional to the input voltage. To arrive at a

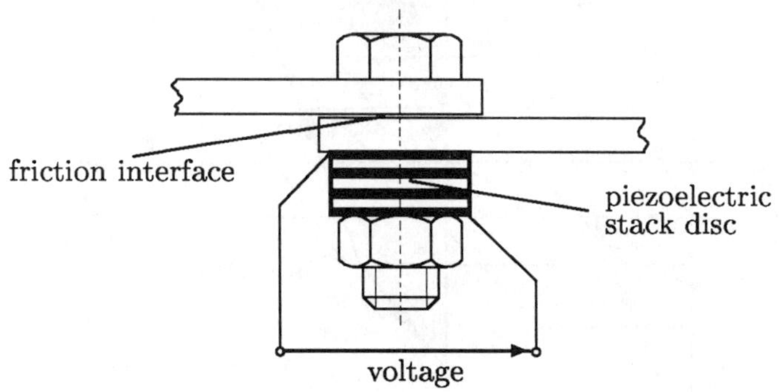

Fig. 8: Semi-active joint.

control law which maximizes the energy dissipation in the joint, a Lyapunov-function representing the system energy is considered [40]. By inspection of its time derivative, one arrives at a 'bang-bang' control law which maximizes the damping contribution of the controlled term. To describe the transfer behavior of the semi-active joint a modification of the dynamic friction model (4) was used in [40]. It is shown that the bang-bang control law applied to normal force of the joint considerably improves the energy dissipation. Moreover the use of a boundary layer around the switching points is needed to prevent 'quasi-sticking' in the friction interface of the joint. Note that a sticking joint, does not dissipate energy. The semi-active joint concept was applied to flexible structures for vibration suppression with considerable success [40, 47, 70].

The design of enhanced friction joints for vibration suppression in flexible structures has also been investigated using semi-active friction elements [60].

6.   Conclusions

The load and moment transfer behavior of joint connections continues to play an important role in the design of structures to predict, measure, and/or enhance vibrational properties. The choice of models describing the transfer behavior of the joints with constitutive or phenomenological approaches depends on the type of application.

If one is interested in a local description of the friction contact derived from processes at the microscale, a constitutive model has to be used.

In general, for control tasks, phenomenological joint models with reduced DOF's are used for designing a model based controller as it was outlined in section 5 for semi-active joints. The research on semi-active friction damping in joints is in its infancy and seems to be a promising concept primarily for vibration suppression in large flexible structures.

References

[1] G. Siegl. *Das Biegeschwingungsverhalten von Rotoren, die mit Blechpaketen besetzt sind.* PhD thesis, TU Berlin, Fachbereich der Physikalischen Ingenieurwissenschaften, 1981.

[2] S.H. Crandall. The role of damping in vibration theory. *Journal of Sound and Vibration*, 11:3–18, 1970.

[3] L. Gaul. The influence of damping on waves and vibration. *Mechanical Systems and Signal Processing*, 13(1):1–30, 1999.

[4] A.A. Ferri. Friction damping and isolation systems. *ASME Journal of Vibration and Acoustics*, 117B:196–206, June 1995.

[5] R.A. Ibrahim. Friction-induced vibration, chatter, squeal, and chaos — Part I: Mechanics of contact and friction. *Applied Mechanics Reviews*, 47(7):209–226, 1994.

[6] R.A. Ibrahim. Friction-induced vibration, chatter, squeal, and chaos — Part II: Dynamics and modeling. *Applied Mechanics Reviews*, 47(7):227–253, 1994.

[7] B. Armstrong-Hélouvry, P. Dupont, and C. Canudas de Wit. A survey of models, analysis tools and compensation methods for the control of machines with friction. *Automatica*, 30(7):1083–1138, July 1994.

[8] C.F. Beards. Damping in structural joints. *The Shock and Vibration Digest*, 24(7):3–7, 1992.

[9] D.I.G. Jones. Damping of dynamic systems. *The Shock and Vibration Digest*, 22(4):3–10, 1990.

[10] I.L. Singer and H.M. Pollock, editors. *Fundamentals of friction: macroscopic and microscopic prcesses*, volume 220 of *NATO ASI series E*. Kluwer, Dortrecht etc., 1992.

[11] J.A. Martins, J.T. Oden, and F.M.F. Simões. A study of static and kinetic friction. *Int. J. Engineering Sci.*, 28:29–92, 1990.

[12] A.P.S Selvadurai and M.J. Boulon, editors. *Mechanics of Geomaterial Interfaces*, volume 42 of *Studies in Applied Mechanics*. Elsevier Scientific Publishers, The Netherlands, 1995.

[13] A.P.S Selvadurai and G.Z Voyiadjis, editors. *Mechanics of Material Interfaces*, volume 11 of *Studies in Applied Mechanics*. Elsevier Scientific Publisher Co., Amsterdam, 1986.

[14] K.L. Johnson. *Contact Mechanics*. Cambridge University Press, Cambridge etc., 1985.

[15] J.J. Kalker. *Three dimensional elastic bodies in rolling contact*. Kluwer, Dortrecht etc., 1990.

[16] P. Lötstedt. Mechanical systems of rigid bodies subjected to unilateral constraints. *SIAM J. Appl. Math.*, 42:281–296, 1982.

[17] J.J. Moreau. *Nonsmooth mechanics and applications*. CISM Courses and Lectures. Springer-Verlag, Wien u.a., 1988.

[18] H. Antes and P.D. Panagiotopoulos. *The boundary integral approach to static and dynamic contact problems*. Birkhäuser, Basel u.a., 1992.

[19] J.P. Den Hartog. Forced vibrations with combined Coulomb and viscous friction. *Transactions of the ASME*, APM-53-9:107–115, 1931.

[20] K.E. Beucke and J.M. Kelly. Equivalent linearizations for practical hysteretic systems. *International Journal of Non-Linear Mechanics*, 23(4):211–238, 1985.

[21] J.R. Anderson and A.A. Ferri. Behavior of a single-degree-of-freedom system with a generalized friction law. *Journal of Sound and Vibration*, 140(2):287–304, 1990.

[22] R.K. Miller. The steady state response of systems with hardening hysteresis. *J. Mech. Design*, 100:193–198, 1978.

[23] F. Pfeiffer. Dynamical systems with time-varying or unsteady structure. *Zeitschrift für angewandte Mathematik und Mechanik*, 71:T6–T22, 1991.

[24] F. Pfeiffer and C. Glocker. Dynamical systems with unsteady processes. In *[5]*, pages 65–74, 1992.

[25] L. Jézéquel. Structural damping by slip in joints. *ASME Journal of Vibration, Acoustics, Stress, and Reliability in Design*, 105:497–504, 1983.

[26] E.H. Dowell. The behavior of a linear damped modal system with a nonlinear spring-mass-dry-friction-damper system attached. *Journal of Sound and Vibration*, 89:65–84, 1983.

[27] A.F. Vakakis and D.J. Ewins. Effects of weak nonlinearities on modal analysis. In *Proc. of the 10th International Modal Analysis Conference*, pages 72–78, San Diego, CA, 1992.

[28] L. Gaul. Zur Dämmung und Dämpfung von Biegewellen an Fügestellen. *Ingenieur Archiv*, 51:101–110, 1981.

[29] L. Gaul. Wellenausbreitung in Strukturen mit Fügestellen. *Zeitschrift für angewandte Mathematik und Mechanik*, 62:T41–T44, 1984.

[30] L. Gaul and S. Bohlen. Vibration of structures coupled by nonlinear transfer behaviour of joints; a combined computational and experimental approach. In *Proc. 5th International Modal Analysis Conference*, volume I, pages 86–91, London, 1987.

[31] L. Gaul and S. Bohlen. Identification of nonlinear structural joint models and implementation in discretized structure models. In *Proc. 11th ASME Conference on Mechanical Vibration and Noise: The Role of Damping in Vibration and Noise*, volume 5, pages 213–219, Boston (USA), 1987.

[32] K. Popp. Non-smooth mechanical systems — an overview. *Forschung im Ingenieurwesen*, 64:223–229, 1998.

[33] B. Friedland and Y.-J. Park. On adaptive friction compensation. In *Proceedings of the IEEE Conference on Decision and Control*, pages 2899–2902, 1991.

[34] D. Ottl. *Schwingungen mechanischer Systeme mit Strukturdämpfung*. Number 603. VDI-Fortschritt-Berichte, Düsseldorf, 1981.

[35] R. Stribeck. Die wesentlichen Eigenschaften der Gleit- und Rollenlager – the key qualities of sliding and roller bearings. *Zeitschrift des Vereins deutscher Ingenieure*, 46(38,39):1342–48,1432–37, 1902.

[36] H. Olsson. *Control Systems with Friction*. PhD thesis, Departement of Automatic Control, Lund Institute of Techonology, April 1996.

[37] B. Armstrong-Hélouvry. *Control of Machines with Friction*. Kluwer Academic Publisher, Boston, Ma., 1991.

[38] J.R. Rice and S.T. Tse. Dynamic motion of a single degree of freedom system following a rate and state dependent friction law. *J. Geophysical Research*, 91:521–530, 1986.

[39] C. Canudas de Wit, H. Olsson, K.J. Åström, and P. Lischinsky. A new model for control of systems with friction. *IEEE Trans. Automat. Control*, 40(3):419–425, March 1995.

[40] L. Gaul and R. Nitsche. Friction control for vibration suppression. *Mechanical Systems and Signal Processing*, 14(2):139–150, March 2000.

[41] P.E. Dupont and A. Stokes. Semi-active control of friction dampers. In *Proceedings of the 34th Conference on Decision and Control*, pages 3331–3336, New Orleans, LA, December 1995.

[42] K. Willner and L. Gaul. A penalty approach for contact description by FEM based on interface physics. In M.H. Aliabadi and C. Alessandri, editors, *Proc. of Contact Mechanics II, Ferrara, Italy*, pages 257–264, Southampton, 1995. Comp. Mech. Publ.

[43] J. Lenz and L. Gaul. The influence of microslip on the dynmamic behavior of bolted joints. In *Proc. of the 13th International Modal Analysis Conference*, pages 248–254, Nashville, Tennesse, 1995.

[44] L. Gaul and J. Lenz. Nonlinear dynamics of structures assembled by bolted joints. *Acta Mechanica*, 125(1–4):169–181, 1997.

[45] K.C. Valanis. A theory of viscoplasticity without a yield surface. *Archive of Mechanics*, 23(4):517–551, 1971.

[46] K.C. Valanis. Fundamental consequences of a new intrinsic time measure. plasticity as a limit of the endochcronic theory. *Archive of Mechanics*, 32:171–191, 1980.

[47] L. Gaul, J. Lenz, and D. Sachau. Active damping of space structures by contact pressure control in joints. *Mechanics of Structures and Machines*, 26:81–100, 1998.

[48] P. Haupt. Dynamische Systeme mit geschwindigkeitsunabhängiger Dämpfung. *Zeitschrift für angewandte Mathematik und Mechanik*, 71:T55–T57, 1991.

[49] J.A. Greenwood and J.B.P. Williamson. Contact of nominally flat surfaces. In *Proc. R. Soc. Lond.*, volume 295, pages 300–319, 1966.

[50] B.B. Mikic. Analytical studies of contact of nominally flat surfaces; effect of previous loading. In *J. Lub. Tech.*, volume 4, pages 451–459, 1971.

[51] T.S. Nguyen and A.P.S. Selvadurai. A model for coupled mechanical and hydraulic behaviour of a rock joint. *Int. Journal for Numerical and Analytical Methods in Geomechanics*, 22:29–48, 1998.

[52] J.A. Greenwood. A unified theory of surface roughness. In *Proc. R. Soc. Lond.*, volume 295, pages 133–157, 1984.

[53] E.S. Levitan. Forced oscillation of a spring-mass system having combined Coulomb and viscous damping. *Journal of the Acoustical Society of America*, 32(10):1265–1269, 1960.

[54] G.C.K. Yeh. Forced vibrations of a two-degree-of-freedom system with combined Coulomb and viscous damping. *Journal of the Acoustical Society of America*, 39:14–24, 1966.

[55] B. Westermo and F. Udwadia. Periodic response of a sliding oscillator system with harmonic excitation. *Earthquake Engineering and Structural Dynamics*, 11:135–146, 1983.

[56] S.W. Shaw. On the dynamic response of a system with dry friction. *Journal of Sound and Vibration*, 108(2):305–325, 1986.

[57] D. Karius. A contribution on forced periodic oscillations with piecewise linear and constant damping. *ASME Journal of Vibration, Acoustics, Stress, and Reliability in Design*, 107(4):383–391, 1985.

[58] N. Makris and M.C. Constantinou. Analysis of motion resisted by friction. I. Constant Coulomb and linear/Coulomb friction. *Mechanics of Structures and Machines*, 19(4):477–500, 1991.

[59] E. Hairer, S.P. Nørsett, and G. Wanner. *Solving Ordinay Differential Equations I: Nonstiff Problems*. Springer-Verlag, Berlin, 1987.

[60] A.A. Ferri and B.S. Heck. Analytical investigation of damping enhancement using active and passive structural joints. *AIAA Journal of Guidance, Control, and Dynamics*, 15(5):1258–1264, September-October 1992.

[61] D.C. Karnopp. Computer simulation of slip-stick friction in mechanical dynamic systems. *ASME Journal of Dynamic Systems, Measurement, and Control*, 107(1):100–103, 1985.

[62] H.J. Klepp. Steady-state and irregular motions of a system with friction under harmonic excitation. *Mechanics of Structures and Machines*, 19(3):327–356, 1991.

[63] F. Pfeiffer and M. Hajek. Stick-slip motion of turbine blade dampers. *Philosophical Transactions of the Royal Society of London, A*, 338:503–517, 1992.

[64] F. Pfeiffer. Complementary problems of stick-slip vibration. In S.C. Sinha and R.M. Evan-Iwanowski, editors, *Dynamics and Vibration of Time-Varying Systems and Structures*, volume 56 of *Proceedings of the ASME 14th Biennial Conference on Mechanical Vibration and Noise*, pages 43–50, Albuquerque, September 19-22 1993.

[65] L. Gaul. Wave transmission and energy dissipation at structural and machine joints. *ASME Journal of Vibration, Acoustics, Stress, and Reliability in Design*, 105, 1983.

[66] L. Gaul and S. Bohlen. Zum nichtlinearen Übertragungsverhalten von Fügestellen. *Zeitschrift für angewandte Mathematik und Mechanik*, 64:T45–T48, 1984.

[67] D.C. Karnopp, M.J. Crosby, and R.A. Harwood. Vibration control using semi-active force generators. *ASME Journal of Engineering for Industry*, 96(2):619–626, 1974.

[68] D.C. Karnopp. Force generation in semi-active suspensions using modulated dissipative elements. *Vehicle System Dynamics*, 16:333–343, 1987.

[69] D.C. Karnopp. Design principles for vibration control systems using semi-active dampers. *ASME Journal of Dynamic Systems, Measurement, and Control*, 112:448–455, September 1990.

[70] L. Gaul, R. Nitsche, and D. Sachau. Semi-active vibration control of flexible structures. In U. Gabbert, editor, *Modelling and Control of Adaptive Mechanical Structures*, number 268 in VDI-Fortschritt-Berichte Reihe 11, pages 277–286, Düsseldorf, 1998. VDI-Fortschritt-Berichte.

[71] L. Gaul. Aktive Beeinflussung von Fügestellen in mechanischen Konstruktionselementen und Strukturen — Active Control of Joints in Members and Structures. German Patent DE 197 02 518 A1, 1997.

NON-STRUCTURAL DYNAMICS

J.K. Hammond
Institute of Sound and Vibration Research
University of Southampton, Southampton, UK.

1. INTRODUCTION

Dynamics is solidly based in Newtonian physics, admitting modelling by equations representing force balancing. Whilst practitioners readily acknowledge that models are 'approximate' in some sense, great effort and faith is put into deterministic models, predominantly linear, unchanging and often finite dimensional - and even with these restrictions, great insight into fundamentals and practicalities has resulted. However, the complexity of even the simplest real systems soon reveals the limitations of these approximations. Accordingly, structural dynamics is evolving along lines that incorporate conceptual, mathematical and probabilistic ideas that allow inclusion of more realism. These are generally referred to in a way that implies exclusion from the accepted and well-developed formalism - and are generally prefixed with - NON - *non*-stationary; *non*linear; *non*-Gaussian; *non*-deterministic. This essay comments on aspects of these *non* developments and speculates on, and questions, their relevance and importance for the future of structural dynamics.

2. NON-STATIONARITY

The wide ranging and topical subject of condition monitoring and fault diagnostics/classification is essentially aimed at quantifying changes in systems. Information on these changes is assumed to be contained in measured signals and particular changes may have revealing signatures. Fundamental to monitoring the condition of a system is the implication that *temporal* changes will manifest themselves in the modes of operation, i.e., if the system changes, then measurements based on system behaviour will be correspondingly non-stationary.

For example, progressive failures (e.g., developing cracks) in mechanical or structural systems can result in changing signatures. These are non-stationary and/or nonlinear features. For example, a cracked vibrating beam exhibits weakly nonlinear characteristics and the resulting dynamical behaviour may be related to crack characteristics [1]. Additionally, crack identification may be perceived through changes in natural frequencies [2]. If changes occur over short time scales then continuous monitoring may be needed. Non-stationarity in signals and systems can be addressed in the time domain or in transform domains. In particular, time-frequency analysis is playing an increasing role in practical dynamics analysis.

2.1 Time-frequency Analysis

The essence of this form of analysis is to consider how the "frequency" of a signal or system "changes with time". This notion is readily accepted but introduces very fundamental considerations.

A starting point is to decide just what is meant by 'frequency' in a varying context. The concepts of instantaneous amplitude, phase and frequency of a signal lead to the analytic signal $\sigma_x(t) = x(t) + j\,\hat{x}(t) = A(t)e^{\phi(t)}$, where $\hat{x}(t)$ is the Hilbert transform of x, $A(t)$ is the instantaneous amplitude and $\dot{f}(t)$ is the instantaneous frequency. This is, for many 'simple' signals (i.e., monocomponent), an intuitively attractive alternative to Fourier methods. The term 'instantaneous frequency' provides the link between frequency and time. Another link between time and frequency is based on Fourier analysis through the concept of group delay, leading to an alternative time-frequency description of a signal. The two descriptions are equivalent (roughly) if a signal is monocomponent and has a large bandwidth-time product. These apparent *alternatives* to linking time and frequency are a basis for considering these more deeply.

Fundamental considerations are given in a text edited by Boashash [3] and a survey paper is [4]. A recent special issue of Mechanical Systems and Signal Processing [5] has been devoted to time-frequency methods with several of the contributions directly related to condition monitoring. The choice of which time-frequency approach is to be used involves deciding the relative merits of linear decomposition (e.g., short time Fourier transform, Gabor forms, wavelet transform) *vs* quadratic forms (Wigner distributions, Cohen class) or nonlinear forms. Presentations in the special issue relate to:

- discussions of providing a rationale for automatic interpretation of time-frequency maps characterising damage which incorporate both physical modelling and pattern recognition.

- The analysis of gearbox faults involving the three stages of obtaining a time-frequency representation with good 'readability', interpretation of the main components with respect to models of gearbox vibration and, finally, validation of the identified models with respect to the original data.

- Sometimes there is a need for clear identification of indicators of early machine degradation when faults may be of such brief duration that they occupy only a small fraction of the analysis interval. The Cohen-Posch distribution is used to locate such events.

- The analysis of speed-dependent vibration signals measured on rotating machinery is accomplished using several sensors giving methods for directional and spatial decompositions using time-frequency plots.

- Gear vibration manifesting itself in amplitude and phase modulation of tooth meshing vibration is analysed using wavelets.

- Higher Order Spectral methods (see Section 3) using the Wigner Higher Order Moment Spectra (WHOMS) are used to detect faults in rotating machinery.

Other time-frequency applications include: Dalianis *et al* [6], who use time-frequency methods to characterise abrupt structural changes in the response of a vibrating beam, and Bonato *et al* [7], who use the cross Wigner distribution to characterise vibrating systems under non-stationary excitation. The class of signals that are termed cyclostationary (i.e., those for which the correlation functions are periodic) are clearly of relevance to rotating phenomena with stochastic elements. There is an extensive bibliography on the subject and we note Hardin and Miamee [8], in which a general class of processes called correlation autoregressive (CAR) is described, which include cyclostationary processes. This is applied to data arising from helicopter noise. Other papers addressing these issues include [9,10] and higher order cyclostationarity has also been discussed [11].

2.2 Time Domain

Purely *time domain* approaches to monitoring changes are essentially aimed at (adaptive) methods for fitting of parametric models to such observations include Prony analysis (where sums of damped exponentials are fitted to ring down data or autocorrelation), maximum entropy spectral analysis, in which AR models are fitted to random time series (assuming white excitation of the system), maximum likelihood estimation using FIR filter forms and others. In general, systems are ARMA in form and the problem of estimating coefficients from noisy output measurements has been given significant attention. The thesis by Kim [12] gives a good overview of the methods. For non-Gaussian inputs, interest has centred on cumulant-based (see Section 3) parametric estimation methods Model-based fault detection methods are described in [13] and [14]. Basseville *et al* [15] consider a system identification approach for damage monitoring in vibrating structures.

2.3 The Future; *Speculation I*

Detection of non-stationarity is vital for condition monitoring, but is it fundamental to the development of structural dynamics? *Specifically*, is modelling of 'short-time' changes in dynamical behaviour important (or even feasible?) for structural dynamics, or should we rely on statistical/time series methods to detect signature changes? In short, is non-stationarity a red-herring for fundamental structural dynamics?

3. NONLINEARITY AND NON-GAUSSIANITY

Nonlinear dynamics is enormous in scope and has a long history - ranging from classic nonlinear differential equation theory to topological methods and chaotic dynamics. The describing differential equations are nonlinear, superposition is not applicable, and multiple solutions can exist. Structures under harmonic excitation can exhibit sub- and superharmonic, quasi-periodic and chaotic responses. There have also been many attempts to generalise the well understood approaches that serve linear systems well. For example, the notions of 'modes' and 'mode shapes' have been re-appraised. The paper by Vakakis [16] is an overview of so-called nonlinear normal modes.

In this discussion, we focus on the generalisation of the linear system input-output form to the Volterra form leading to higher-dimensional impulse response functions and corresponding frequency response forms. This formalism is well-established but limited in its more general practical utility in dynamics. (Why?) It is also appropriate to note the inclusion of the concept of non-Gaussianity. The response of a nonlinear system driven by a Gaussian input is in general non-Gaussian. Accordingly, the two concepts are sometimes linked - that is, the detection of non-Gaussianity *may* be considered as evidence of nonlinearity.

The full probabilistic structure of a random process is often too difficult to compute but moments of the process can sometimes be useful and more easily obtained. Second order properties, namely, auto- and cross-correlation functions and spectra, are used for Gaussian processes. Higher order moments, cumulants and spectra are used for non-Gaussian processes. This area of study is referred to as higher order statistics (HOS), e.g., [17]. The linking of non-Gaussianity to nonlinearity was emphasised by Rao [18]. The use of either moments or cumulants is one of choice. The key characteristic of cumulants is that they are blind to Gaussianity whilst moments are not. Accordingly, if cumulant-based methods are applied to signals composed of a non-Gaussian process and additive Gaussian noise, then the cumulants are (asymptotically) unaffected by the noise.

Furthermore, cumulants and higher order spectra contain amplitude and phase information about the process in contrast to auto-correlation and power spectra. This additional information contained by HOS is vital to system identification (e.g., identification of non-minimum phase systems), signal reconstruction (of non-minimum phase signals) and inverse problems.

The first of the higher order spectra is the third order spectrum, the bispectrum. However, the bispectrum only yields information in cases where the random process has a skewed distribution. In a significant number of physical problems, systems are symmetrical and yield unskewed output signals; in these circumstances the bispectrum is an uninformative measure; see [19] which shows the concepts associated with the bispectrum carried over to the fourth order spectrum, referred to as the trispectrum, and discusses how the trispectrum can be used to analyse symmetric nonlinearities.

Higher Order Statistics are relevant in the exploitation of non-Gaussianity even in the context of linear systems in relation to system identification and in inverse problems. In surveillance, extraction of the 'cause' of a measured signal is often necessary. This area of signal processing is called inversion and in the linear case may be expressed as

$$\mathbf{y} = A\mathbf{x} + \mathbf{n}$$

with the aim of recovering **x** from measurement **y** in the presence of noise n. The operator A may or may not be known. The problem may be expressed in the time or frequency domain. This involves the inversion (in some sense) of the operator A such that the recovered signal (say $\hat{x}$) is a good approximation to **x**. Generally, inverse problems are ill-conditioned, in that small perturbations in y or A lead to large changes in the recovered estimate of **x.** The aim therefore is to find methods to create inverse operators that are not unduly sensitive to these perturbations. The procedure for achieving this is 'regularisation'.

This is a widely studied problem resulting in various approaches to the construction of the pseudo-inverse to matrix A, e.g., using Tikhonov regularisation, singular value decomposition, generalised cross validation, etc; see [20]. However, if the operator A is unknown, this is the *blind* inversion problem and now higher order statistics again come into play, in which the success of the procedure is dependent on the non-Gaussianity of the input to be recovered. The use of cumulant based objective functions in the design of linear inverse filters for impulsive type excitation is described in [21].

3.1   The Future; *Speculation II*

Nonlinearity can occur in unlimited forms. However, many real systems are an assemblage of essentially linear systems interconnected through *local* nonlinear connections.

Is there an effective and practical approach to this problem, and what role does HOS have to offer/play here?

4.   NON-DETERMINISM

By a 'deterministic' structure we mean one that is described by a mathematical model that (subject to the fact it is an approximate description in some sense) we treat as exact. For *forward prediction* of the response, we solve the equations for the particular inputs and boundary conditions; for *determination of unknown systems* we use input and output data to estimate *parameters* of the model; for the *determination of the input* we use the measured output and invert the system in some way. This section discusses the practical issue of lack of knowledge of the system and the measured signals, i.e., non-determinism. We distinguish between two types of system non-determinism.

Non-determinism from Complexity

Distributed systems are infinite dimensional and are often approximated by a high order finite dimensional model. Detailed consideration of such systems becomes impractical and approximate methods have been introduced and are currently of great interest. These include Statistical Energy Analysis (SEA), fuzzy structure theory, and approaches combining modal analysis and SEA.

Non-determinism from Variety

When there are many possible configurations of a system, writing down a set of equations *a priori* may not be definitive. Indeed, a variety of models may be candidates.   Under such circumstances, system identification and parameter estimation is resorted to, i.e., experimental 'verification' of the model. *But, which model?* The problem of fitting parametric models to data is well worn, including aspects of model order determination, but more formalised approaches to model selection in addition to order determination and parameter estimation are needed. Recently, Bayesian methods have been demonstrated to offer great insight into signal processing/time series problems and new computational methods for numerical integration and optimisation have made the procedures increasingly relevant.

The remainder of this contribution is aimed at suggesting that these concepts arising from dynamics and from probabilistic/time series methods could come together to provide a powerful and effective approach to dealing with real dynamical systems.

4.1 Complexity

It is often the case that determination of the individual modes is unrealistic owing to the 'high modal density' arising with increasing frequency. This has led to Statistical Energy Analysis, in which a system is modelled as a collection of coupled subsystems characterised by their vibrational energy; each subsystem must contain a number of resonant modes in the analysis bandwidth. The Proceedings of a recent Symposium on SEA gives a current overview of theory, experiment and applications [22]. Langley and Bremner [23] point out that this implies that the wavelength of subsystem deformation must be less than or of the same order of the dimensions of the subsystem. This is not necessarily always the case, e.g., bending motion may satisfy this, but in-plane motion may not. Accordingly, they propose an approach that combines conventional modal analysis for modelling long wavelengths and SEA for modelling short wavelengths. This approach complements the fuzzy-structure approach in which a master-structure has attached to it a set of uncertain or fuzzy attachments. Use of this formulation allows representation of the dynamics in matrix partitioned form for global response and (coupled) local response. The global analysis essentially follows the 'usual' procedures and are linked in with the local responses (described using SEA methods) with due allowance for the coupling existing between the two types of response. An important general result of this is that if the local modes have a high degree of modal overlap then the main effect of the local mode dynamics is to add damping and an effective mass to the global co-ordinates, in line with fuzzy-structure theory.

Fuzzy structure theory was introduced by Soize in 1986 and described in [24]. The aim of this theory is to predict the medium-frequency local response of a 'master' structure coupled to a large number of subsystems or secondary structures. Owing to their complexity these subsystems are called fuzzy substructures. [*Note:* [24] fuzzy structure theory does not correspond to a classical dynamics problem with random uncertainties nor does 'fuzzy' have anything to do with the mathematics of fuzzy sets and logic. Fuzzy structure theory introduces a random boundary impedance operator to model the effects of the fuzzy substructures on the master structure.]

4.2 Variety

We now describe further non-determinism by including within the mathematical framework the possibility of different models, each with different parameters of unknown order, being candidates to describe a data set. This formalism is set within a time series and probabilistic context and recent theoretical developments, matched by major steps in algorithm development have led to some impressive practical applications - reported mainly in the signal processing literature. This is the Bayesian approach led by the Cambridge University Signal Processing Group over a number of years. The book by O'Ruanidh and Fitzgerald [25] has arisen from this work.

Bayesian Analysis [25]

Two basic problems exist in data analysis. First, we need to choose (from a set of possible models) that which is supported by the data - called model selection. Second, we wish to determine the values of the model parameters - called parameter estimation. The essence of the method may be briefly summarised as follows.

We start with the likelihood function. For an observed data set, $\mathbf{d}$, $p(\mathbf{d}|q, M_k)$ denotes the probability of realising data given $\mathbf{d}$ given parameters $\square$ (of model and noise) and model M_k ($k = 1,...,M$ denotes candidate models). The maximum likelihood approach estimates parameters $q\square\square$by locating the supremum of p, i.e., by maximising the probability of observing the data by selecting q, but (i) this does not use prior information that may be available; (ii) the likelihood on its own does not limit the *number* of parameters that may 'fit' the data, i.e., it does not control complexity of the model. Ways of controlling complexity are well known, e.g., Akaike's Information Criterion (AIC) and the Minimum Description Length (MDL), amongst others [26].

The Bayes' approach builds in *prior* information expressed as

$$p(\theta|\mathbf{d}, M_k) = \frac{p(\mathbf{d}\mid\theta, M_k)\,p(\theta\mid M_k)}{p(\mathbf{d}\mid M_k)}$$

or in words
$$\text{posterior} = \frac{\text{likelihood}\infty\text{prior}}{\text{evidence}}$$

i.e., $p(\theta|M_k)$ is the prior probability summarising knowledge of the parameters prior to observing the data. The denominator $p(\mathbf{d}|M_k)$ is called the *evidence* which is of interest in model selection. The quantity $p(\theta|\mathbf{d},M_k)$ is the posterior density summarising the state of knowledge about the parameters after the data is observed, i.e., Bayes provides an approach by which prior information is updated on the basis of new data. The posterior density may be used for parameter estimation, i.e., Maximum a Posteriori (MAP) estimation.

The Bayesian approach offers much for model selection, parameter estimation *and* a rational basis for compromise between accuracy of fit to the data and parametric complexity (Ockham's razor), but it introduces difficulties too. What form should priors take? How can we compute the conditional probabilities and find optima? The book [25] and associated and more recent results from the Cambridge Group are interesting demonstrations of the power of this approach.

4.3 The Future; *Speculation III*

Can the Bayesian approach and the existing methods of SEA, fuzzy structures, uncertain parameters, etc., be brought together to provide a unified approach to model determination and parameter estimation for identification of classes of structural systems?

5. CONCLUDING REMARK

This essay has noted three speculations and they should be ranked in importance. They have been presented in the sequence: non-stationarity; nonlinearity and non-Gaussianity; non-determinism. Whilst not diminishing the intrinsic importance of each, it is the author's *speculation* that their importance to the future development of structural dynamics is in the reverse order - what do you think?

REFERENCES

1. J.N. Sundermeyer and R.L. Weaver 1995 Journal of Sound and Vibration 183(5), 857-871. On crack identification and characterisation in a beam by nonlinear vibration analysis.

2. P. Cawley and R.D. Adams 1979 Journal of Strain Analysis (I.Mech.E.) 14(2), 49. The location of defects in structures from measurements of natural frequencies.

3. B. Boashash (ed.) 1992 Time-frequency Signal Analysis - Methods and Applications. Longman-Cheshire.

4. J.K. Hammond and P.R. White 1996 Journal of Sound and Vibration 190, 419-447. The analysis of nonstationary signals using time-frequency methods.

5. Mechanical Systems and Signal Processing, 11(4), Special Issue: Time-frequency Methods. 1997.

6. S.A. Dalianis, J.K. Hammond, P.R. White and G.E. Cambourakis 1998 Journal of Vibration and Control, 4, 75-91. Simulation and identification of non-stationary systems using linear time-frequency methods.

7. NATO-Advanced Study Institute on 'Modal Analysis and Testing', May 3-15 1998, Sesimbra, Portugal, 1-47. Ed. J.M.M. e Silva and N.M.M. Maia.

8. J.C. Hardin and A.G. Miamee 1990 Journal of Sound and Vibration 142(2), 191-202. Correlation autoregressive processes with application to helicopter noise.

9. P.J. Sherman and L.B. White 1995 J.Acoust.Soc. Am. 98(6). Improved periodic spectral analysis with application to diesel vibration data.

10. A.V. Dandawaté and G.B. Giannakis 199 IEEE Trans. on Signal Processing. Reconstruction of almost periodic signals using cyclostationarity.

11. C.M. Spooner and W.A. Gardner 1992 An Overview of the Theory of Higher-Order Cyclostationarity in Nonstationary Stochastic Processes and their Application (ed. A.G. Miamee) World Scientific.

12. D.H. Kim 1998 Ph.D. Thesis, University of Southampton. Identification of Nonstationary Parametric Models using Higher Order Statistics.

13. S. Godshill 1993 Ph.D. Thesis, University of Cambridge. The Restoration of Degraded Audio Signals.

14. R. Isermann 1995 Proc. of the 2nd International Symposium on Acoustical and Vibratory Surveillance Methods and Diagnostic Techniques, CETIM, Senlis, France. Fault detection and diagnosis - methods and applications.

15. M. Basseville, A. Benveniste, D. Gach-Devauchelle, M. Goursat, D. Bonnecase, P. Dorey, M. Previsto and M. Olagnon. 1992 Proceedings of the International Symposium on Recent Advances in Surveillance and using Acoustical and Vibratory Methods, Senlis, France. Statistical methods for in situ damage monitoring and diagnostics of vibrating structures.

16. A.F. Vakakis 1997 Mechanical Systems and Signal Processing 11(1), 3-22. Nonlinear normal modes and their applications in vibration theory: an overview.

17.	C.L. Nikias and J.M. Mendel 1993 IEEE Signal Processing Magazine. Signal processing with higher-order spectra.

18.	T.S. Rao and M.M. Gabr 1984 Lecture Notes in Statistics 24, Springer Verlag: New York. An introduction to bispectral analysis and bilinear time series models.

19.	W.B. Collis, P.R. White and J.K. Hammond 1998 Mechanical Systems and Signal Processing 12(3). HIgher-order spectra: the bispectrum and trispectrum.

20.	S.H. Yoon 1998 Ph.D. Thesis, Institute of Sound and Vibration Research, University of Southampton. Reconstruction of acoustic source strength distributions and their interactions by inverse techniques.

21.	J.Y. Lee and A.K. Nandi 1998 Mechanical Systems and Signal Processing 12(2), 357-371. Deconvolution of impacting signals using higher-order statistics.

22.	IUTAM Symposium on Statistical Energy Analysis. 1999 Ed. F.J. Fahy and W.G. Price. Kluwer Academic Publishing.

23.	R.S. Langley and P. Bremner 1999 A hybrid method for the vibration analysis of complex structural-acoustic systems. In proof; Journal of the Acoustical Society of America.

24.	R. Ohayon and C. Soize 1998 Structural Acoustics and Vibration,. Mechanica lModels, Variational Formulation and Discretization. Academic Press Ltd.

25.	J.J.K. O'Ruanidh and W.J. Fitzgerald 1996 Numerical Bayesian Methods Applied to Digital Signal Processing. Springer-Verlag.

26.	F. Gustafsson and H. Hjalmarsson 1995 Automatica, 31(10), 1377-1392. Twenty-one ML estimators for model selection.

Multifunctional Structures of the Next Millennium

Daniel J. Inman
Center for Intelligent Materials and Structures
Department of Mechanical Engineering, MC 0261
Virginia Polytechnic Institute and State University
Blacksburg, VA 24061 USA

This paper is an essay on my view of one possible way forward in structural dynamics. Smart materials or material transducers are materials that contain sensors, actuators, and control systems that allow structures to respond or adaptively change as the result of external conditions. Such materials form transducers that are able to convert electrical or magnetic energy into mechanical motion or force and vice versa. Structures made of these materials present new opportunities for the structural dynamics community. Piezoelectric patch and stack actuators, shape memory alloys and electrorheological fluids, when incorporated into structures, can provide a myriad of benefits and solutions to vibration problems such as vibration suppression, aeroelastic control, shape control, and diagnostics. It is clear that new technologies and materials form a key to creating improved structures. It is also clear that our user community is extremely conservative and resists change. A discussion of smart structures is presented followed by comments on dealing with change, with the hope of provoking action and response from the structural dynamics community.

INTRODUCTION

In the 1990's many of those working in structural dynamics started to seek solutions to various vibration and design problems using unusual materials (Rogers, 1993, and Tzou, 1998). These material systems and structures, known as smart structures, consist of a variety of components that have the ability to change one or more of their physical properties as the result of an externally applied field (usually magnetic or electric). Conversely these systems can produce currents as the result of mechanical forces or strains. Such devices have been used for years as accelerometers and shakers, but when fully integrated throughout a structure in some composite fashion, form many new possibilities for structural dynamics. For example, currently several air frame manufacturers are looking into the possibility of removing flaps from an aircraft wing in favor of an "adaptive wing" which is capable of changing its shape. This "new" technology has forced the integration of the field of structural dynamics with the fields of control, materials and circuit theory.

In a similar way, MEMS (Micro Electro Mechanical Systems) have produced the ability to integrate sensing and actuation functions into structures, forming new possibilities. An example of a MEMS device is an accelerometer made at the molecular level. Both smart structures and MEMS technologies require knowledge of mechanical and electrical components, highlighting the importance of the field of Mechatronics, which has emerged in the last decade. Mechatronics is the discipline of integrating mechanical and electrical components at an initial design stage rather then treating them separately. This trend to integrate the functions of sensing, actuation and electronics into the structure itself has raised interest in defining Multifunctional Structures with the hope of advancing the performance and utility of structures. Multifunctional Structures may be loosely defined as the integration of devices into structures and structural design. A simple example of a multifunctional structure is integrating the antenna functions of an aircraft into the skin of the airframe. All of these technology areas have important consequences for structural dynamic engineers.

The hope of modern structures is application driven. The goal is to increase the performance of structures by reducing their weight, increasing strength and utility. This is not profound. An examination of the history of engineering shows that most great breakthroughs in design, modeling and analysis came as the result of a focused, application driven projects: flight, reaching the moon, spanning rivers, etc. The exception may be computer technology which has developed under its own steam, yet provided advances in structural dynamics analysis and measurement by making new tools available.

The concept of working in structural dynamics is an evolving one. Between the first and second World Wars, it simply met that you knew how to compute dynamic safety factors from static calculations. At one time, the forefront of structural dynamics research implied the ability to write large FORTRAN codes. Important sub sets of structural dynamics required an understanding of fluid dynamics which eventually led to the field of aeroelasticity. An engineer working in structural dynamics for the space program soon found a need to couple an understanding of structural dynamics with thermodynamics. As applications changed, performance demands increase and the abilities required of structural dynamics engineering have evolved into numerous related areas. The goal of this article is to suggest that the next generation of structural dynamics engineers will need expertise in smart materials and generally be adept at integrating electric, thermal, fluid and magnetic properties into their skills.

Several examples of the use of smart materials integrated into structures to form smart material systems or smart structures are presented to provide the flavor of some possible uses of this technology in the structural dynamics community. Many other examples may be found in the literature. Hopefully these examples will motivate the integration of these materials into standard practice in the structural dynamics community.

SMART DAMPING

This is the notion that many of these new materials can provide passive damping by changing mechanical motion into an electrical field, then using a circuit, or shunt, to dissipate electric energy as heat. The smart material that has received the most attention is the piezoceramic shunt that has come to the attention of vibration experts in the early 90s. Specifically the piezoceramic effect may be used along with a passive circuit to produce a damping mechanism with unique capabilities comparable to the damping properties of viscoelastic based constrained layer damping treatments commonly in use.

Passive piezoceramic materials have illustrated the ability to modify the resonant response of a structure. Forward (1981) demonstrated the use of a resistive shunt placed across a piezoceramic to provide damping and verified this result experimentally. Since then, a number of researchers have modified and developed this simple idea into a variety of applications (Lesieutre, 1998).

Rather then applying a viscoelastic layer to a host structure, a piezoelectric device (usually a piezoceramic) is layered into or on a host structure and shunted to a resister or resister and inductor. As the host vibrates, the piezoelectric effect changes the induced strain into a voltage that is then dissipated as heat through the resistive shunt circuit. The result is a system that produces a loss factor versus frequency curve much like that of a viscoelastic material. Only loss factors of about 0.45 for longitudinal vibration and just 0.08 in the transverse direction are obtained when practical values of resistance are used. However, the peak value of the loss factor can be easily change from one value of frequency to another providing increased design flexibility.

If an inductor is added to the shunt, it behaves like a vibration absorber and can produce larger loss factors at resonance than a comparably sized constrained layer damping treatment, but over a narrow band. In addition, the shunted piezoceramic system is not as temperature dependent as the viscoelastic counter part, and is much stiffer then a viscoelastic material (VEM). Piezoelectric materials also do not suffer from creep, common to viscoelastic materials. While shunted piezoceramics do not directly compete with VEM for adding damping, they do offer more design flexibility, increased stiffness and temperature stability. This approach may also provide some advantages in terms of temperature response, mass location and design parameters.

One of the technical difficulties with designing a vibration suppression system with a shunt is that the size of the inductance required increase with the lever of damping added. This requires the use of synthetic inductors. The use of additional capacitors can reduce the required inductance. This sort of problem points directly to the importance of the interactions between structural dynamics and circuit theory.

A variety of other smart materials, such as shape memory alloys or magnetostrictives, may provide improved damping solutions in passive settings when combined with a shunt of some sort. This is largely unexplored, but several researchers are working in this direction. Already those who work on building dampers and vehicle shock absorbers have exploited use of more exotic materials and this has resulted in improved damping.

ADAPTIVE WING PROGRAMS

One of the more exciting prospects for the use of smart materials is in structural dynamics of airfoils. Here the goal is to remove traditional control surfaces in favor of wings that change shape and other properties to provide improved performance as well as to provide maneuver control. Some of the motivation for this is the desire to produce an all-electric aircraft. Adaptive wings use various active materials (shape memory alloys, piezoceramics, elecetrostrictives, etc.) to bend, twist and change the surface of a wing in flight in order to obtain more favorable lift and drag, to replace flaps, prevent flutter, etc. Most of the world's airframe manufacturers have some sort of adaptive wing program under way and some have even moved into the flight test stages (Kudva, et. al 1997). However, the most probable use in practice will come from unmanned aircraft, the development of which is supported by the US Air Force.

Current wing and control surface design may eventually give way to wings that change shape to replace hydraulically activated flaps or to provide multiple wing profiles. Airfoil design is always a compromise and the concept of adaptive wings might allow less compromise. For example, through proper heat treatment and processing, a shape memory alloy 'memorizes' a configuration. After large plastic deformation, greater than 1% strain, the alloy will return to its memory shape upon heating above a characteristic transition temperature. This memory effect is a result of a phase transformation from a low modulus martensitic phase, at low temperatures, to a high modulus austentic phase, at high temperatures. The change in modulus can be as much as a factor of 2.4 (e.g. from 35 GPa to 83 GPa). These materials when properly integrated into the wing design may allow for new flight regimes and drastic reductions in limitations on fighter and commercial aircraft.

One of the major factors in airfoil design is the wing chord line: the line between leading and trailing edges. Flaps were invented to provide a way to change the chord line of an airfoil during flight. In a sense, flaps are really a lumped parameter approximation to changing the chord line, based on the technology available in the 20s. With the new technology of smart materials and structures, it may be possible to perform this same task in a distributed fashion and greatly enhance the performance and capabilities of flight control systems by having an advanced and distributed mechanism for changing camber.

Tab Assisted Control (TAC) is a technique for avoiding lock up of control surfaces on submarines. Here again, the concept of using an active material may be ideal for performing the same task without adding another mechanical surface to the structure. The use of shape memory alloy (SMA) in particular may provide a variety of solutions to engineering problems which require actuators which can deliver high force, high stroke, high energy densities, and, most significant for this research, high force-to-volume (or weight) ratios. As noted above, recent studies have shown the suitability of SMA actuators to aerodynamic problems that require large torque from actuators that must be housed within an airfoil's cross-section. These studies used SMA torque tubes as actuators, and were able to achieve high torque, 500--2000 in lbf, due to the high recovery stress available from SMAs. Similarly, large stroke actuators are possible due to the high recoverable strains, as much as 4 to 8%. The extension of that research to tab-assisted control (TAC) surfaces is natural.

There are several possible designs that a distributed TAC may use. Two torsional SMA actuator designs: *torque tubes* and *torque rods* may be suitable for this application. If these actuators are twisted resulting in 'plastic' deformation, then upon heating the deformation is recovered due to the memory effect of the material. Torque tubes have been used in a variety of aerodynamic applications to effect the twist of lifting and control surfaces. Torque rods are a recent development and are attractive because they can be designed to exhibit two-way memory. Normally, shape memory actuators return to their original state at an uncontrolled relaxation time making it difficult to provide actuation in two directions. Torque tubes are constructed from thin-walled alloy stock; this ensures plastic deformation of all of the material. These actuators exhibit one-way memory and are used in an agonist-antagonist configuration; in this configuration, as the heated actuator cools, the cold actuator supplies a necessary bias stiffness restoring the active actuator to its initial position. Torque rods, however, can be designed for two-way memory. If properly designed, the central section remains elastic and acts as a bias stiffness for the actuator. The size of this elastic region, and hence the bias stiffness, is dependent upon the applied load. As a torque rod actuator begins recovery upon heating it cannot fully recover because of the internal bias. Because the center section has stored some potential energy, the stored energy will be released upon cooling and the actuator will rotate back to its initial position.

Two examples of applications using smart structures have been mentioned here: submarines and aircraft. However, many other application areas are possible. The point is to illustrate that current technology in structural dynamics has traditionally focused only on structural properties. Going forward, it appears that the integration of control and sensing functions into structures and hence into the discipline of structural dynamics is a viable possibility.

SELF-DIAGNOSTIC STRUCTURES

Smart materials are keenly suited for integrating into health monitoring and fault detection systems based on vibration signals. They are small and unobtrusive and come in a variety of sizes and abilities, allowing them to be placed almost anywhere. The most interesting aspect of smart materials, however, is that they can also serve as actuators to provide driving signals as well as sensing, for systems that do not contain natural excitation forces, or for diagnostic algorithms that require a known, well controlled excitation. Thus by combining smart structures technology with diagnostics, one can imagine structural systems in the future that have self contained and self-diagnostic components, minimizing maintenance and inspection cycles. Three examples are summarized here, one which is based solely on signal processing and involves no model of the structure ending with one which requires a substantial model of the structure. A summary of recent health monitoring and damage detection methods can be found in Doebling, et. al. (1998).

In general the diagnostic or damage detection methods seek to answer the questions:

Is their damage?
Where is it?
How big is it?
What is its orientation?
What should be done about it?

The first of these questions can be answered without having much of a model of the structure or part of interest. The need to know more about the nature of the defect, however, requires an increasing knowledge of the model of the structure that is not always possible. Thus methods that strive to answer a number of these questions quickly become intertwined with modeling issues. Modeling in structural dynamics is often not compatible with actuation, nonlinearity, damping, interaction with acoustic or fluid phenomena. Our traditional modal model is not good beyond a certain frequency for most structures. What we need is something (test and analysis) in the mid-range such as Statistical Energy Analysis, wave propagation models, impedance modes, etc.

First consider the idea often used in damage detection that a defect will produce a small change in stiffness and/or mass, and hence frequency. Early vibration based damage detection methods often looked at frequency response function (FRF) data for a small change in frequency. However, small changes in frequency are generally difficult to measure using FRF data. One method that avoids this difficulty is to continuously compare the healthy time signal of the structure under study to the current time signal of the structure. If a small difference in frequency exists between these two signals, then when they are combined they will produce the beat phenomena that serves to magnify small differences in frequencies. This effect has been used successfully to detect the presence of small amounts of damage (less than 0.1% change in mass) in plates and in helicopter blades by using internal piezoceramic materials to both excite and sense the various time histories.

This approach illustrates a time-domain procedure that is totally self-contained in a moving structure, capable of self diagnostics using embedded piezoceramic sensing and actuation at relatively low power costs. The procedure depends only upon subtracting two signals, does not require any modeling of the structure and hence is simple enough for onboard use (Cattarius and Inman, 1997).

A second example of a diagnostic procedure is impedance based and may also be used in a self-diagnostic configuration by incorporating local, embedded and/or surface-mounted piezoceramic sensors and actuators. This approach is a high frequency impedance-based method that looks for a shift in electrical impedance measurements as an indicator of damage. In the impedance-based qualitative health monitoring technique, the real-time implementation relies on a simple scalar damage index that can be easily interpreted. Using this damage index in conjunction with a damage threshold value, the approach can warn the operator in a green/red light form, whether or not the threshold value has been reached. This approach has been used on numerous bolted joint structures and in large concrete structures. Again, this system uses the integration of sensors and actuators into the structure forming a smart structure capable of self-monitoring (Park, et. al., 1998).

Banks, et.al. (1996) presented a method which takes advantage of a detailed and well identified model of the structure. The results are mentioned here to provide experimental proof that a piezoceramic based diagnostic system cannot only determine the existence of damage, as can the previous methods, but is also able to determine the size and location of the damage. The method is based on using a partial differential equation model of a structure that is partially layered with piezoceramic patches which again forms a "self-diagnostic" structure. The algorithm uses a spline-based approximation of the equations of vibration and successfully identifies the existence, size and location of holes in a beam. The method works by estimating functions of the longitudinal direction of the beam corresponding to the damping parameters (both Kelvin-Voigt and air damping), the modulus and the density. Each of these is allowed to be discontinuous in order to allow holes to be included in the solution set of functions.

It is important to note that while this method works very well, it has the disadvantage of requiring a very detailed model of the structure including a model of the damping mechanisms. The parameters are allowed to have some measure of uncertainty as they are estimated in the inverse procedure used to identify the damage. However, the form of the governing differential equation must be known in substantial detail. With this noted, the method very effectively identifies the damage, where it is and possibly its orientation. Furthermore, the results are consistent across several different types and sets of measurements.

Another important feature of the method is that it does not use modal data, but rather is based on time domain measurements. The experiments were repeated with traditional excitation means (an instrumented hammer) and response measurement (accelerometer, and position probe) as well as with the internal self-sensing actuation scheme offered by the integrated piezoceramic patch. Besides verifying a damage detection algorithm, the test shows conclusively that it is possible to use piezoceramic materials to form self-monitoring devices.

Almost any diagnostic or monitoring system can be integrated into a structure to make it a self-diagnostic system. Here, three examples have been mentioned which make the point that the integration of sensors and actuators directly into a structure provides interesting new problems for the structural dynamics community. In addition, the three examples illustrate that increased knowledge of the structure allows more information regarding the health of the structure to be determined.

SELF-REPAIRING STRUCTURES

The area of smart materials also holds some promise for carrying diagnostics one step farther by adding an ability to perform self-repair. Because smart materials are able to change their stiffness and damping properties according to an applied voltage, it is thought that they may be able to compensate for some types of limited damage, allowing the continued safe use of a damaged structure. One possibility under consideration is the use of woven fabric-like composites with shape memory alloy fibers. When damaged, the fibers can be activated to increase strength.

Another example of a self-repairing structure made possible by smart materials is the smart bolt. A smart connection, or bolt, consists of structural members joined together by bolt and nut combinations fitted with piezoceramic elements as washers. These combinations can be used to monitor bolt tension and connection damage. Suppose that torque is used as a measure of the health of the system. Then weak impedance signals can be sent out from the piezoceramic patches and sensed by the opposing washer to determine the level of torque on the bolt. In some configurations, temporary adjustments of the bolt tension can be achieved actively (and thus remotely). Experimental results to date clearly show that changes in the mechanical characteristics of bolted connections can be identified through localized impedance measurements of connection components. Initial experiments indicate that changes of the order of an inch pound can be determined in the Coulomb region. In particular it may be possible to use a washer system made from an active material such as piezoceramics. Then as the bolt becomes out of torque, changes in the impedance of the system will result. It may then be possible to tighten the bolt by applying a voltage to cause the piezoceramic to expand. This idea incorporates the aspects of self-diagnostics with active control to improve the behavior of a bolted structure under dynamic load. Details of such systems are being worked on by a small company in the US and by Prof. Gaul of the University of Stuttgart who has invented the concept of a semi-active bolt.

BIOLOGICALLY INFLUENCED STRUCTURES

One approach to formulating smart structures has been to mimic nature. In fact many of the first attempts at flight sought to mimic nature. Of course artificial flapping wings were not to succeed, but birds did provide motivation and eventually pointed the way to successful ideas. Perhaps more appropriately then, biological systems have been used to motivate design of new smart structures and material systems as well as uses of these materials. This has led to a number of different mechanisms. Some have used shape memory alloys in an attempt to mimic muscle. Others have attempted to use piezoceramics to form miniature bugs that walk about (smart ants) with miniature video cameras in hostile environments.

The need for large solar panels and antennas in space coupled with the need to have very small packages to send into orbit has produced a focus on inflatable structures and a class of structures known as tensegrity structures. Tensegrity, is an example of structural advantage gained by biological motivation. Tensegrity refers to a repeated structure that stabilizes itself because of the way in which tension and compression forces (stresses) are distributed and balanced within the structure. It is possible for a system of wires and rods to be configured in a plane such that they will deploy to a three dimensional configuration that is rigid (stable).

Inflatable structures are also under consideration for use in expanding components in modern satellites. The idea is to use a collapsed Mylar like material stowed for launch, then expanded in orbit to provide the structural framework for satellites and antennas. Both types of configurations are motivated by nature's unfolding of planar systems into three-dimensional formations. Both types of structures will also benefit from the inclusion of smart materials to serve as actuation or muscle imitating functions and as sensors. Such structures offer a challenge to structural dynamics in terms of providing predictive models that may be used in design and analysis.

The examination of biological functions may motivate others in structural dynamics to invent new structural forms that solve existing problems or create new products. In thinking about combining structures with sensing and actuation it is important to match requirements with abilities. Actuators are usually designed in terms of force, stroke and time constant. Sensors are designed in terms of band width and sensitivity. Locations of sensors and actuators inside a structure are also extremely important.

HISTORICAL PERSPECTIVE

In looking to the new fields of structural dynamics it is important to review the past. Defining the elements of success and failure is best tried in a historical setting. The history of aerospace structures provides a nice setting for this. Much of the success of the aerospace business is summarized in an excellent collection of articles edited by Flomenhoft (1997). Many of the successes in structural dynamics in the early days were driven by failures in practice. In particular, flutter of aircraft drove the discipline of aeroelasticity and many innovations related to the structural dynamics areas. This greatly improved our understanding of structures. However, as applications became more complex, knowledge became more specialized and the discipline of structural dynamics became essentially narrow-minded. To advance, the discipline has to return to an interdisciplinary state.

Another important lesson from history is the role of computation, simulation and experiments. Today we have intense computational abilities. Often one can simulate hundreds of designs before actually building anything. On the other hand it is very important to perform experiments to verify theories. It may not be enough to develop good models, techniques and designs that work. In addition they must also be user friendly and this often means simple. Not all problems have simple solutions however, but we can as a group develop better techniques for communicating with practicing engineering professionals.

History points out that ideas are often discarded prematurely. Think, for example, of steam-powered automobiles. Critiques of the time used the problems of steam power to cast stones at the concept of an automobile. Yet a few years latter the solution appeared in the form of gas-powered engines. Thus some previous ideas should be considered again in view of current technological advances. This is particularly true with regard to computational methods in structural dynamics.

THE WAY FORWARD

The optimal way forward is not clear, but it must begin with educational aspects, short courses, funding agencies and corporate tolerance of high-risk research. Our society has evolved into a stage of immediate gratification. Corporations now want to support research only if it will turn a profit by the next stockholders report. This is like a farmer with a four month growing season, plowing up his plantings after the first quarter because he hasn't sold any vegetables.

How can we change the current industrial "research" mode? We need to follow the model of the drug and computer industries where research directly drives their products and hence enjoys much greater support. A tough mind set exists in industries using structural dynamics when it comes to new materials and designs: if we haven't used it yet, we don't want to! If it's a new idea, competition with existing technology is often a problem of importance to consider, meaning that new ideas are often suppressed because they do not offer an immediate advantage over existing systems.

What we need for structural dynamics to grow is to create challenge problems, benchmarks and to pursue crazy ideas. Those of us in university settings have to be creative and exotic in our approach because no other institutions (corporate research or government labs) are willing or able to take the risk of failure. The need for funding is obvious and critical, yet the university community needs to carry the ball even without support from external sources.

As engineers we often are not the best at communicating or reading. We need to refocus our concept of information dissemination. We have new tools (i.e., the WEB) and we need to learn to take advantage of this. We tend to reinvent the wheel in a cyclic fashion. How do we avoid this? How do we make it more efficient? Realize that some things need to be revisited on purpose (eigensolvers and parallel computing). However, we do need to guard against blindly ignoring previous results. Net searches should help this situation as do review articles. We have much more information at our disposal, but we also have efficient ways of searching for this information.

SUMMARY

The message or goal of this contribution is to convince the reader that the next generation of structural dynamicists will have to have skills that allow them to comfortably work with electronics, controls, and materials as well as traditional topics in structural dynamics. We need to replace the idea that stiffness and damping properties are all that is important with the notion that structures are integrated components of a total system.

While several successful examples of the use of structural systems integrated with smart materials have been presented, it is important to note that these devices all have limitations and one must examine the force, deflection and time constant needed to form a successful and useful system. The inclusion of smart structures into the discipline of structural dynamics offers new opportunities and responsibilities. It requires the integration of a more multidisciplinary role for the structural dynamics community.

Smart materials and structures should provide motivation for structural dynamics engineers to expand their expertise to include electromechanical devices. Many improvements are possible with these new materials as well as many new opportunities, and it is reasonable to view them as one possible advancement in structural dynamics for the future.

ACKNOWLEDGEMENT

The author thanks the following agencies of the United States. The Defense Advanced Research Projects Agency, Army Research Office, the Air Force Office of Scientific Search, the Office of Naval Research, the National Science Foundation and the National Aeronautics and Space Administration for sponsoring various aspects of our efforts in smart structures.

REFERENCES

Banks, H. T., Inman, D. J., Leo, D. J. and Wang, Y., 1996, "An Experimentally Validated Damage Detection Theory in Smart Structures', *Journal of Sound and Vibration*, Vol. 191, No. 5, pp. 859-880.

Cattarius, J. and Inman, D. J., 1997, "Time Domain Analysis for Damage Detection in Smart Structures*", Mechanical Systems and Signal Processing*, Vol. 11, No. 3, pp. 409-423.

Doebling, S. W., Farrar, C. R. and Prime, M. B., 1998, "A Summary Review of Vibration-Based Damage Identification Methods", *The Shock and Vibration Digest*, Vol. 30, No 2., pp. 91-105.

Flomenhoft, Hubert, I., Editor, 1997, *The Revolution in Structural Dynamics*, Dynaflo Press, Palm Beach Garden Florida.

Forward, R.L., 1979, "Electronic Damping of Vibrations in Optical Structures," *Journal of Applied Optics*, 18(5), pp. 690-697.

Ingber, Donald, E., 1997, "Tensegrity: the Architectural Basis of Cellular Mechanotransduction", *Annual Review of Physiology*, Vol. 59, pp. 575-599.

Park, G, Kabeya, K, Cundey, H. H., and Inman, D. J., 1998, Dynamic Impedance Based Structural Health Monitoring for Temperature Varying Applications, *International Journal of the Japanese Society of Mechanical Engineers*, to appear.

Kudva, J., Appa, K., Jardine, P., and Martin, C., 1997, "An overview of Recent Progress on the DARPA/USAF Wright Laboratory Smart Materials and Structures Development- Smart Wing," In *SPIE-Smart Structures and Materials*, vol. 3044, pp. 24-32.

Lesieutre, G.A., 1998, "Vibration Damping and Control using Shunted Piezoelectric Materials", *The Shock and Vibration Digest*, Vol. 30, No 3, pp. 181-190.

Rogers, C. R., 1993, "Intelligent Material Systems: The Dawn of a New Material Age", *Journal of Intelligent Material Systems and Structures*, Vol. 4, No 1, pp. 4-14.

Tzou, H. S., 1998, "Multifield Transducers, Devices, Mechatronic Systems, and Structronic Systems with Smart Materials", *The Shock and Vibration Digest*, Vol. 30, No. 4, pp. 282-294.

COMPUTATIONAL AND EXPERIMENTAL ANALYSIS OF COUPLED PROBLEMS

Roger Ohayon

Structural Mechanics and Coupled Systems Laboratory
Conservatoire National des Arts et Métiers (CNAM)
2, rue Conté – 75003 Paris, France

1. INTRODUCTION - MOTIVATIONS OF THIS ESSAY

In this essay, it is proposed to investigate the various methods of modeling the vibrations of real systems with arbitrary geometries, with various complicated boundary conditions, which do not relate to "analytical" solutions.

Those complex systems are generally heterogeneous, for example systems with many components, multilayered composites or sandwich structures, structures connected with complex interfaces, structures coupled with fluids (liquids or gas) for structural acoustic problems (internal and external noise reduction), aeroelasticity of structures embedded into an external flow, vibrations of structures containing liquids.

We are therefore dealing with systems which are characterized by the presence of different wavelengths due to those various heterogeneities, within the context of weak or strong interactions between physical - multiphysic analysis - or spatial subsystems.

Some parts (substructures) of the system can be very complicated to model and experimental inputs are then required, the general philosophy being to carry out "less costly experiments", "experiments with higher measurements quality" and one must " never carry out an experiment without a companion predictive model".

2. STRUCTURE - STRUCTURE COUPLING

We consider hereafter, within the linearized elasticity theory, the coupling between heterogeneous structures of arbitrary shapes connected along complex interfaces.

Let us first consider the coupling of two substructures A and B connected along a perfect geometric interaction surface interface S. If the substructure B is dissipative, a classical result in structural dynamics occurs, namely a linear relationship can be established between the surface forces exerted by substructure B on substructure A along interaction interface S and the values of displacement field in B on the interface S, through a frequency-dependent dynamic stiffness operator. (When using the velocity field, the involved operator is the frequency-dependent impedance operator; when using the acceleration field, the involved operator is the frequency-dependent dynamic mass operator.)

Those operators are in essence boundary integral operators which can be constructed using appropriate discretization methods (finite element discretization of the classical displacement variational formulation of the problem or finite element discretization of a frequency-dependent boundary integral formulation of the problem leading to so-called boundary element methods).

In some particular cases, those operators can be frequency-independent. This is the case when the behaviour of substructure B relates to Guyan condensation modeling. This means that substructure B has a non resonant behaviour, therefore static condensation can be applied. A static, frequency-independent linear relationship allows the elimination of interior degrees of freedom as a function of the degrees of freedom (displacement field u) on the interface S.

Of course, the frequency behaviour of the displacement field must be considered and a distinction must be made between the modal domain (low modal density) and the high frequency domain (relating to Statistical Energy Analysis methodologies, through the intermediate medium frequency range which needs appropriate numerical methodologies but where the boundary conditions still play an important role).

2.1 Complex Structures with Complex Connections

The heterogeneities of the systems can introduce many local modes. Therefore, the problem of local – global, weak or strong interaction between those modes occurs. The discussion is connected to the geometry of the substructures (large or small substructures) and to the local mechanical characteristics of those substructures. We will analyze briefly those complex situations.

Of course, in the absence of appropriate experimental informations, one way is to carry out, for instance, a detailed finite element analysis on heavily parallel computers. The finite element software must give reliable results, namely the displacement field and their gradients (leading to the stress field) at any point of the substructures.

To attain this objective, *a posteriori* error estimates methodologies must be included in those finite element codes. Various error estimates have been proposed in the computational mechanics literature, mainly for static problems with local high displacement gradients. Those analysis are *a posteriori* in the sense that they utilize the finite element solution and starting from this approximate solution for a given mesh of the domain, a refined solution is obtained using the local equilibrium equations (see the works of O.C. Zienkiewicz, I. Babuska, T. Oden and several others in the appropriate literature).

Many attempts to extend those methodologies in structural dynamics are the subject of deep investigations but the problem is much more difficult than in linear static problems. In effect, in structural vibrations of complex heterogeneous structures, local modes can occur among the global modes and some substructures behave as if they were in the low density (low frequency modal domain), and others behave as if they were in the higher modal density regime (medium and high frequency domains). So further investigations are needed from the computational mechanics community on those aspects.

Alternative methodologies of computation can be carried out, for instance, dynamic substructuring procedures using appropriate vector basis of the admissible space C of the unknown field - the displacement field u - of the variational formulation of the problem (see Hurty and Craig-Bampton basic methodologies within finite element discretizations of the substructures).

A presentation of the method, at the continuous level, consists in an appropriate decomposition of the admissible space C into direct algebraic sum of two subspaces C_1 and C_2 of C. A Ritz-Galerkin projection is then applied to the variational formulation of the global problem leading to two variational formulations of each substructure by restriction of the admissible space C to the two subspaces C_1 and C_2 of C.

Within the dynamic substructuring context, there exists several possible appropriate decompositions of a given vector space into two (or more) algebraic direct sum of vector subspaces and it constitutes an efficient way for constructing various dynamic substructuring procedures. Hybrid dynamic substructing procedures can also be applied using a standard finite element procedure for one of the two substructures. If, for instance, the interface is defined as a third physical interface medium and not as a geometric surface interface, then such a hybrid dynamic substructuring procedure may be applied using, if necessary, a refined finite element modeling of the interface physical medium.

Those matrix model reduction techniques are useful, in order to avoid the use of a prohibitive number of degrees of freedom, for sensitivity analysis purposes and for correlation between experiments and computations (parametric updating procedures). An important aspect concerns also the reduction of vibrations (using for example smart materials and structures technologies). A huge number of degrees of freedom creates strong difficulties in the active control analysis of such systems, which is why, in many cases, the active/passive control is carried out from an experimental point of view only. In those cases, appropriate reduced models must be constructed in order to have a predictive computational model of the system.

An alternative procedure to reduce the size of the problem can be derived using asymptotic appropriate analysis prior to any numerical techniques. Those procedures are classical in computational fluid mechanics and in mathematical derivation of beam, plate and shell theories starting from the three dimensional partial differential equations of the boundary value problem (see the works of Antman, Ciarlet and several others).

Difficulties are expected in asymptotic analysis if several small parameters are simultaneously involved in the problem with the same order of magnitude.

2.2   Spatial Localization

Due existing heterogeneities, wavelengths in each substructures composing the global structure can have different orders of magnitude. Specificities such as the sizes of each substructure, their relative rigidity and mass can affect the dynamic behaviour of the system. At first glance, a local mode can be defined as a mode

74

which is localized within a local subdomain of the global system. A global mode can be defined as a mode with a wavelength of order of magnitude of a predominant specified characteristic length of the system. Those definitions are gross definitions as the real physical situations are much more tricky to deal with.

One of the most "classical" examples is the case of space trusses composed of beams with appropriate junctions. Those trusses belong to the category of almost-periodic structures and some comments are necessary. A space truss system, depending on the overall characteristic length compared to those of the generic cell, on the mechanical characteristics of the beam members, and of course on the joints, may exhibit high modal frequency density with local and global modal interactions, specially for very "long" space trusses.

If we consider a periodic structure (periodic in one dimension for the sake of brevity), this periodic structure is by definition infinite. The dynamic behaviour of such systems is relevant to wave propagation analysis in periodic structures which exhibits classical dispersion curves (wave number as a function of the frequency) with so-called Brillouin zones. Those analysis are classical in the vibration analyses of crystals in solid state physics. Introduction of small dissymmetries in the system (almost-periodic infinite structures) leads to a much more complicated analysis.

In this case, the system can exhibit localization phenomenon, called Anderson localizations (see the works of C. Pierre on those subjects). If the system is bounded and almost-periodic, then the problem is much more complicated. In general, the conditions of weak or strong interactions between local and global modes remain an open problem. Space antennas (two dimensional structures) exhibit similar complicated dynamical behaviour. The case of shells are also interesting from those point of views.

For axisymmetric shells, varying mechanical behaviour may occur, such as flexural-membrane-torsion effects due to different wavelengths present in the system. For cylindrical shells, for instance, the vibration behaviour is governed by the circumferential Fourier index, which appears as powers of $1/n$ in the flexural-membrane-torsion equations. Weak or strong interactions between those effects can then be discussed using asymptotic appropriate analysis (considering $1/n$ as as small parameter).

The previous examples show that for structures with complicated arbitrary geometries, the problem is much more difficult to handle.

2.3 Physical Localization

Let us consider now the cases of composite and sandwich structures. If we take the example of a composite plate, several basic parameters appear such as the thickness h, the global characteristic length l and the wavelength L of the phenomenon.

The standard homogeneous beam, plate and shell theories, usually valid for low modal density cases, i.e. low frequency modal domain, corresponds to the situation where h is much smaller than L which is of the order of magnitude of l. The introduction of higher order theories in order to take into account shear effects (which are of prime importance in structural vibrations even in low frequency cases) does not change the above hypothesis as those effects are considered as corrections to the usual basic Kirchhoff-Love assumptions, leading to Mindlin-Reissner theories.

In literature, even for high frequency analysis, such beam, plate and shell theories are used. Those situations correspond to the case where h is much smaller than L which is in turn much smaller than l. In this respect, the partial differential equations obtained from the three dimensional analysis (through a priori displacements/stresses physical assumptions or through asymptotic analysis - it should be noted that shear effects are difficult to obtain through direct asymptotic considerations) are valid from a mathematical point of view. The most complicated situation corresponds to the case where h is of the order of magnitude of L which is much smaller than l. In this last case, local through-thickness vibrations may occur and a three dimensional analysis must be carried out.

If we consider now the cases of heterogeneous systems, such as composite multilayer plates or honeycomb sandwich structures, for low frequency cases, some homogenization mathematical methods give good results, but for higher frequencies, there is a major problem, specially for sandwich structures, as the mechanical characteristics can depend upon the frequency (damping viscous effects). A micro/macro multiscale analysis may then be necessary to catch the local phenomenon which can in turn affect the global behaviour of the system. A direct analysis will introduce a huge number of degrees of freedom - so-called MegaDOF systems - and some specific analysis is then required.

2.4   Micro-Meso-Macro Multiscale Modeling

A way to the above considerations is to start from a three dimensional theory and to incorporate some hypothesis on the microstructure. Of course, the level considered to define a "microscale" is strongly dependent on the user and on the degree of refinement of the desired results. If appropriate experiments or highly refined and detailed computational models show that local effects are of prime importance on the dynamics of the global complex system, then it is necessary to model those local effects.

Starting from the classical Euler-Bernouilli homogeneous beam theory which relates, through a priori mechanical assumptions, generalized quantities (mean-values and first order moments) defined on a reference line to local internal displacements and stresses, higher order theories can be derived by adding local effects, i.e. additional field variables or "internal" physical or generalized degrees of freedom.

Another example concerns acoustic coatings, such as porous (fluid/structure) medium, which are used for internal noise reduction (passive classical systems). Refined mixture constitutive equations, classically known as Biot theories, are derived taking into account the local microstructure.   Adaptive structural "intelligent systems" such as a hybrid passive/active complex multilayered composite with embedded thin viscoelastic layers and active piezoceramics wafers which need multiphysic detailed modeling is also relevant to the above discussion. It should be noted that for such coupled systems, independent analysis carried out for each physical subsystems considered alone does not mean that the behaviour is the same when those subsystems are coupled. A "coupling" physical analysis is necessary to evaluate the weak or strong interactions between those subsystems. For example, within the above considered coupled system, the piezoelectric thin wafer has local vibrations which do not interfere with the vibrations of the composite structure itself in the usual low frequency range of interest. At this stage, one should mention the growing research activities on multiscale analysis in computational mechanics area (see the works of T. Hughes, W.K. Liu, T. Belytschko and several other investigators) in order to catch the local behaviour at the global macroscopic level. The global field describing the medium - the displacement field for instance - is decomposed as a sum of quantities which will describe the local phenomenon at each level. The evaluation and the importance of such procedures in structural vibrations remain to be investigated.

The emerging research activities on nanotechnologies are an example of such investigations in order to improve the feasability of constructing computational models of complex coupled systems taking into account local phenomena.

3.   FLUID - STRUCTURE LINEAR INTERACTION

In many cases, the structures are coupled with an internal and/or external environment such as linear structural acoustic problems (noise reduction), vibrations of structures containing liquids.

3.1   Internal Structural Acoustics

If the internal fluid is a gas, then the ratio between the mass density of the gas with respect to the mass density of the structure can be considered as a small parameter (with respect to 1). If the internal fluid is a liquid, then the ratio between the velocity of sound in the liquid with respect to the velocity of sound in the structure (supposed for instance homogeneous) can be considered as a small parameter. For complex structures of arbitrary shapes, the problem consists in taking into account those facts prior to any huge computations.

The answer to this question is rather difficult. In effect, if the structures are complex (multilayered composites or sandwich structures), the velocity of sound (scalar) is replaced by a tensor and several small parameters are present simultaneously. A rational methodology consists, taking as an example the low frequency domain (low modal density situation), in deriving from an appropriate variational formulation of the local boundary value problem, a modal matrix reduced model of the system. The construction uses a Ritz-Galerkin projection after a decomposition of the admissible vector space of the variational formulation into an algebraic direct sum of two admissible vector subspaces (physical dynamic substructuring approach).

An open problem still exists and is of prime importance: how to quantify a weak or a strong coupling between subsystems in a predictive way.

3.2   External Structural Acoustics

For exterior problems (vibrations of structures submerged into an external fluid environment), the usual approach consists in the derivation of a boundary element discretized description of the exterior boundary value problem (radiation Sommerfeld conditions at infinity).

Much care must be taken in order to derive a formulation valid for every value of the real frequency and which is symmetric with respect to the chosen unknown variables. An other approach has also been derived using a doubly asymptotic approximation (DAA), due to T. Geers, which takes into account, in the boundary value problem, prior to any numerical resolution, the physical behaviour of the external fluid for short and long time scales. This analysis is not a real asymptotic analysis (matched asymptotic developments as in fluid mechanics) but it takes into account two extreme cases (an incompressible fluid approximation for long time scales and a geometric acoustic short wavelength approximation for short time scales).

It should be noted that the far field solution is very sensitive to the degree of refinement of the vibrations of the structures (sensitivity to a small normal displacement of the fluid structure external interaction interface).

Some recent ideas, from T. Hughes and W.K. Liu, based on variational multiscale approach by decomposition of a primal field into the sum of two unknowns, one leading to a resolvable problem - large scale -, and the other one connected to a microscale level in order to catch the local behaviour of the acoustic medium in the far field, so-called unresolvable scale, are under investigation.

3.3 Vibrations of Structures Containing Liquids

Let us consider vibrations of elastic structures containing incompressible liquid. If gravity effects on the free surface are neglected, then the classical concept of frequency-independent added mass is obtained. Two questions may arise concerning the possible effects of gravity (sloshing) and of compressibility of the liquid. Here again, the questions of weak or strong interaction arise and the wavelengths of the various phenomena are of prime importance. In those cases, construction of appropriate reduced matrix models constitutes a rational way to solve the problems.

4. CONCLUSION - OPEN PROBLEMS

The role played by wavelength considerations for complex systems of arbitrary shapes has been clearly put into evidence. However, the *a priori* quantification of possible weak or strong interactions between geometric subsystems or between physical phenomena is an open area of research. The interest of the construction of a family of appropriate reduced order models adapted to specified frequency domains has been clearly stated (see references 1 and 2) and the limit of validity of such reduced models must be specifically analyzed.

5.   REFERENCES

1. Morand, H. and Ohayon, R., *Fluid Structure Interaction*, Wiley, 1995
2. Ohayon, R. and Soize, C., *Structural Acoustics and Vibration*, Academic Press, 1998

What can Mechatronics do for Structural Dynamics?

Gerhard Schweitzer
Institute of Robotics, ETH Zurich, 8092 Zurich, Switzerland

ABSTRACT
Mechatronics, as a combination of mechanics, electronics and computer science opens new ways of designing and shaping the dynamics of mechanical systems. Concepts, methods and examples will be presented, and an outlook on potential future developments of intelligent machines will be given. Comparisons and extensions to structural dynamics, active dynamics or smart structures will be drawn.

1.   WHAT IS MECHATRONICS?

Let us begin with a paradigm! In classical mechanics our standard question is: which motion will a body perform when a given force, for example gravity, is acting upon it? To answer this question we write down the equations of motion in a well-known way, and solve them. It is a problem of *analysis*.

This kind of question can be inverted, too, and then it runs: which force has to be exerted on the body such that it actually performs a certain motion? For example, how do the braking forces for a car have to be tuned so that the car can negotiate a curve safely even during braking? This is a problem of *synthesis*, and for the car example this actually has led to the Anti-Blocking System (ABS), a safety feature installed in most cars today.

For solving such a synthesis task we of course need engineering mechanics, but in addition we need knowledge in systems theory and in control techniques. Further, if we want to realize the suggested solution we additionally have to use knowledge and methods from electrical engineering, electronics and computer science. The term mechatronics for such a synthesis task came from Japan in about 1980, having been coined, it is said, by an employee of the Yasukawa Company. This interconnection of disciplines is actually not new: in aerospace engineering especially it has been well known for a long time and has been successful. For some years now, however, this interdisciplinary field has been growing rapidly and it has developed a weight of its own. This has become possible by the availability of relatively cheap computational power and it is further supported by the rise of versatile power electronics.

Mechatronics characterizes a general trend resulting from the increasing importance of information processing in machinery and other products. Software has become a machine element and an integral part of the product.

In recent years a number of different *definitions for mechatronics* have been suggested. At the ETH we have defined mechatronics in a way that clearly brings out the novel possibilities of combining different disciplines and the potential for machine intelligence [SCHWEITZER 96]. It is the extension and the completion of mechanical systems with sensors and microcomputers which is the most important aspect. The fact that such a system picks up changes in its environment by sensors, and reacts to their signals using the appropriate information processing, makes it different from conventional machines (Fig. 1).

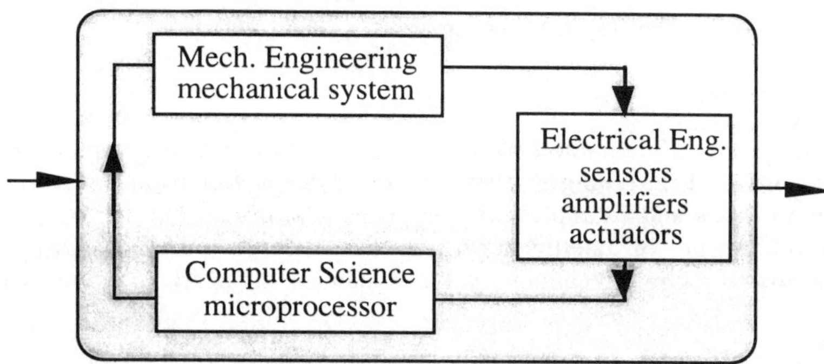

Fig. 1 Mechatronic system.

The system picks up signals from its environment, processes them in an intelligent way and reacts, for example, with forces or motions. Methods for connecting the various areas of knowledge - mechanical, electrical engineering and computer science - are provided by the basic engineering sciences, system theory, control techniques and information processing

Examples of mechatronic systems are robots, digitally controlled combustion engines, machine tools with self-adaptive tools, contact-free magnetic bearings, automated guided vehicles, etc. Typical for such a product is the high amount of system knowledge and software that is necessary for its design. Furthermore, and this is most essential, software has become an integral part of the product itself, necessary for its function and operation. It is fully justified to say software has become an actual "machine element".

2.   RELATIONS OF MECHATRONICS TO OTHER FIELDS

The approach, to regard information processing and actuation as an integral part of a product, in addition to its obvious mechanical structure, may have been a kind of revelation in the technical area and a good reason to coin a special term for it such as mechatronics. In biology, however, there was never any doubt that a biological system needs more than just its physical-mechanical structure to make it work. Thus, an extension of mechatronics, or a technical derivation of biology are bio-inspired systems, characterizing artificial organs, insect-like robots, or the artificial eye or retina. Comparing technical and biological systems may lead to another interesting argument concerning the aspect of reliability. Mechatronics, it is said, will lead to very complex and therefore unreliable systems, an argument, which certainly could be backed up by numerous case histories. On the other side, information processing in mechatronic devices is increasingly being used to make them safer and more reliable, just think of the dynamics control of modern cars. And this certainly is only the beginning towards intelligent machines [SCHWEITZER 96a]. Biological systems, even the simplest ones, are of a tremendous complexity, probably necessary to give them the capability for survival, or to support their ability to operate under uncertain conditions.

The number of new research fields and application areas with the appendix "...tronics" or with the attributes "active, smart, intelligent" is growing. In this way they are indicating the ability to process information within the system under consideration, and to react in an intelligent manner. Some of these new names are known as biotronics, structronics, adaptronics, thermotronics, active structures, active acoustics, active fluid-structure interaction, smart materials, smart products. Of course, this indicates as well, that the original fields are shifting their emphasis in research to new centers, they are adding new methods and they are opening new fields of applications [FULLER, et al. 96], [BOLLER 98]. What this may mean for structural dynamics will be discussed in the next section.

3.   MECHATRONIC APPROACHES IN STRUCTURAL DYNAMICS

Structural Dynamics is very strong in modeling techniques such as FEM and in simulation. Through modal analysis techniques experimental results can be related to simulations. The structure is characterized by modal parameters, the FE model is validated or refined by updating the parameters. In the first place all these techniques are serving the analysis.

This objective may be shifted under mechatronic aspects or even replaced. A major objective in mechatronics is to control the structure in such a way that the dynamics follow given specifications. The model which the control design is based on can be a non-parametric model, i.e. a set of frequency responses without detailed knowledge of the physical parameters involved. Section 4 will give an example of this. Of course, for the design phase of the structure itself an insight into the physics of the structural dynamics is still desirable, and for design modifications even a detailed model may be necessary. Methods for setting up structural models that are useful for control purposes and where control elements such as sensors and actuators can be integrated optimally have been discussed, for example by [SKELTON 95]. The observability and controllability of structural motions by placing sensors and actuators in suitable locations or distributing them optimally is being investigated.

Another mechatronics-inspired objective would be self-calibration and self-diagnosis of dynamic behavior, a feature that goes beyond mere monitoring. While the monitoring of structural properties such as condition monitoring in composite materials or in reinforced concrete is an actual research topic already, the step to actively investigate a structure by means of its control system has still to be taken. Once a control system has been incorporated into a structure, being an active structure now, a self-diagnosis consists in coming up with a suggestion for a fault, and to generate an internal control signal to verify the suggestion or to modify it.

This would require a strategy to compare expected behavior with measured behavior. One way to describe the expected behavior would be through a detailed model, which can accommodate the needs of such an active self-diagnosis.

4. EXAMPLE FOR IDENTIFICATION AND CONTROL OF A FLEXIBLE ROTOR

In a recent BRITE/EURAM project [MARS 96], part of the project was to support a flexible rotor with Active Magnetic Bearings [GAEHLER et al 96]. The test rig, which has been built at Darmstadt University, is shown in Fig. 2, two characteristic modes are presented in Fig. 3, and the main principle of an Active Magnetic Bearing (AMB) is indicated in Fig. 4.

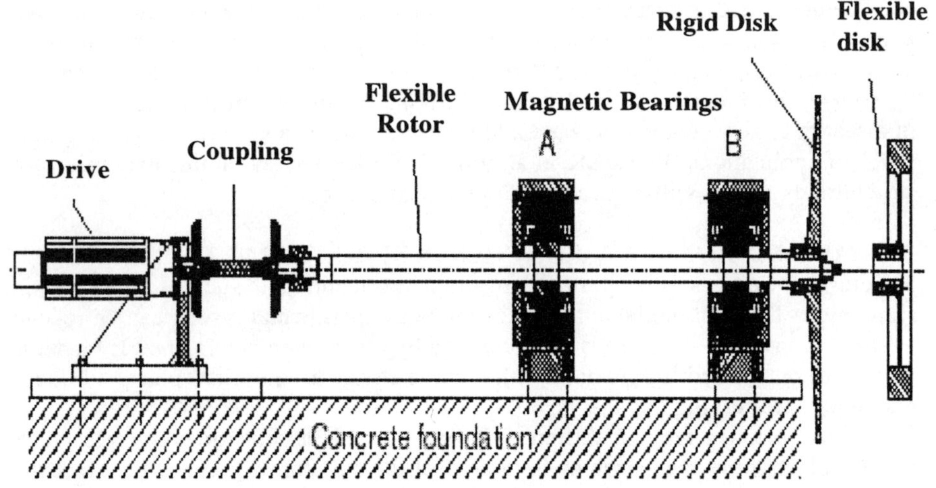

Fig. 2 Test rig configuration

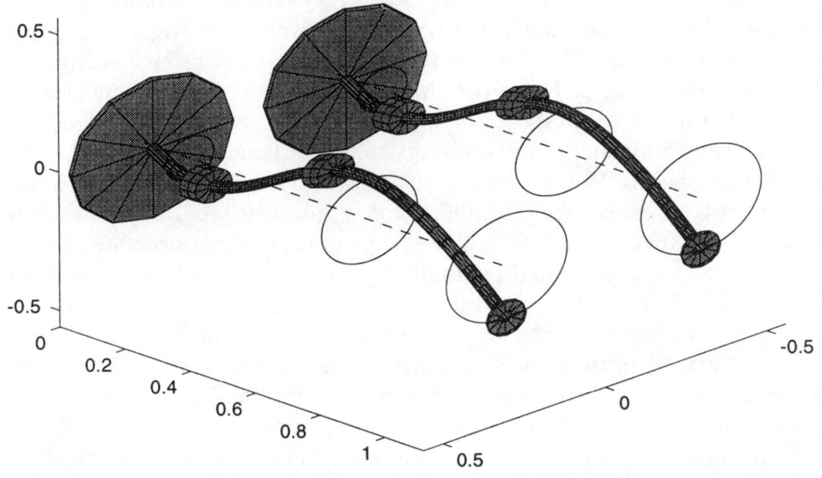

Fig. 3 A forward mode (left) and a backward mode (right) of the elastic rotor

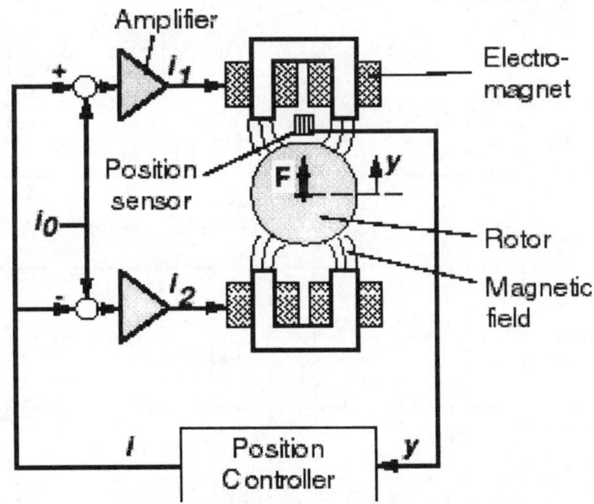

Fig. 4 Principle of a double-sided AMB for one degree of freedom

84

First, a simple model of the rotor bearing system was derived, and a robust control was designed that made the rotor just hover. Two ways to improve the performance were available. The model could have been refined using a modal analysis approach with updating the finite element model. This, however, would not have taken into account the needs of modeling the elements of the control loop as well, the sensor dynamics, the actuators and amplifiers. Therefore, a more direct approach was chosen. Within the closed control loop of the AMB the open loop frequency responses of the flexible rotor were measured. They describe its input/output behavior without prior knowledge of the internal structure of the rotor. The test signals were generated internally by the AMBs themselves and not by an external exciter. For the design of the controller a precise model is required and to this end the measured frequency responses were parameterized by rational functions. This identification was done in an iterative way. The controller determines to a large extent how large a model error can be accepted at a certain frequency. This means that not only the controller design depends on the model, but also the modeling depends on the controller. Thus, modeling and identification are no longer considered as two independent problems, but rather as one joint problem. The results are very satisfying. One element of the 4 by 4 transfer matrix is shown in Fig. 5.

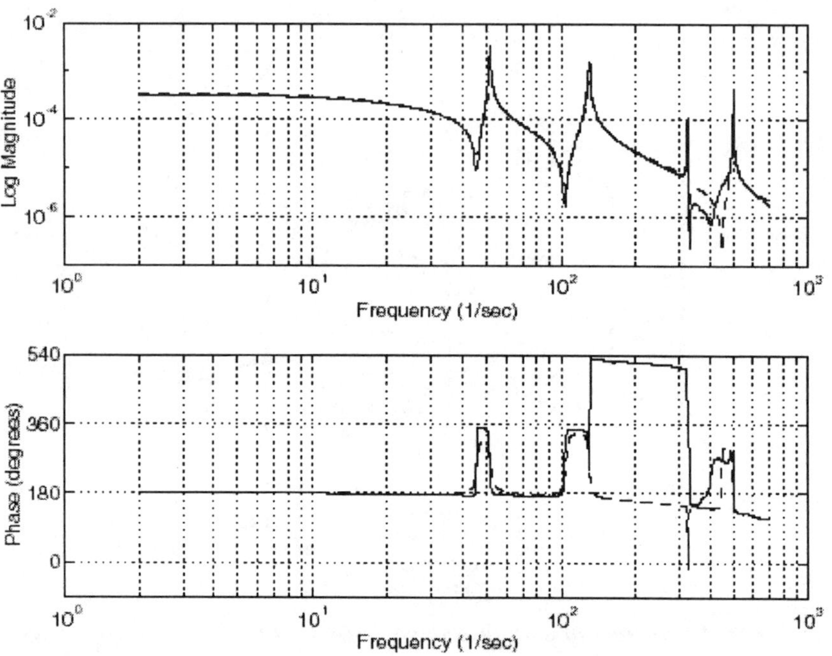

Fig. 5 Identification results. Solid: measured FRF; dashed: identified model (the larger differences in the phase are multiples of 360 degrees)

5. CONCLUSIONS

Mechatronics has successfully developed concepts, methods and tools, which are already useful in other fields as well. It is expected that a strong relation between Structural Dynamics and Mechatronics will evolve, based on the following statements:

- The objective to build a system, product or structure to perform motions in a specified way leads to the task of synthesis. This goes beyond the classical task of modeling and understanding a system by proper analysis.
- The objective of specifying motions, in general, requires the introduction of a control loop, with means for sensing, information processing, and actuating. In particular, the integration of information processing will lead to intelligent products.
- The system models have to be extended and modified to accommodate the needs of control, too.
- By making use of the control, the procedures for monitoring and identification can be developed further into self-diagnostics and other features.

One of the consequences of this emerging trend may be that educational curricula in Structural Dynamics will have to consider control and related information processing as a basic method.

REFERENCES

[BOLLER 98] Boller, C.: State of the Art and Trends in Using Smart Materials and Systems in Transportation Vehicles. J. Systems and Control Engineering, Proc. Instn Mech. Engrs Vol. 212 Part I, 1998

FULLER et al 96] Fuller, C.R., S.J. Elliott and P.A. Nelson: Active Control of Vibration. Academic Press, 1996

[GAEHLER et al. 96] Gaehler, C., M. Mohler and R. Herzog: Multivariable Identification of Active Magnetic Bearing Systems. Proc. IUTAM Symp. on the Interaction between Dynamics and Control in Advanced Mechanical Systems, Eindhoven, The Netherlands, April 21-26, 1996

[MARS 96] EURAM/BRITE Project MARS BRE2-CT92-0223, Development of validated structural dynamic modeling and testing techniques for vibration prediction in rotating machinery (Ewins, D., R. Nordmann, G. Schweitzer), 1996

[SCHWEITZER 96] Schweitzer, G.: Mechatronics – Basics, Objectives, Examples. J. Systems and Control Engineering, Proc. Instn Mech. Engrs Vol. 210, 1996

[SCHWEITZER 96a] Schweitzer, G: Mechatronics for the Design of Human-Oriented Machines. IEEE/ASME Transactions on Mechatronics, Vol.1, No. 2, June 1996

[SKELTON 95] Skelton, R. E.: Integrated Structure and Controller Design. ACC, 1995

[SPENCER 96] Spencer, B.F.: Recent Trends in Vibration Control in the USA. Proc. Third Internat. Conf. on Motion and Vibration Control (MOVIC '96), Chiba, Sept. 1-6, 1996

Vibroacoustics Beyond 2000: Looking for the Sound of Silence

Aldo Sestieri
Dipartimento di Meccanica ed Aeronautica
Univerità di Roma La Sapienza

1. INTRODUCTION

The subject of vibroacoustics concerns the interaction between structural motion and acoustic field. As such it involves a large range of frequencies, i.e. the bandwidth 20 - 20000 Hz of the human hearing. It is a custom and a necessity to divide vibroacoustics into two different problems - external and internal, and subdivide each of them into three different parts: the low, medium and high frequency regions. The need for that relies on the different approaches that are used for the solution of each particular problem: the external problem is related to an infinite medium, typically characterized by fluid wave propagation, whilst the internal one is commonly described by acoustic eigenmodes or stationary waves. With respect to frequencies, the low frequency region involves large wavelengths or low eigenmodes, while the contrary is expected for high frequencies. And it is obvious that, from a numerical point of view, large wavelengths imply a coarse discretization while short wavelengths impel to use fine meshes. It should be evident that a strict bound among low, medium and high frequencies does not exist, in that these regions are related to the ratio between the dimensions of the acoustic medium and the involved structural-acoustic wavelengths.

Although the solution of each of these different problems required a great amount of research until the end of the 1980s, nowadays it can be stated that both the internal and external problems at low and low-medium frequencies can be quite easily solved by classical Finite Element Methods (FEM) or integral formulations (Boundary Element (BEM) [1-3] or Succi's method [4]. The same statement cannot be stressed for the high frequency problems, for a number of reasons. Therefore the actual research in vibroacoustics is concentrated on this high field, as it begun to be over thirty years ago. It is worthwhile to point out that we are here referring to the modal or wave approach to the problem. The ray approach, so far widely used for the internal acoustic problem of large closed spaces, is not considered here in that it cannot account for the interaction between vibrating structures and fluid, while it is commonly used for the design of large rooms, such as auditoriums, conference rooms, etc.

In the context of high frequency structural-acoustic problems there are two different arguments that deserve high consideration for the development of appropriate solution techniques. The first one is obviously connected to the definition of a suitable fine mesh that implies heavy numerical computations.

While one could claim that the development of computers will make it possible to solve problems of larger and larger dimensions, it should be stressed that the larger are the problem dimensions the lower is the solution accuracy, and the lower is the possibility of performing appropriate structural modifications to improve the acoustic response. The second item is related to the statistical behavior of structural and/or acoustic systems at high frequencies: because of inherent uncertainties about the geometrical and physical parameters of the system, on the joints and/or end constraints, the vibrational-acoustic behavior of any system differs unpredictably from similar ones when the modal density or modal overlap increases over a certain amount [5-8]. Therefore at high frequencies the attempts to obtain accurate results by large computational models of complex systems will be in vain or illusory when many modes participate to the response. Figure 1, after Kompela et al. [9], clearly focuses this concept.

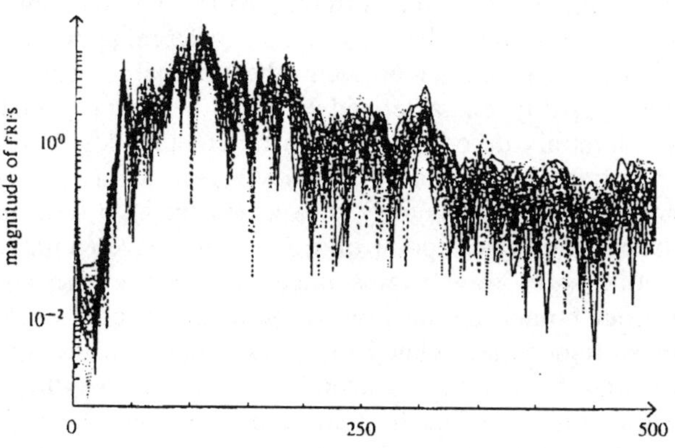

Fig. 1   *A series of structure-borne FRF amplitudes of pickup trucks for the driver microphone*

One would keep in mind that an appropriate model for high frequency structural-acoustic problems should consider both these aspects: the need for decreasing the computational burden, related to a reduction of the computational degrees of freedom, and the opportunity to create a model capable of providing a response in the statistical sense.

It appears now incredible that the pioneers of the Statistical Energy Analysis (SEA), the most developed (philosophical) technique so far available for the solution of vibroacoustic problems, had the intuition of accounting for both these aspects over thirty years ago when formulating the SEA approach. In fact SEA is supposed to be a statistical approach not accounting for the vibroacoustical behavior of a particular system but rather for the "ensemble-average" behavior of an entire population. Although it is questionable that SEA is an actual statistical approach because neither an average response is in fact determined nor any kind of variance is provided nor some probabilistic distribution is required or assumed to solve the problem in SEA terms, there are smart elements in SEA capable of providing some kind of statistical response. SEA is a *"minimum system descriptor necessary for a prediction of ensemble-average behaviors ...and a gross parameter model. ...SEA does not provide the ensemble-mean energy response function for a population but rather it estimates the frequency-average value of the energy response functions of individual archetypal subsystems over interval of frequencies}"* [8]. Then, which are the "statistical" elements in SEA?

- the evaluation of modal densities by general geometrical descriptors such as the area (volume) of the structural (acoustic) subsystems, instead of its real dimensions, without accounting for the specific boundary conditions;
- the evaluation of damping and coupling loss factors without referring specifically to the subsystem under analysis;
- the use of a global space descriptor (mean-pressure or mean square velocity averaged in space) instead of a local descriptor which smoothes away unavoidable uncertainties and space fluctuations;
- the use, as global descriptor, of the time-averaged energy that is phase independent;
- the possibility of overcoming one of the basic and critical SEA assumptions, the weak coupling between the subsystems in the model, thus averaging and lowering the inter-coupling effects among subsystems. This justifies the increasing use of SEA in many transport industries: a large number of subsystems are used in recent applications of SEA, and the results, that previously were almost unreliable, begin to match better with the experimental results [10].

However, notwithstanding a much larger comprehension of the SEA basic assumptions, the large amount of theoretical and experimental work developed on the subject in over thirty years, the availability of more friendly user and less expensive SEA codes, the miss of other valid alternative methods, SEA is not still considered a reliable approach to use at a design stage.

Therefore, since the end of the 70s, but more systematically since the end of the 80s, an important piece of work has been developed to overcome some SEA limitations. Some studies were particularly devoted to fill the frequency gap between the high frequency limitation of deterministic finite element methods related to mesh refinement (implying a high computational burden), necessary for a description of the short wavelengths involved, and the low frequency bound of Statistical Energy Analysis, due to an insufficient number of modes in the modal subsystems. However, most of the work was addressed to the appealing possibility of providing a more rich description of the vibroacoustic response along the considered systems. In this framework the energy flow formulations and the envelope formulations gain a particular role: the first because, in recent years, it was often assumed that the high frequency vibrations could be modeled by a vibrational conductivity equation analogous to the steady equation governing the heat flow, which is an extension of SEA laws into differential terms. On the contrary, the envelope formulations disregard the thermal analogy while trying to use a new description of the dynamic response by a suitable transformation that would be capable, as the thermal methods, of providing a smooth or quasi-static solution for the high frequency oscillating field.

In section 2 an overlook of the SEA state of the art with regard to its developments, limitations and perspectives is presented.

In section 3 some interesting formulations are discussed: they provide a deep insight into some basic SEA assumptions and throw a bridge toward some recent and important developments that promise to be very useful for the practical solution of structural-acoustic problems. In the same session the energy flow formulations and envelope formulations are presented.

In figure 2, a block diagram on the connection among these different approaches is shown, with reference to some researchers that contributed to their developments.

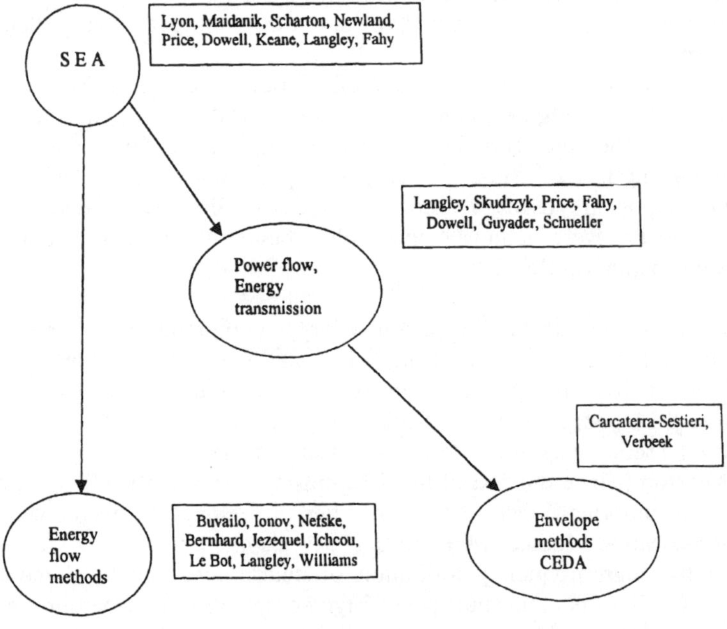

Fig. 2 Block diagram representing the high-frequency vibroacoustic research.

2.   STATE OF THE ART ON SEA

The pioneering works on SEA were developed by a group of scientists (**Lyon, Maidanik, Smith, Heckl, Noiseux**) that met together in 1961 to discuss how to predict the rocket noise and vibrations of satellite launch vehicles ([11]). In 1962 **Lyon and Maidanik** published a first systematic work on SEA [12]: they consider the average power flow between two simple harmonic oscillators excited by random force, discovering that the power flow is proportional to the difference of kinetic energy between the two oscillators, i.e. the energy (more precisely the modal energy) of each system is equivalent to the temperature in thermal systems. Then they extend this result to two coupled multimodal subsystems and introduce the first restrictive assumption on the type of coupling between subsystems, that is requested to be weak, in the sense that the interaction forces between them must be much smaller than the internal forces within each subsystem. In that paper they provide the concept of modal groups as the systems interchanging their energy, and that of equipartition of energy to specify that all the modes of a group contribute equally to the total energy of the subsystem. Each modal group can be regarded as a collection of simple oscillators. Furthermore, they assume that the energy exchange between the two multimodal systems can be obtained by superposition of the energy exchange among the elementary oscillators of each group.

Further developments in this area are due to **Lyon and Scharton** [13], **Kakar** [14], **Lyon and Eichler** [15], that contributed, until 1975, to a better comprehension of SEA bases and limitations.

An important work to understand the limits of validity of SEA is due to **Woodhouse** in 1981 [16]. He disregards the mathematical formulation developed by Lyon and Maidanik for the coupled system and proposes a general approach for studying the power flow among mechanical systems by using a technique similar to a previous one developed by Lord Rayleigh. In this way he reaches important and original results. He finds a condition of energetic reciprocity that is a necessary condition so that modes belonging to two different groups be coupled in the way suggested by Lyon. This condition is not easily verified, excluding the case of two elementary single degrees of freedom oscillators coupled together and the case of a set of homogeneous resonators (**Lyon and Scharton** in 1968). With this condition he proves quite simply most of the previous obtained results, but shows that, in general, the Lyon's approach is not rigorous for three or more coupled oscillators. Finally he proves that the necessary condition is also sufficient under the assumptions of *i*) statistical independence of the random forces applied to each oscillator, *ii*) weak coupling and *iii*) small damping.

In 1983-85 **Dowell and Kubota** [17] developed an original and efficient approach to study high frequency vibration problems. It is a systematic simplification of modal analysis and it can be considered valid when the structural (or acoustic) response is characterized by a high modal density: thus their approach is called Asymptotic Modal Analysis. Either white random forces or harmonic forces can be used, the last case not being permitted by SEA. Two types of averages are performed on the response: the first one is a spatial average on the whole structure, the second one is a frequency average on the band of excitation. In this way Dowell obtains several classical results of SEA for both the structural coupling and the structural-acoustic coupling, providing a rigorous interpretation of SEA assumptions and, possibly more important, presenting a new systematic approach for the theoretical analysis of high frequency problems.

Further insight into the SEA theoretical bases was provided by **Keane and Price** in 1987 and **Langley** in 1989.

Keane and Price [18] give a number of restrictive hypotheses under which the proportional link between power flow and energy exchange is valid:
- the forces acting on each modal subsystem must be statistically independent;
- power flow and energy must be averaged on rather narrow frequency bands, provided that they contain several modes;
- the subsystems' damping must be proportional;
- the coupling must be weak and conservative.

However, they also stress that the weak coupling assumption can be eliminated provided that the overall response of each subsystem is not dominated by a single mode of that group: in fact this general assessment can be obtained by either a weak coupling or by assuming the presence of many interacting modes.

Langley's approach [19] provides a convincing theoretical analysis on SEA bases. He starts from the motion equations of a continuous elastic system and assumes that the exciting forces are random and spatially uncorrelated, while the coupling is conservative. He finds that the general expression of the power flow can be reduced to the proportional form proposed by Lyon when the following conditions hold:
- uniform mass density;
- damping proportional to the mass;
- energy averaged over a population of random systems;
- weak coupling, where by weak coupling Langley means that the Green function of any isolated component does not vary sensibly when it is coupled to the other subsystems.

In 1990 **Maidanik** reformulated the theoretical bases of SEA [20], redetermining the governing SEA equations by means of a wave approach to the dynamic problem. The basic assumptions are:
- the subsystems energies must be averaged over space and statistically, i.e. with reference to a population of structures;
- the total energy must be considered as a simple superposition of energies associated to each elementary wave propagating within the medium.

Although limited to continuous one-dimensional systems (simple coupled waveguides) Maidanik's approach is more general and an extension to complex system can be likely expected. His rigorous formulation clearly emphasizes the assumptions that must be established to obtain the SEA formulation from the general formulation of energy exchange. However, if this is true from a mathematical point of view, his formal assumptions cannot be easily interpreted in physical terms.

As a conclusion of this contribution, it is worthwhile to point out some general comments on the past and future developments of SEA.

When thinking to the governing equations and even accepting the validity of the fundamental relationships between energies and power flows (although it was shown that this result cannot be considered valid in any situation), it is very difficult to provide a valid SEA model of the dynamic system. If the SEA equations are quite simple, the effective unknowns are not only the subsystems' energies but also the coupling and internal loss factors, that are dependent on the frequency band excited by the input(s), modal densities and input powers. With regard to such quantities the analyst does not know anything a priori and SEA does not provide a general criterion to determine these factors, although they can be evaluated experimentally or theoretically by different methods related to modal analysis, wave approach or mobility functions, or even using finite element techniques. Moreover the input power is never practically known, and its estimate open complex problems that were partly analyzed by **Pinnington and White** [21]. Finally the definition of appropriate modal groups is not necessarily a simple task, in that modal groups do not generally coincide with structural components. The effective simplification of SEA relies on the elimination of the space dependence (through the spatial average) and the frequency dependence (by the ensemble-average that is in fact substituted by the frequency-average).

Another point that is critical and has been seriously considered in several alternative methods to SEA is the frequency limitation of the SEA approach. In fact the limitation concerns the frequency bandwidth where the number of modes is too high for finite element analysis but too low to provide accurate averages for SEA.

To conclude it can be stressed that while the input requirements are rather heavy (input power, coupling loss factors, loss factors, modal densities), on the contrary the output is rather poor since any local information is lost as well as any resonant behavior. Although this is inherent in the same soul of SEA that, in this way, would provide a "statistical" response, no doubt this is felt as a limitation, and, possibly, it can explain why SEA has not yet reached the important role that it would deserve in structural dynamics and vibroacoustics.

However, it is important to note in the last five years a renewed interest toward the generalization of SEA assumptions (equipartition of energy, rain on the roof load, etc.): a special contribution to this subject must be acknowledged especially to **Heron**, **Guyader** and **Orefice**. This work will certainly pay in the future.

3.   FROM SEA TO ENERGY FLOW AND ENVELOPE FORMULATIONS

After the definition of the concept of power flow or structure-borne intensity by **Noiseux** in 1969 [22], related to the product of force and velocity (or stress and velocity), many basic experimental and theoretical studies were developed by several authors.

Based on these power flow concepts, especially during the 80s, new approaches were studied and new formulations proposed to provide a deeper

94

insight into SEA philosophy, fundamentals and implications, to avoid the uncertainties of SEA results and to provide a vibrational energy distribution rather than an overall mean energy as in SEA. Although only a few of these formulations are really convenient for the solution of the problem, they gave an important stimulus to new research in high frequency vibroacoustics.

3.1 Alternative approaches to SEA

Certainly the Asymptotic Modal Analysis of Dowell, described in the previous section, belongs to this category, being contemporary a procedure capable of explaining many of the SEA assumptions, while providing a systematic analysis for problems at high frequencies.

Important contributions to the comprehension of some SEA results, but not necessarily related to this goal were given by **Newland** who, as one of the maximum experts in random vibrations, developed studies on coupled oscillators subjected to random forces. In particular, in [23] a perturbation method is proposed for calculating the statistics of energy transfer between weakly coupled oscillators, showing that the first order approximation for the mean power flow coincides with the classical SEA result. Moreover it is stressed that this approach can provide important statistics and more accurate results than SEA when second and higher order approximations are considered.

Skudrzyk in 1980 [24] studied an interesting and basic approach to predict the dynamic response of complex coupled systems. Unlike SEA, the mean-value method of Skudrzyk is not a statistical theory, although it shares with SEA a basic task in that it tries to eliminate unnecessary details to describe the dynamic response of complex and coupled structural systems. By a limited knowledge of few structural parameters, such as the mass and the density of resonances, and by some information on the excitation and the response location, the mean-value method predicts a mean-line through the frequency response curve of the vibrating system and the envelopes of the resonant and antiresonance peaks in almost the whole frequency range of interest, thus not showing the classical limitations of FEM and SEA.

In trying to eliminate the frequency gap between FEA and SEA, and proceeding along the line shown by Skudrzyk, **Cuschieri** uses the power flow concept as complementary to FEA and SEA [25], by considering the mean level of the transfer function mobilities: he shows that the mean responses are independent on the exact geometry but depend only on the general structural characteristics. He also stresses that, at high frequencies, the power flow method converges to SEA results, provided that frequency averages are suitably performed.

A very important work in this framework is the Wave Intensity Analysis (WIA) proposed by **Langley** in 1992 [26]. In the WIA Langley regards the displacement field as the result of a superposition of waves travelling along any direction, with proper amplitudes and phases. If the phase dependencies are neglected, the effects of resonances and antiresonances, related to wave interactions in correspondence to in-phase and out-of-phase effects, are eliminated. From a modal point of view, this corresponds to perform modal averages on the response. Once the phase dependence is eliminated, only an energetic beam is associated to each wave, corresponding to the mean energy of the travelling wave. The general beam which is dependent on the spatial coordinates, is expanded into Fourier series. An energy balance equation is then written for the interacting beams and the coefficients of the Fourier series are determined by a Galërkin procedure. If the series expansion is arrested to the first linear term, Langley shows that the energy balance exactly corresponds to the SEA equations. In this context WIA is a generalization of SEA, and SEA becomes an approximate formulation in which:

- the phase effects of the travelling waves are eliminated, while the reflection, dissipation and transmission effects are still maintained;
- the energetic field is approximated by the superposition of the mean energy (first term of the series expansion) transported by each wave. Thus Langley determines more general equations than the classical SEA and proves that, in this way, better results than SEA are usually obtained.

3.2 The thermal analogy and related approaches

Particular mention is deserved, in the framework of methods alternative to SEA for the analysis of vibroacoustic systems, by a set of similar procedures generally known as heat conductivity methods or thermal analogy approach. In fact, they represent a very attractive development, promising to overcome SEA limitations while providing results of more informative content. However, recently several arguments have been discussed that contrast the validity of the thermal analogy bases.

In 1977 three soviet researchers, **Belov, Rybak and Tartakovski** [27]), and later on in 1979, **Buvailo and Ionov** [28, proposed an extension of the SEA laws into differential terms. More precisely they assumed that elemental volumes within any vibrating elastic medium exchange energy in direct proportion to the difference of their energy levels. Thus, by using the same relationship proposed in SEA for the dissipated power, they establish a local energy balance as:

$$\frac{\partial E}{\partial t} = -\nabla \bar{q} - \Pi_{diss}$$

where E is the energy and $\bar{q}$ the power flux obtained as the product of the stress tensor by the velocity vector. Then, thanks to the thermal assumption $(\bar{q} = -\mu\, grad(E))$ and to the hypothesis that the dissipated power $\Pi_{diss} = \alpha E$, where μ and α are constants depending on the material and excitation frequency, they obtain, for a stationary process:

$$\nabla^2 E - \beta^2 E = 0 \qquad (1)$$

This equation is formally equivalent to the heat diffusion equation, as it is expected in that it is obtained by an extension of the macroscopic thermal SEA law into differential terms.

With respect to the classic equation of motion described by some form of hyperbolic differential wave equation, the previous equation presents outstanding advantages from a numerical point of view. In fact, the heat equation is a parabolic equation describing a diffusion phenomenon and admits solutions exponentially decaying from the source, without oscillations. On the contrary, the wave equation describes a propagation phenomenon and has oscillating solutions in space, whose wavenumbers increase in direct proportion to the exciting frequency. This difference permits to solve the parabolic equation with a coarse mesh that is usually frequency independent while the wave equation requires a mesh that becomes more and more demanding as the frequency increase and the space passes from one to two and three-dimensions. This finally implies a heavy computational burden for any computer.

Probably as a result of being published only in Russian, the works by the soviets were almost ignored until **Nefske and Sung** [29] reformulated the problem in 1987, and proposed, for the first time, some interesting applications on beam structures excited by harmonic concentrated loads. The comparison with modal analysis results show that the solution of the thermal equation provides a kind of average trend of the energy density along the beam (figure 3, after Nefske and Sung [29]). The agreement is good, even for coupled system of beams, although it represents an heuristic solution providing acceptable results for one-dimensional systems only. Notwithstanding the poor theoretical assumptions of this approach, some groups begun to work on it, by proposing interesting applications to coupled beams (**Palmer, Williams, Fox** [30, 31]) that confirm the good results for one-dimensional systems but show severe limitations when extended to two-dimensional plates. In this framework, **Bouthier, Wolhever and Bernhard** developed theoretical considerations related first to one-dimensional systems, then to plates [32, 33].

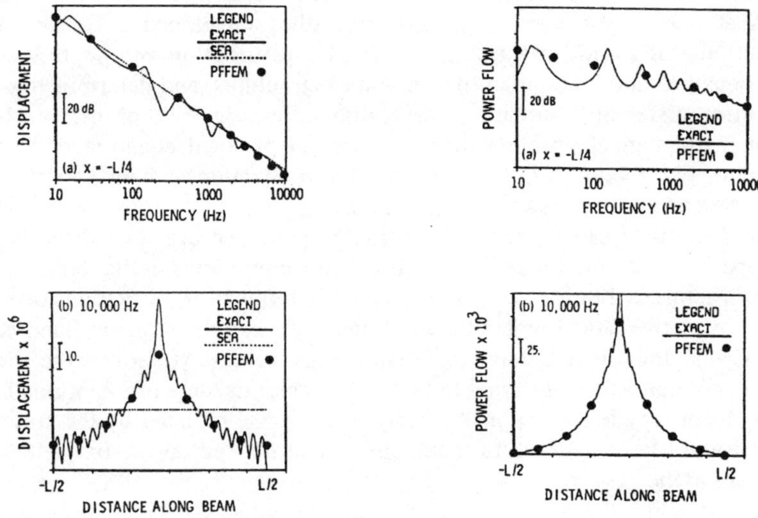

*Fig. 3 Comparison of Nefske-Sung results with the exact (modal)
and SEA solutions*

They state that the diffusion equation can be considered valid under the hypotheses that:

- the excitation is concentrated and harmonic;
- the reference energy is the total energy per unit length (sum of kinetic and potential energy) averaged in time over a period of oscillation; moreover some kind of local average must be performed to eliminate harmonic terms appearing in the reference energy;
- the near field contribution is neglected, i.e. the energy field must be considered far away (more than a wavelength) from the source.

Under these assumptions the results they obtain are identical to those of Nefske and Sung, although they attribute different meanings to the energy variable in the thermal equation. Nefske and Sung interpret it as local energy averaged on time and then on some appropriate frequency band (SEA inspiration); on the contrary Wolhever and Bernhard [32] give to it the meaning of local energy averaged on time and space, in this way eliminate the harmonic energy terms. Although their analysis refers rigorously to one-dimensional systems only, successively Bouthier and the previous authors assume the general conviction that if time and spatial energy averages are performed, the resulting energy variable satisfies a diffusive equation. Under this subjective assumption they extend the thermal approach to two-dimensional systems [33].

98

At the beginning of the 90s **Le Bot, Jezequel** and others begun to study energy propagation in structures with the aim of developing alternatives procedures to SEA at high frequencies. Originally [34] they developed a General Energy Formulation (GEF) for describing the spatial distribution of the time-average energy densities and lagrangian of continuous structures, and determined a pair of exact linear differential equations of eighth order. Because of the much more difficult mathematical problem than the original physical equation of motion in terms of displacement, it is obvious that these equations are meaningless for a solution procedure of vibroacoustic problems.

Incidentally it can be stressed on this purpose that although the energy is a quite appealing variable for studying vibroacoustic problems in that both the sound and the structural vibrations can be expressed in terms of it, difficulties arise when looking for mathematical matching conditions between the structural and acoustic fields because they usually have different energy values. Moreover, the structural analyst is not generally interested to know the energetic state of vibrations because it is not the energy level responsible only of the noise radiated by the structure or of the fatigue life of a structural component, that depend on the frequency or the stress level in the system.

To overcome the numerical difficulties related to the GEF formulation, **Le Bot and Jezequel** [35] produced, for the total energy a simpler equation (Simplified Energy Formulation: SEF) by performing spatial averages on the single wavelengths. At least for one-dimensional systems, it corresponds to the heat equation, giving more credit to the thermal approach.

Successively **Le Bot and Luzzato** [36] try to extend the SEF model to two-dimensional systems, but they do not succeed in determining a heat equation for them. The non validity of the thermal analogy for two-dimensional structures was previously obtained by Le Bot and Jezequel [34], when examining the case of circular membranes excited by a harmonic force, by numerical simulations and theoretical arguments.

In 1996 **Ichchou, Jezequel and Lase** [37, 38] propose a wave approach to build up an energy description of some dynamic systems, in contrast with the approach followed by Bernhard *et al.* in that the concept of space average used by them is not considered to be generally acceptable. However, using the same hypotheses proposed by the previous authors and adding the assumption of neglecting the interference among propagation waves (by means of a statistical average between two general propagation directions, as explained later in this section) to generalize the space average concept, Ichchou and Jezequel derive and confirm the same heat diffusion equation, that they stress to be valid for one as well as for two-dimensional systems. In particular the energy model derived by them for plane wave fields is written in terms of the group velocity of the plane waves as:

$$-\left(\frac{c_g^2}{\eta\,\omega}\right)\nabla^2 <\bar{e}> + \eta\,\omega <\bar{e}> = 0$$

where $<\bar{e}>$ is the time and space average of the energy density, and η is the loss factor. This result is quite important because it would confirm that the thermal analogy can be justified by different approaches, by introducing only a few new assumptions: in this case the elimination of wave interference in the high frequency range. Wave interference is an important concept considered also by Langley in developing his wave intensity technique, but also by Carcaterra and Adamo to justify the non-validity of the thermal approach, especially for two-dimensional systems: this point will be discussed later on. In any case, these authors, by their wave approach, claim for a generalization of the thermal analogy, and in fact in successive papers they present their model for one-dimensional systems (bars and beams) [37] and then for plates and general two-dimensional structures [38].

3.3 Contributions to a better understanding of energy flow in structures

An important group of papers in the 90s deeply analyze the problem of energy flow in structures: most of these papers originate from a critical review of the thermal analogy.

A first analysis on the subject was given by **Carcaterra and Sestieri** and later on by **Langley**, **Smith**, **Xing and Price**, **Carcaterra and Adamo**, **Le Bot**, **Orefice-Cacciolati and Guyader** by theoretical and experimental developments. A work that can be appropriately inserted in this context, although having a different background derived from the dynamics of stochastic systems, has been recently presented by **Pradlwarter and Schuëller.**

As already stressed, the thermal analogy tries to describe the mechanism of mechanical power propagation within a structure, by assuming that the power flow is only proportional to the gradient of the energy density. In [39] the following results are shown by theoretical considerations.

- For longitudinal rods, the thermal analogy is valid provided that no dissipation effects are introduced into the system. In the presence of dissipation effects the thermal analogy can be obtained only if some kind of spatial average is performed so that the time-average potential energy is equal to half the total time-average energy.
- For flexural beams, the heat equation as determined in [29] is not valid even if the near field contributions are neglected: in fact, the time-average far field energy has two terms that are not thermal. Moreover, in the presence of dissipation effects, it is not possible to derive a time-average energy equation from the displacement equation of motion.
- For flexural plates, the time-average energy equation differs from the heat equation.

  On the bases of these results, further arguments can be developed to contrast the validity of the thermal analogy.

While SEA states its thermal propagation law for the mechanical energy with reference to systems of finite dimensions, the thermal analogy tries to extend this law into differential terms, thus referring to elemental systems. Although SEA equations can be obtained by different approaches, the SEA transmission law relies on the following points: *i*) each subsystem is equipped with modes; *ii*) the number of modes in the bandwidth must not be too small; *iii*) the modal groups must be weakly coupled. On the contrary the elemental elements of an elastic medium do not present a modal behavior and are strongly coupled because the interaction forces are of the same order of magnitude of the internal forces. Therefore there seems to be a methodological error in extending the SEA laws into differential terms, and it would be inappropriate to describe the power flow in structural problems by the thermal analogy: a more complex mechanism would be required.

Another important contribution to a better understanding of the energy propagation mechanism is given by **Carcaterra and Adamo** in [40]: this work gives a special insight into the thermal wave approach, and can be conveniently correlated to the works by Langley [26, 41] and Ichchou and Jezequel. In [40] the authors highlight different scale laws in the mechanical energy transmission, controlled by a suitable μ parameter related to the ratio between a characteristic space-average length and the excitation wavelength. For $\mu < 1$ (small scale) the vibrational conductivity fails, while for $\mu > 1$ the asymptotic thermal analogy is valid for rods and beams but not for plates or other two-dimensional systems. The difference is explained by considering two different types of energy interaction between wave energies: the coincident wave energy (c.w.e.) determined by waves propagating in the same direction and the interference wave energy (i.w.e.) depending on waves propagating along different directions. It is shown that their asymptotic features are totally different: while the c.w.e. tends to satisfy asymptotically the energy balance in thermal form for $\mu > 1$, the i.w.e. exhibits a complex behavior showing a non thermal contribution even for $\mu > 1$. It is focused by asymptotic considerations that the non thermal component is largely dominant in the whole medium scale range. The authors clearly show by theoretical and experimental tests performed by a scanner laser that while the one-dimensional systems tend to behave thermally in the large scale because of the c.w.e mechanism, two and higher dimensional systems are dominated by the i.w.e. mechanism, that inhibits the thermal behavior.

Actually the interference problem is often considered by a statistical point of view. Ichchou and Jezequel in the mentioned papers, but also Langley in his WIA, assume that the expected value between two wave phasors associated to two general directions of propagation θ_i and θ_j are uncorrelated, i.e.:

$$E \left\{ A(\theta_i)\, A^*(\theta_j) \right\} = 0 \qquad (2)$$

Carcaterra and Adamo suggest in [40] that equation [2] would hold for reverberant fields while, for a direct field, an "isotropic condition" should rather be considered, i.e. ($E \left\{ A(\theta_i)\, A^*(\theta_j) \right\}$ = cost). Therefore the thermal analogy would be valid for reverberant fields but not for direct fields: thus the sum of these two contributions, total field, would not be thermal.

Similar concepts are considered by **Langley** in [41]. Particularly he shows that:

-    for axi-symmetrical wave fields, the wave interference cannot be neglected so that the thermal analogy fails. More specifically he states that the conditions under which the vibrational conductivity equation holds are not met by the direct field arising from a point load: the heat equation predicts a far field energy density which decays in proportion to $1/\sqrt{r}$, while the exact analysis yields $1/r$

- for two-dimensional structures such as plates, the vibrational conductivity approach yields an energy distribution which is more spatially uniform than the true result (see figure 4 after Langley, [41]).

Addressed to the same goal is the paper by **Smith** [42], who tries to reformulate the thermal analogy by separating the direct and reverberant fields, so obtaining better results than those related to the standard thermal analogy.

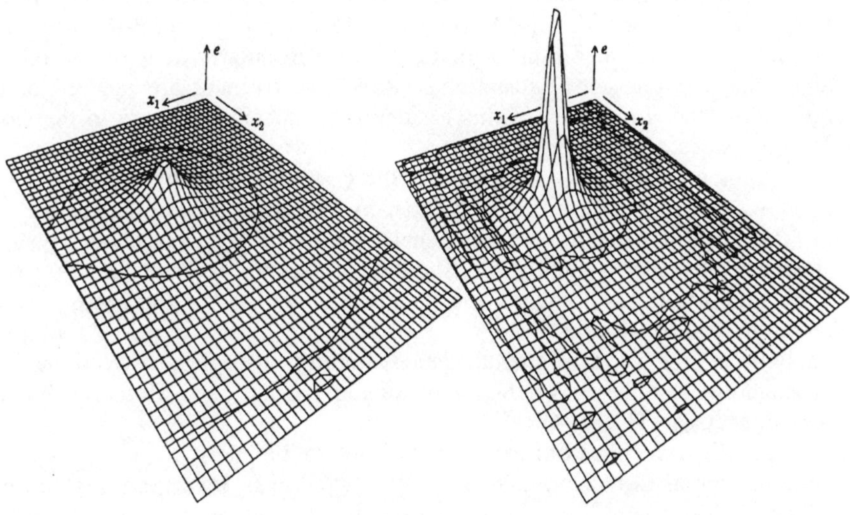

Fig. 4 Comparison between the surface plot of the energy density obtained by the vibrational conductivity equation (left) and the exact analysis

In the same line of the previous papers is the work by **Xing and Price** [43] that develop a mathematical model describing the energy flow associated with the dynamics of a continuum: in this way they derive the energy flow equations of dynamically excited systems, and analyze the particular cases of rods, beams and plates, under the point of view of the heat conductivity. They state that necessary and sufficient condition for the existence of an energy flow potential, that can justify the proportionality of the energy flow to the gradient of the energy density, is that the energy flow vector is irrotational. By analyzing the case of a simple rod, they show that a similarity between the energy transmission along the rod and the flow of thermal energy in a heat conduction problem does not exist, and conclude by stressing that the development of any hypothesis or modeling based on such analogy is of limited value.

A convincing work for the study of high frequency problems based on the energy formulation seems to be a recent paper by **Le Bot** [44]. It represents the last evolution of previous works by the same author, Jezequel and Guyader, and summarizes many of the works on the energy flow by Bernhard. It could be said that it keeps the wealthy part of all the energy formulations, while eliminating most of the controversial parts. He considers the wave field generated by a point source in a free field and determines a relationship between the intensity vector and the energetic field: in this way he finds a simple energy balance equation that, in general, is not thermal anymore, thus confirming several results determined in [39, 40].

Some analogies with this last work can be found in a paper by **Orefice, Guyader and Cacciolati** [45]. They introduce a power/energy transfer function, and discuss some important properties of it. Particularly, it is discussed whether and when the energetic cross-terms (wave interference) are negligible, making quite clear the limitations of this assumption, especially related to the stochastic characteristics of the input power.

Another work [46], addressed to the comprehension of energy transmission phenomena in vibroacoustics, is a paper by **Pradlwarter and Schuëller**. It has a different origin from those so far analyzed in that it refers to the dynamics of stochastic systems. Actually this should not wonder because SEA itself finds a basic support on this branch of mechanics: **Morse, Ingard** and **Lyon** have several common connections, and Morse, a well known physicist, developed a lot of work on both atomic physics and acoustics, finding analogies between the acoustic propagation and the wave theory in atomic physics (see Morse: Vibration and Sound, McGraw Hill, 1948).

Pradlwarter and Schuëller in [46] determine energy equations for dynamic systems similar to those obtained by SEA, valid even for strong non-conservative coupling. They start their analysis from the Lyapunov equation describing the evolution of the output covariance matrix for a linear system under random excitation. This seems to be a correct starting point for the energy analysis because the covariance matrix represents:

- the simplest statistical indicator for the system's response;
- the simplest indicator of energy distribution among the system's degrees of freedom. In particular its diagonal terms represent the expected values of the energetic levels for each degrees of freedom.

The analysis developed in [46] shows how the energy is stored and transmitted within mechanical systems. While the method is not useful to predict the structural behavior at high frequencies because the knowledge of the eigenparameters participating to the response is required, however, it works at low frequencies where the number of participating modes are few: in this case the energy distribution of the system is determined as well as the coupling loss factors.

3.4 The envelope methods

The envelope models developed by **Carcaterra and Sestieri** try to overcome the theoretical limitations encountered by the thermal methods. In the years three different methods were successively proposed: the envelope energy model (EEM) [47], the envelope-phase energy model (EPHEM) [48] and the complex envelope displacement analysis (CEDA) [49].

In the EEM an energy variable is used (the kinetic energy density), obtained by an envelope energy definition that uses the Hilbert transform averaged over time and space. In the EPHEM an energy variable is still used that is exactly defined as in the EEM, but the structural response is also characterized by a second variable, the phase, introduced to recover energy jumps at the discontinuities. With this model the physical dynamic response can be reconstructed combining together the envelope energy and the local phase: a main drawback is related to its nonlinear character.

Completely different is the CEDA approach. The new variable is not the energy anymore but rather a variable directly related to the physical displacement. The dynamic response is determined by the local displacement without performing any kind of average. Unlike EEM and EPHEM, neither the knowledge of transmission or reflection coefficients nor the input power are necessary, because the forcing term is a complex envelope force directly related to the excitation force only.

The complex envelope displacement theory relies on a suitable variable transformation, achieved through the action of the envelope operator $\mathbf{E}$ on the displacement w, defined as:

$$\mathbf{E}(\cdot) = [\mathbf{I} + j\mathbf{H}(\cdot)]\, e^{-jk_0 x}$$

where $\mathbf{H}$ and $\mathbf{I}$ are the Hilbert and identity transformations, respectively, $k_0 = \omega_0/c$ is the carrier wavenumber, corresponding to the harmonic excitation frequency ω_0, and c is the phase wave speed in the considered system. Therefore for the new variable - complex envelope displacement $\tilde{w}$ - one has the relationship:

$$\tilde{w} = \mathbf{E}[w] = [w + j\bar{w}]\, e^{-jk_0 x} = \hat{w}\, e^{-jk_0 x}$$

where $\hat{w}$ is the analytic displacement $\hat{w} = [w + j\bar{w}]$, $\bar{w}$ the Hilbert transform of w and j the imaginary unit.

By this definition a very convenient transformation is obtained, at least for one-dimensional systems. It can be easily shown that, if the physical displacement spectrum is band limited around the carrier wavenumber k_0, the complex envelope displacement is band limited around the wavenumbers' origin. In fact, for a band limited spectrum around k_0, the Fourier transform $W(k)$ of the physical displacement is concentrated within two limited regions around $\pm k_0$ of bandwidth Δk (figure 5a). The Fourier transform of the analytical displacement $\hat{w}(x)$ is then given by:

104

$$\hat{W}(k) = W(k) + j\tilde{W}(k)$$

and it can be verified that it provides:

$$\hat{W}(k) = W(k) + sign(k)W(k)$$

i.e. the negative wavenumber contribution of $W(k)$ is deleted, and the positive one is doubled (figure 5b). Finally the Fourier transform of the complex envelope displacement provides $\overline{W}(k) = \hat{W}(k + k_0)$, implying a shift of the positive wavenumber contribution of $\hat{W}(k)$ toward the origin of wavenumbers (figure 5 c). According to the Nyquist criterion, this property suggests that, unlike the physical displacement requiring a space sampling that increases with the excitation frequency ω_0, the complex envelope displacement, which is a low wavenumber

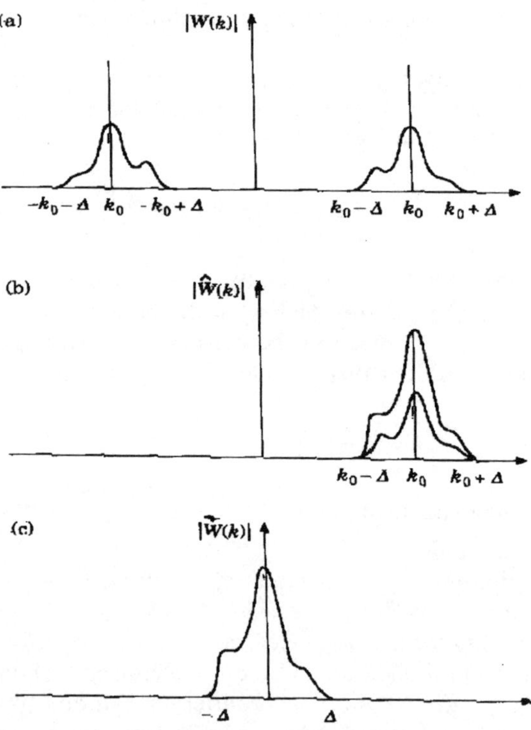

Fig. 5 Fourier transform of the physical displacement (a), analytic displacement (b), and complex envelope displacement (c)

function, can be correctly described by a limited number of samples. Therefore, provided that a suitable governing equation is obtained in terms of this new variable - which is actually available [49], the dynamic response of the structure can be obtained by a coarse set of sampled points even for high frequency excitation.

The physical displacement rapidly oscillating in space is transformed into a slowly oscillating solution that presents the further advantage of admitting an inverse transformation so that, if needed, the physical displacement can be recovered from the envelope displacement. Similar developments can be obtained for the pressure field in acoustic problems.

In figures 6a and 6b the complex envelope displacement and the recovered physical displacement for the flexural beam of figure 7 are presented, showing the advantages of this approach.

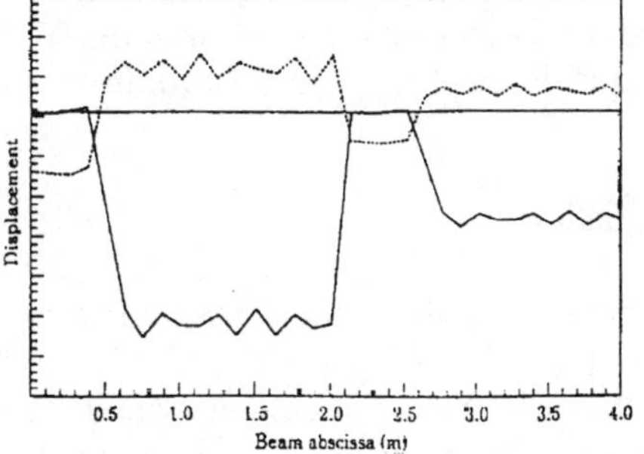
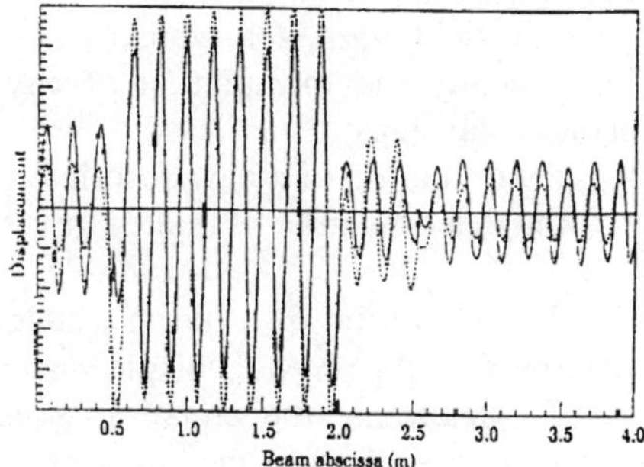

Fig. 6 Complex envelope displacement - real (---) and imaginary (- - -) parts, and comparison between physical (---) and recovered solution from CEDA (- - -)

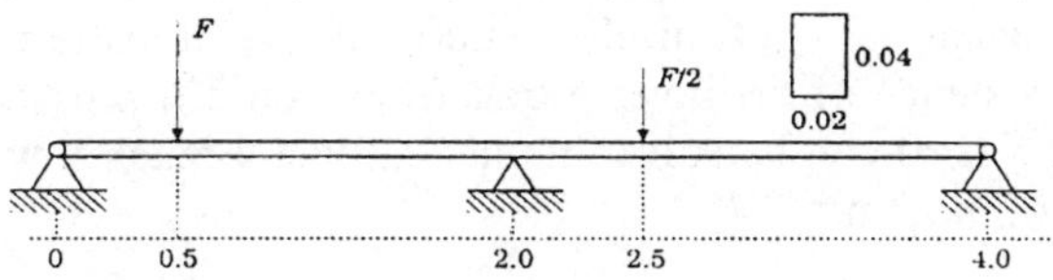

Fig. 7 Flexural beam

Notwithstanding these promising results, it must be stressed that some difficulties must be overcome before a complete extension to two-dimensional systems and to coupled structural-acoustic problems can be reached. In fact the extension of this approach to two-dimensional systems is not immediate in that it implies a more complex treatment of the wavenumber spectrum shift.

In fact, it must be underlined that, for two-dimensional systems, the spectrum of the analytic signal is not anymore concentrated around the wavenumber k_0 but rather it occupies an approximate ring-shaped zone in the two-dimensional wavenumber spectrum around the radius k_0, that is much more difficult to shift into the wavenumbers' origin. Different approaches to overcome this difficulty have been studied and compared, and a definite answer is expected shortly.

Momentarily a hybrid approach has been developed to provide a solution to two-dimensional problems (**Adamo, Sestieri** and **Carcaterra** [50]). Although the original ideas of this method rely on the CEDA procedure, it is not anymore an envelope approach, so that it cannot be considered in the framework of the envelope methods.

3.5 Comments on the field variables and governing equations

Any high frequency approach uses, instead of the physical displacement, a different field variable: typically some kind of energy average is introduced. Consequently the formulation of any technique is implicitly performed in two fundamental steps:
-	the definition of a new descriptor;
-	the determination of related governing equations.

The definition of a new variable ξ is obtained by the action of a transform operator $\mathbf{T}$ on the physical displacement w: $\xi = \mathbf{T}w$.

To determine the equation governing the new field variable it would be necessary to start from the equation of motion $\mathbf{L}[w] = p$, where $\mathbf{L}$ is a structural dynamic operator and p the external load. The new equation is determined as:

$$\mathbf{G}(\xi, p) = 0$$

Two operators are then introduced: a transform operator $\mathbf{T}$ and a governing operator $\mathbf{G}$. To understand advantages and limitations of each formulation it is of paramount importance to confirm the nature of the transform operator and, particularly, the existence of the inverse transformation $\mathbf{T}^{-1}$, provided that the new governing equation $\mathbf{G}=0$ can be solved numerically at a much lower cost than the original equation of motion.

When the existence of the inverse operator fails, serious drawbacks arise:
-	the use of ξ instead of w produces a loss of information. This is a typical situation with any energy approach. The knowledge of an average energy does not allow to recover the physical displacement, and any information on the local response is lost;
-	the existence of $\mathbf{G}$ itself is not guaranteed, and, even if it exists, its determination is not simple;
-	the evaluation of the boundary and joint conditions for the new variable is a difficult task. Very often, as in SEA and thermal approaches, different

experimental and numerical techniques are required to determine appropriate coefficients that are necessary to assemble the systems together;

- even when **G** exists, the forcing term cannot be determined in function of the physical load p alone, but rather depends on both p and w. This is typical of all energy approaches, where the forcing term is the input average power.

These limitations seem to be the price to pay for the use of the new variable ξ, presenting a lower information content than w, although, for the same reason, the solution of $\mathbf{G}(\xi, p) = 0$ implies a lower computation cost than $\mathbf{L}[w] = p$.

However this is not a general rule: provided that the inverse of **T** exists, sometimes it is possible to obtain a simple solution for the problems mentioned above.

Although a common definition of the energy variables is not encountered in the thermal methods, so that a non unique definition of the transform operator can be provided, in general for them the inverse of **T** does not exist, with consequent problems for the joint conditions, the impossibility of recovering the physical solution and the difficulty of computing **G**.

On the contrary, for the envelope methods it is shown that the envelope operator admits an inverse, i.e. $\mathbf{E}^{-1} = \mathrm{Re}\{(\cdot)\,e^{jk_0 x}\}$. In particular for the CEDA, the transform operator is the envelope operator, which is invertible, and the form of **G** is easily determined by the relationship $\mathbf{G}=\mathbf{E}\mathbf{L}\mathbf{E}^{-1}$. These circumstances would suggest that the CEDA approach can provide reliable results under good general conditions: it is possible to recover the physical displacement, it is easy to express the boundary and joint conditions in terms of the physical conditions and, last but not least, the forcing term is a simple function of the physical load.

4. COMMENTS AND CONCLUSIONS

Vibroacoustics covers the human hearing bandwidth between 20 and 20000 Hz. Only in the low frequency range will reliable and efficient solutions exist that can be obtained by classical finite element methods or boundary elements or the Succi's approach. For the medium or high ranges, SEA, energy methods and envelope methods have been proposed to provide solutions that can be used at a design stage. Unfortunately SEA has not reached full acceptance nor are the other methods yet to be considered valid alternatives.

Notwithstanding the large amount of work developed, there is still a real theoretical and practical interest - and a large space - on this field that should push researchers to turn their interest toward this argument. Industrial requirements would be a convincing reason, but I would say that the subject per se is really fascinating involving interdisciplinary topics ranging from theoretical physics, statistics, acoustics and classical structural vibrations.

ACKNOWLEDGMENT

Most of the figures in the text are drawn from other papers by other authors. Permission for reproducing these figures has been obtained from the editors of the different journals and publications: this permission is gratefully acknowledged.

REFERENCES

[1] Nefske D.J., Wolf J.A., Howell L.J., "Structural acoustic finite element analysis of the automotive passenger compartment: a review of current practice", *J. Sound and Vibration*, vol. 80, 247-266, 1982.

[2] Shenk H.A., "Improved integral formulation for acoustic radiation problems", *J. Acoustic Society of America*, vol. 44, 41-58, 1968.

[3] Sestieri A., Del Vescovo, Lucibello P., "Structural-acoustic coupling in complex shaped cavities", J. *Sound and Vibration*, vol. 96(2), 219-233, 1984.

[4] Succi G., "The interior acoustic field of an automobile cabin", *J. Acoustic Society of America*
 vol. 81(6), 1688-1694, 1987.

[5] Carcaterra A., Sestieri A., " Sensitivity analysis for statistical characterization of uncertain systems", *Proc. IMAC 15*, Orlando, 1997

[6] Ibrahim R.A., "Structural dynamics with parameter uncertainties", *Applied Mechanics Review*, vol. 40(3), 309-328, 1987.

[7] Fahy F., "Statistical Energy Analysis: a wolf in sheep's clothing?", *Proc. Internoise*, Leuven (Belgium), 1993.

[8] Fahy F., "Statistical Energy Analysis: a critical overview", on *Statistical Energy Analysis Ed. A.J. Keane \& W.G. Price*, Cambridge Univ. Press, 1994.

[9] Kompella M.S., Bernhard B.J., "Measurement of the statistical variation of structural-acoustic characteristics of automotive vehicles", *Proc. SAE Noise and Vibration Conf.*, Warrendale, USA, 1993.

[10] Plunt J., Personal communication.

[11] Maidanik G., "Cursory historical commentary on SEAmanship: then (1961) through now (1990)", *Proc. Int. Conf. on Air and Structure-Borne-Sound and Vibration*, Auburn, USA, 1990.

[12] Lyon R.H., Maidanik G., "Power flow between linearly coupled oscillators", *J. Acoustic Society of America*, vol. 34(5), 623-639, 1962.

[13] Lyon R.H., Scharton T.D., "Vibrational-energy transmission in a three-element structure", *J. Acoustic Society of America*, vol. 38(2), 253-261, 1965.

[14] Kakar M.P., "Power flow between linearly coupled oscillators", Ph.D. Thesis, Sheffield Univ., 1969.

[15] Lyon R.H., Eichler E., "Random vibration of connected structures", *J. Acoustic Society of America*, vol. 36(7), 1344-1354, 1964.

[16] Woodhouse J., "An approach to the theoretical background of Statistical Energy Analysis applied to structural vibration", *J. Acoustic Society of America*, vol. 69(6), 1695-1709, 1981.

[17] Dowell E.H., Kubota Y., "Asymptotic Modal Analysis and Statistical Energy Analysis of dynamical systems", *J. Applied Mechanics*, vol. 52, 949-957, 1985.

[18] Keane A.J., Price W.G., "Statistical Energy Analysis of strongly coupled systems", *J. Sound and Vibration*, vol. 117(2), 363-386, 1987.

[19] Langley R.S., "A general derivation of the Statistical Energy Analysis equations for coupled dynamic systems", *J. Sound and Vibration*, vol. 135(3), 499-508, 1989.

[20] Maidanik G., Dickey J., "Wave derivation of the energetics of driven coupled one-dimensional dynamic systems", *J. Sound and Vibration*, vol. 139(1), 1990.

[21] Pinnington R.J., White R.G., "Power flow trough machine isolators to resonant and non resonant beams", *Shock and Vibration Control Course*, ISVR, 1982.

[22] Noiseux D.U., "Measurement of power flow in uniform beams and plates", *J. Acoustic Society of America*, vol. 47, 238-247, 1969.

[23] Newland D.E., "Calculation of power flow between coupled oscillators", *J. Sound and Vibration*, vol. 3(3), 262-276, 1966.

[24] Skudrzyk E., "The mean-value method of predicting the dynamic response of complex vibrators", *J. Acoustic Society of America*, vol. 67(4), 1105-1135, 1980.

[25] Cuschieri J.M., "Power flow as a complement to SEA and finite element analysis", *ASME NCA-vol. 3, Statistical Energy Analysis,* 1987.

[26] Langley R.S., "A wave intensity technique for the analysis of high frequency vibrations", J. *Sound and Vibration*, vol. 159(3), 483-502, 1992.

[27] Belov V.D., Rybak S.A., Tartakovski B.D., "Propagation of vibrational energy in absorbing structures", *Soviet Physics Acoustics*, vol. 23(2), 1977.

[28] Buvailo L.E., Ionov A.V., ''Application of the finite element method to the investigation of the vibro-acoustical characteristics of structures at high audio frequencies'', *Soviet Physics Acoustics*, vol. 26(4), 1980.

[29] Nefske D.J., Sung S.H., ''Power flow finite element analysis of dynamic systems: theory and application to beams'', *ASME NCA-vol. 3, Statistical Energy Analysis*, 1987.

[30] Palmer J.D., William E.J., Fox C.H., ''Application of the energy flow approach to vibrational analysis of real structures'', *Proc. 5th Int. Conf. on Recent Advances in Structural Dynamics*, Southampton, U.K., 1994.

[31] Palmer J.D., William E.J., Fox C.H., ''High frequency power flow in structures'', *Proc. IMAC10*, San Diego, USA, 1992.

[32] Wolhever J.C., Bernhard R.J., ''Mechanical energy flow models in rods and beams'', *J. Sound and Vibration*, vol. 153, 1-19, 1992.

[33] Bouthier O., Bernhard R.J., Wolhever J.C., Energy and structural intensity formulation of beam and plate vibrations'', *Proc. 3rd Int. Conf. on Intensity Techniques*, Senlis, France, 1990.

[34] Le Bot A., Jezequel L., ''Energy formulation for one-dimensional problems'', *Proc. Acoustica 93*, Southampton, 1993.

[35] Le Bot A., Jezequel L., ''Energy methods applied to transverse vibrations of beams'', *Proc. 4th Int. Conf. on Intensity Techniques*, Senlis, France, 1994.

[36] Le Bot A., Luzzato E., '' Smooth energy formulation for multi-dimensional problems'', *Workshop on Methods in medium and high frequencies: the alternatives to SEA*, Clamart (Electricité de France), France, 1994.

[37] Lase Y., Ichchou M.N., Jezequel L., ``Energy flow analysis of bars and beams: theoretical formulations'', *J. Sound and Vibration*, vol. 192(1), 281-305, 1996.

[38] Ichchou M.N., Jezequel L., ``Comments on simple models of the energy flow in vibrating membranes and on simple models of the energetics of transversely vibrating plates'', *J. Sound and Vibration*, vol. 195(4), 679-685, 1996.

[39] Carcaterra A., Sestieri A., ``Energy density equations and power flow in structures'', *J. Sound and Vibration*, vol. 188(2), 269-282, 1995.

[40] Carcaterra A., Adamo L., ``Wave energy exchange in dynamical systems: theoretical and experimental analysis'', in print on *J. Sound and Vibration*, 1999.

[41] Langley R.S., "On the vibrational conductivity approach to high frequency dynamics for two-dimensional structural components", *J. Sound and Vibration*, vol. 182(4), 637-657, 1995.

[42] Smith M.J., " A hybrid method for predicting high frequency vibrational response of point-loaded plates", *J. Sound and Vibration*, vol. 202(3), 375-394, 1997.

[43] Xing J.T., Price W.G., "The energy flow equation of continuum dynamics", *IUTAM Symposium on Statistical Energy Analysis*, Southampton, 1997.

[44] Le Bot A., "A vibroacoustic model for high frequency analysis", *J. Sound and Vibration*, vol. 211(4), 537-554, 1998.

[45] Orefice G., Guyader J.L., Cacciolati C., "The energetic mean mobility approach (EMMA)", *IUTAM Symposium on Statistical Energy Analysis*, Southampton, 1997.

[46] Pradlwarter H.J., Schüller G.I., ``Statistical Energy Analysis in view of stochastic modal analysis", *IUTAM Symposium on Statistical Energy Analysis*, Southampton, 1997.

[47] Sestieri A., Carcaterra S., ``An envelope energy model for high frequency structural problems", *J. Sound and Vibration*, vol. 188(2), 283-295, 1995.

[48] Carcaterra A., Sestieri A., ``Envelope versus envelope-phase energy model for high frequency vibrations", *Proc. IMAC 13*, Nashville, USA, 1994.

[49] Carcaterra A., Sestieri A., ``Complex envelope displacement analysis: a quasi-static approach to vibrations", *J. Sound and Vibration*, vol. 201(2), 205-233, 1997.

[50] Adamo L., Sestieri A., Carcaterra A., ``Vibroacoustics: extension of the envelope approach to two dimensional structures ", Proc. ISMA 23, 1998

Warping the time Axes or, Non-linear is what we call that which we do not understand

Håvard Vold
Vold Solutions, Inc., 1716 Madison Road, Cincinnati, OH 45206

This author has butted his head into many a ceiling in trying to fit elaborate parametric models to test. In his younger days he looked towards colleagues and mentors for help, and more often than not, the source of the trouble was labelled non-linearity. As the author grew older, wiser, but not as smart as before, he realised that his kind of non-linearity has a very pragmatic definition: Non-linear is what we call that which we do not understand. This definition is also subjective; your linearity may be my non-linearity.

The basic rule in statistics and empirical science is that the more that we invest in up front assumptions and parameters, the more resolution our estimation models will have. This is called statistical degrees of freedom, and shows itself, for example, in the increased spectral resolution of a maximum entropy spectrum over a discrete Fourier transform spectrum. The problem is that as we add a priori assumptions, our models become less robust, and we often have severe problems explaining the lack of fit of models to experimentally acquired data. What our disciplines really need are good ways of visualising test data without requiring restrictive assumptions, such as linearity.

One of the more challenging aspects of experimental dynamics is the description of transient noise and vibration phenomena. Transient here means that the system, loads, or boundary conditions change with time. The usual approaches to analysing such data are to perform successions of partially overlapped myopic analyses, using tools that assume stationary behaviour. The usual effects are smearing and poor resolution. This may be acceptable if the non-stationarity is mild.

The author proposes a general approach for an important subclass of transient phenomena, namely those that can be reduced to a quasi-stationary case by suitable monotonic transformations of the time axis. Quasi stationary means here that the apparent variations of systems properties are more benign than the original formulation, and that hence, stationary analysis tools have a wider range of applicability.

The prime example of such a transformation is in the author's opinion the resampling of rotating machine data from the time domain into even angle increments around the primary shaft. In this angle domain, frequency analysis tolls may be applied, because all rotation-induced phenomena seem to happen at fixed frequencies; all significant frequencies separated by a common step. Another application is to resample data from a moving microphone source according to an instantaneous Doppler shift correction factor. Frequencies on the ground become the same as frequencies at the moving source, and, to compound this, another resampling may be performed to transform to the crankshaft angle domain. Phase lock between moving data acquisition stations may be achieved by tracking out of band reference tones.

The point is that by choosing appropriate transformations of the time axes, transient and bewildering phenomena may become tractable with classical tools. Linearity has nothing to do with this, we have just taken some tricks from relatively theory and applied them to classical mechanics. Much insight can be gained by elementary means by investing into an accurate determination of the time axis transformations.

This author believes that a ruthless exploitation of time axis transformations, coupled with classical data analysis methods will enable a large class of phenomena to be visualised and quantified. The challenge is to define standard procedures for identifying such transformations and then applying them.

Part III

SD2000 Review –
Experimental Modal Analysis

David L Brown
University of Cincinnati

INTRODUCTION

This conribution is a review of the current state-of-the-art technologies in the area of modal analysis. Modal analysis refers to the process of estimating either analytically or experimentally a set of basic characteristic functions or solutions (modal parameters) which can be used to describe the static and dynamic response of a system. All possible responses of the system can be estimated by a superposition of these characteristic solutions. These characteristic functions are the eigenvalues, eigenvectors and scale factors of the equations of motion of the system. In this contribution the precise mathematical description of the area of modal analysis will be sacrificed and a more conceptual presentation will be made for brevity.

This review will be presented with a historical perspective because the evolution of a technology often will fold over on itself. Methods which were discarded are often rediscovered in the presence of evolving technologies.

ACRONYMS AND NOMENCLATURE

$[A]$	Residue Matrix
$\{f\}$ or $\{F\}$	input vector (forces)
$[h(t)]$	unit impulse response matrix
j	square root of minus one
$[H(\omega)]$	frequency response matrix
$[L]$	modal participation matrix (right hand eigenvectors–input characteristic functions)
p	index for response point
q	index for input point
t	time
$\{x\}$ or $\{X\}$	response vector (displacements)

116

$[\lambda]$	diagonal matrix whose elements are functions of time or frequency (eigenvalues–temporal characteristic functions)

$$\lambda_{rr}(t) = e^{\lambda_r t}$$

$$\lambda_{rr}(\omega) = \frac{1}{j\omega - \lambda_r} + \frac{1}{j\omega - \lambda_r^*}$$

r	index for eigenvalues
$[\Psi]$	modal matrix (left hand eigenvectors–response characteristic functions)
ADC	Analog Digital Converter
CEA	Complex Exponential Algorithm
CMIF	Complex Mode Indicator Function
DAC	Digital Analog Converter
DFT	Discrete Fourier Transform
DOF	Degree-of-Freedom
eFRF	Enhanced Frequency Response Function
EMIF	Enhanced Mode Indicator Function
ERA	Eigenvalues Realization Algorithm
FFT	Fast Fourier Transform
LSCE	Least Squares Complex Exponential
MAC	Modal Assurance Criteria
MDOF	Multiple Degree-of Freedom
MIMO	Multiple Input Multiple Output
MRIT	Multiple Reference Impact Testing
IMAC	International Modal Analysis Conference
ITD	Ibrahim's Time Domain
PC	Personal Computer
PFD	Polyreference Frequency Domain
PTD	Polyreference Time Domain
SDOF	Single Degree-of-Freedom
SV	Singular Values
SVD	Singular Value Decomposition
UMPA	Unified Matrix Polynomial Approach

CONCEPTS

In this section the concepts of modal analysis will be addressed. Modal parameters are characteristic functions which are the building blocks for all possible solutions to the equations of motion of a system. For mechanical systems, all possible input-output (force-displacement) solutions are simple superpositions of these characteristic functions. Think of these characteristic functions as Lego (children's building toy).

Complicated structures can be built from a small set of different Lego elements. These Lego elements are the building blocks, and superposition of the blocks can be used to construct complicated structures. Since the geometry of these elements is known, it is easy to design a structure. This leads to one of the important assumptions of modal analysis. The system must be linear. For the Lego case, if the geometry of the elements depends upon their position in the structure or a point in time, it would be difficult to design the structure. This would be the case if the system was nonlinear or non-stationary. The characteristic functions would not be constants of the system and superposition would not work.

The characteristic functions are functions of input-output locations and time/frequency. In other words, they describe the spatial and temporal information about the system. There is one set of characteristic functions which describes the input space, a second set which describes the output space, and a third set which describes the temporal space (time-frequency). The temporal characteristic function is represented by the system eigenvalues, and the input-output functions by sets of left hand and right hand eigenvectors. A three dimensional characteristic space can be defined with the input coordinate along one axis, the output coordinate along the second axis, and the temporal information (time/frequency) along the third axis. This characteristic space corresponds to the three dimensional frequency response matrix or unit impulse matrix. Figure 1 is a graphical representation of the characteristic space. Any point in the frequency response matrix has the value of the frequency response function at a given frequency for a given input coordinate and given output coordinate. At first glance, this characteristic space would appear to contain nearly infinite information, but in reality it can be generated with only a small number of eigenvalues and eigenvectors.

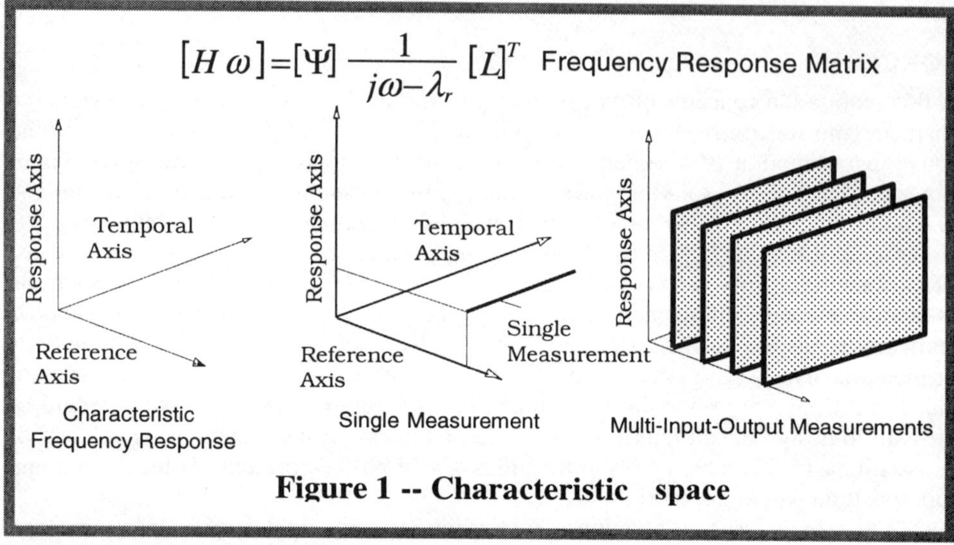

Figure 1 -- Characteristic   space

Mathematically, the response of the system can be computed from the system characteristic functions by the following relationships:

Time Domain

$$\{\chi(t)\}=[h(t)]*\{f(t)\} \qquad \text{Convolution} \qquad (1)$$

Frequency Domain

$$\{X(\omega\}=\{H(\omega)]\{F(\omega)] \qquad (2)$$

where

$$[h(t)]=[\Psi][\lambda(t)][L]^t \qquad \text{Unit Impulse Response Matrix} \quad (3)$$

and

$$[H(\omega)] = [\Psi][\lambda(\omega)][L]^t$$

Frequency Response Matrix   (4)

In equations (3) and (4), the columns of the $[\Psi]$ matrix are the left hand eigenvectors which are the output space characteristic functions. The columns of the $[L]$ matrix are the right hand eigenvectors which are the input space characteristic functions. These functions are spatial characteristics and are only functions of geometry. The $[\lambda]$ matrix is a diagonal matrix whose diagonal elements are determined by the eigenvalues and are functions of time or frequency. In order to experimentally determine these characteristic functions, it is important to observe all three dimensions of the characteristic space.

This means that a set of measurements would have to be made by exciting at all potential input points and measuring the response at all output points. These measurements would also have to span the temporal space. In other words, the system has to be observable. Experimentally, this would be a formidable if not impossible task. Fortunately, the equations of motion for most mechanical systems are symmetric. This results from the fact that the effective stiffness and damping which act between point p and point q are the same as what acts between point q and point p. In other words, most mechanical structures obey reciprocity. Analytically, this means the left and right hand eigenvectors are the same or transposes of each other, depending upon how the eigenvectors are expressed mathematically. As a result, it is only necessary to completely span either the input space or the output space.

Theoretically, this means that when experimentally testing a structure, it is only necessary to excite at a single input point with the following exceptions:

- The excitation point is not at a node of one of the eigenvectors.
- There are no repeated eigenvalues.

This reciprocity attribute makes modal testing a conceivable process.

Practically, many eigenvectors are non-observable in the presence of measurement noise or are so closely coupled that they appear to be repeated. Therefore, it is generally necessary to use multiple excitation points.

The above concepts are used to explain three of the important assumptions of experimental modal analysis:

1. The system has to be linear.
2. The system has to be time invariant.
3. Experimentally, the system has to be observable.

Another conceptional point of interest is that modal parameters that can easily be generated analytically are often very difficult to measure experimentally. The modal parameters are the result of an eigenvalue-eigenvector solution technique. This method performs a coordinate transformation, which uncouples the equations of motion. The modal parameters are solutions in this transformed coordinate system (modal coordinates). When testing a system, all forces are applied in the physical coordinate system. In a physical coordinate system, the forces are applied at discrete input points on the structure, while in modal coordinates the forces are distributed over the structure. Distributing forces proportional to the eigenvectors is difficult to do experimentally.

For modal coordinates, the transformation vectors are the eigenvectors. In this coordinate system, the forcing function is proportional to the given eigenvector of the system. In order to measure one of the characteristic solutions, the system has to be excited with a modal forcing function. As result, directly measuring the modal parameters is difficult.

When the structure is excited at a discrete point, a large subset of the modal characteristic functions are present in the data set. The modal parameters must be extracted from this data set by a parameter estimation process. In recent years, the largest research effort in experimental modal analysis has been the development of parameter estimation algorithms to extract modal parameters from the measured input-output responses.

The problem of matching analytical and experimental conditions is an important consideration that has historically created problems between the analyst and the experimentalist. It is important when developing a test that the limitations of the testing be understood. For example, a common testing requirement is to specify that a test object be tested in a fixed mounting condition. The test object, most likely, is not mounted in this configuration in practice but is analytically modeled in this configuration. In the laboratory it may not be possible to support this test article in a fixed condition, therefore, the test results will not reflect a fixed boundary condition. Analytically, the test object can easily be constrained in another manner which can be more easily tested. It is important for the analyst and the experimentalist to understand the limitations of both technologies. Of course, these considerations are important for other parties that are involved in planning a test. An example where this can be a problem is when an interested party insists that the object be tested in an operating configuration. If there are large nonlinear components in this configuration, then it can be difficult to perform a modal test of the complete test object. By removing these components and replacing them with linear simulators, the test can be performed. The linear characteristics of the complete test object can be verified. Auxiliary testing can be performed to identify the nonlinear characteristics of components which were removed.

EXPERIMENTAL TESTING PROCEDURES

Modal testing has become a fairly mature technology and there are numerous testing methods which can be used to perform a test. There are many reasons for performing a modal test, and some test methods are better suited to certain tests. However, in most cases, even for a given test there are several choices that are available. In the next several sections, modal testing procedures will be reviewed. Detailed information on each technique will not be included. Only a general description of the technique will be given.

GENERALIZED PROCEDURES

In experimental modal testing, the response to external forcing functions is measured along with the external forcing functions when possible. In most cases the input forces can be measured, but there are cases where unmeasurable operating forces or free decays may be used. The modal parameters are estimated from the output response data and the input data. Historically, two basic techniques have been used. In the first technique, a modal force is applied and the response corresponds to one of the system eigenvectors. In the second technique, forces are applied at discrete input points so that the response is the summation of a subset of the modal parameters. For this case, the modal parameters are extracted by curve fitting to an analytical expression, or by fitting a system model to the input-output measurements and solving this system model for the modal parameters.

A typical modal test can be divided into the following general steps:

1. Test Setup–Pretest Analysis
   a. reason for performing test
      i. trouble shooting
      ii. finite element model (FEM) verification
      iii. finite element model updating
      iv. generating an analytical model based upon experimental modal parameters
      v. condition monitoring
   b. pretest analysis
      i. Finite Element Analysis
         (1) target modes
      ii. excitation locations
      iii. response locations
      iv. frequency range
         (1) frequency resolution
   c. pretest testing
      i. nonlinear testing
         (1) is the structure nonlinear
         (2) what kinds of nonlinearities
         (3) can nonlinearity be removed
      ii. impact testing to check pretest analysis

                (1) are exciter locations good?
                (2) initial estimates of eigenvalues

   d.   testing methods
      i.   normal mode method
      ii.   parameter estimation method
      iii. operating inputs (ambient testing)
      iv. free decays

   e.   excitation methods
      i.   roving input
      ii.   fixed inputs
      iii. operating inputs

   f.   boundary conditions
      i.   fixed
      ii.   free-free
      iii. hybrid
      iv. operating

   g.   calibration of sensors
   h.   cable management

2.   Data Acquisition
   a.   real time
   b.   post processed
   c.   forces
      i.   single inputs
      ii.   multiple inputs

   d.   responses
      i.   sequential
      ii.   simultaneous

   e.   type of inputs
      iii. periodic
      iv. random
      v.   transient
      vi. operating

3.   Signal Processing and computations of measurements
   a.   FFT analysis
      i.   windows
      ii.   frequency response estimation
            (1) H_1
            (2) H_2
            (3) H_v

      b.   time domain
          i.   windows
          ii. correlation matrices
          iii. random decrement

4.   Parameter Estimation
      a.   curve fitting analytical expression
          i.   time domain
          ii. frequency domain
          iii. SDOF
          iv. MDOF
      b.   estimating system model
          i.   time domain
          ii. frequency domain
          iii. spatial domain

5.   Modal Model Assimilation and Validation
      a.   assembling modal model from various estimation processes
      b.   validation of modal model

The above list addresses many, but not all, of the aspects of performing a modal test. These are the items that would normally be covered in a survey or complete review of modal testing. Obviously, several complete books could be written on this subject, for example, a book on just the parameter estimation aspects of modal testing. There have been hundreds of papers written exclusively on the subject of modal parameter estimation. In fact, the best reference for the complete modal testing may be the proceedings from the International Modal Analysis Conference (IMAC) where most of the modal analysis technology has been published in recent years. In the following sections, selected items taken from the above list will be discussed.

TEST SETUP

The first question is, What is the purpose of the test? The test purpose could be to solve a simple troubleshooting problem, in which case, the test could simply mean grabbing a few accelerometers, an instrumented hammer and a simple portable multiple channel analyzer off the shelf, mounting the accelerometers on a test object, beating the test object with a hammer to measure a set frequency response functions, and using simple parameter estimation procedure to get a quick estimate of the modal parameters. In other words, performing a multiple reference impact test to solve the trouble-shooting problem. In this case, very little pretest analysis is required if the test object is already a known item. This contrasts with a very large modal test for finite element verification of updating. For this type of test, pretest planning can be extensive. Sometimes these tests are scheduled a year or more in advance. FEMs are built of these test structures to determine which modes are important to identify, the best locations to excite these modes, and the best response locations to identify the target modes. For the largest of these tests, 10 to 20 exciter positions and up to 1500 response positions may be used and a number of configurations tested.

Afterwards, the pretest analysis is performed. A pretest testing program can be performed to determine the degree of nonlinear response of the system. If the system has significant nonlinear response, then additional testing is performed to better understand the location and the characteristics of the nonlinearities. Can the system be linearized? The pretest testing can also be used to determine the excitation and response locations and to check on the modal density and give an initial estimate of the system eigenvalues. These initial measurements are useful in selecting a test method.

MODAL ANALYSIS APPLICATIONS

Experimental Modal Analysis has been used for a number of important applications:

Troubleshooting

Historically, troubleshooting has been by far the biggest application of experimental modal analysis. Solving vibration, noise, and to a lesser extent controls problems have been the major troubleshooting applications. A typical application might be solving a self-excited vibration or chatter problem on a machine tool. This application is a field testing application because the machine tool is normally tested in the manufacturing plant. Another application may be testing a bridge or another type of infrastructure to determine if it has deteriorated or suffered damage. Again, this is a field testing situation.

For troubleshooting applications, a quick and dirty testing procedure can very often be used. Multiple reference impact testing is one of the more modern testing methods for troubleshooting. In this testing method, a number of accelerometers are mounted on the test specimen and a roving instrumented hammer is used for input. This method can easily be set up in field testing applications. The preferred data acquisition system is a small portable multi-channel analyzer connected to a notebook computer. Multiple reference data is obtained with this method and the modal parameters can be extracted using any of the commercial multiple reference parameter estimation algorithms. However, Complex Mode Indicator Function (CMIF), which is one of the newer spatial domain algorithms, is preferred. This algorithm is very simple to use and minimizes the problems of sorting computational modes from real modes. The parameter estimation algorithms will be covered in more detail in later sections of this review.

Testing methods which use operating input as the excitation are also good for troubleshooting. In these methods the actual operating forces excite modes of interest. If the inputs are periodic, then order tracking can be used to extract operating mode shapes at problem frequencies. If the operating inputs can be measured or estimated indirectly, then ordinary frequency response functions can be measured. These frequency response functions can be used to estimate modal parameters. For large structures such as bridges, the operating inputs are often broadband random inputs caused by the wind or traffic flow. In these cases, power spectra, cross spectra, correlation matrices, or random decrement methods can be used as inputs into parameter estimation algorithms for extracting the modal parameters.

Finite Element Verification

Finite Element Modeling (FEM) is becoming a commonly used design tool for analyzing the static and dynamic characteristics of mechanical systems. FEM is used to develop an initial design of a structural component. After a prototype of the component is manufactured, it is important to verify the FEM model experimentally so that FEM can be used to make design modifications to the initial design. Also, the verification process builds up experience so that the FEM can be used with confidence on future design projects.

Finite element verification requires the most sophisticated modal testing procedures. There have been two general methods for performing these types of tests. The first method is the normal mode testing method, and the second method is the broadband multiple exciter testing method. The first method has been used since the early '60s for testing aerospace structures. The second method has been the standard testing method over the past ten years.

In both of these methods, the number and location of the exciter positions are determined in the pretest analysis. Likewise, the number and location of the response points are also determined. The finite element model is used to define these points and to identify target modes. The pretest testing program is used to identify potential testing problems such as nonlinearities.

For the more standard broadband excitation method, typically 200 to 500 accelerometers are mounted on the test structure and two to four exciters are used to excite the structure. A multiple input data acquisition system is used to simultaneously measure the input and response matrix. Uncorrelated random excitation signals are used to excite the structure. Multiple Input Multiple Output (MIMO) signal processing is used to compute the frequency response matrix. Advanced parameter estimation algorithms (PTD, ERA, etc) are used to extract the modal parameters.

Finite Element Updating

The updating of a FEM model is an important process. This involves locating and correcting errors which may exist in the original FEM model. This is a difficult process since the FEM has a large number of Degrees Of Freedom (DOF). The experimental database is normally limited to the first 10 to 100 eigenvalues, depending upon the nature of the test article. Often this is not enough information to locate and characterize the errors in the FEM model. Another point that complicates the testing is that often the test object is a component of a larger system which connects into the system at a number of points. It is normally impossible to experimentally match the boundary conditions at these connection points. This means that the 10 to 100 modes measured may not include the important information at these connection points. Auxiliary measurements need to be taken at these connection points and this data has to be incorporated into the updating process.

The testing methods used for the finite element updating are the same as those used for the verification process.

Several new techniques are being investigated for the updating cases. The purpose of these methods is to increase or supplement the experimental database. The first method supplements the modal data with driving point frequency response measurements at the connection points. This data are used to update the model at the connection points. The second method is the Perturbed Boundary Condition (PBC) testing method. The PBC test procedure measures a larger database for verification and/or updating of Finite Element models. The test article is tested in a number of configurations where the boundary conditions are drastically perturbed. These additional configurations expand the database that is used for the updating process. Each configuration can generate 10 to 100 additional eigenvectors that can be used in the updating process. Since all of the transducers are premounted on the test article, the data for each configuration can be taken in approximately 30 minutes.

Experimental Modeling

Very often a FEM model for a component does not exist, but the hardware for the component does exist. An example of this might be an air conditioner compressor used to cool an automobile. This air compressor may contribute to a noise problem. If a model of the engine-accessory package is developed, an analytical model of the air compressor can be developed by experimentally testing the existing air compressor and determining either an impedance model or a modal model of the compressor. This model can be used as a sub-component in the engine-accessory system model.

There are two common types of analytical models based on experimental data. The first is an impedance model where the measured frequency response matrix is used to model the relationships between the input and outputs of the system. The second is a modal model. The testing methods for the component could use an impacting or an exciter test method. This decision depends on the size and complexity of the test item. For the impedance model, measurements have to be made only at the connection points, but a complete impedance matrix must be measured. In other words, measurements must be made between every input and output point of the structure. This would be an unimaginable task if the number of points was large. However, for cases where the number of connection points is small, this is a viable method. The impedance functions contain all of the modal information for the frequency range of interest. This method is frequently used to design mass dampers that can be applied to a structure to reduce the influence of troublesome modes of vibration.

The second method is to measure the eigenvalues and eigenvectors of the component and to generate a modal model. This model is only as good as the measured modal database. Very often, modes that are not in the frequency range of interest are important. It is important to include the influence of these modes in the model. This is done by measuring residual terms at the connection points or by using a testing method such as the PBC method.

Condition Monitoring

In more recent years there has been increasing interest in using modal parameters for evaluating the health of infrastructures. In general, these are large structures that are very often located in remote locations. Two testing techniques are currently being pursued. The first is the Multiple Reference Impact Testing (MRIT) method and the second is ambient testing methods. For smaller structures such as overpass bridges, wind turbines, large industrial fans, etc., the MRIT impact testing has been used with success. For very large structures such as dams, large bridges and large buildings, ambient testing is being used. In both of these testing methods, it is almost a requirement that the transducers be premounted.

For some of the infrastructure testing applications, it is important not to interfere with the function of the structure. Generally, this is true for the testing of bridges since interruptions of traffic flow are undesirable. This means that the MRIT has to be performed in a short period of time and that any ambient testing must be done in the presence of the operating forcing functions.

TESTING METHODS

There are four basic testing methods:

Normal Mode Testing

The normal mode testing method is one of the earlier testing procedures. It dates back to the '50s. In this time period, computing capabilities were very limited. Measurements were performed in an analog sense. ADCs were nonexistent or very expensive. The testing was performed with sine testing and analog filters to exclude noise or harmonic distortion components. For this time period, about the only choice for performing a large scale modal test was to use modal forcing vectors. These forcing vectors would excite one modal characteristic function at a time. The deformation vector at the excitation frequency corresponded to an estimate of the eigenvector. The frequency and damping of the free decay, after the forcing function terminated, corresponded to an estimate of the eigenvalues. The forcing vector that was applied was also designed to remove the damping or to make the damping proportional to the mass or stiffness distribution so that a pure normal mode could be measured. Over time, more sophisticated methods were developed to estimate the eigenvalues. The modal forcing function(vector) was swept over a small frequency band around the eigenvalue which corresponded to the eigenvector being measured. The amplitude of the response of selected points was plotted as a function of frequency. The eigenvalues could be determined from the half power points measured in this plot. Later, parameter estimation algorithms were used to estimate the eigenvalues.

The problem of using this method was that it took great skill in being able to adjust the forcing function so that a single normal mode was excited. The adjustment included controlling the frequency and distribution of forces.

Over the years, this method has been refined by incorporating some of the parameter estimation technology into methods for tuning the modes and estimating the eigenvalues.

If used properly, this method gives excellent results.

Parameter Estimation Testing Methods

In the late '50s and early '60s, the only practical way to perform a big modal test was to use the normal mode testing method. This method required extensive equipment which was very expensive at the time. The only practical method for performing a normal mode test was to premount a large number of sensors to the structure. These sensors and the signal conditioning required by the sensors were very expensive. An approximate method that was used which required minimum equipment was to rove a single sensor or a small number of sensors over the structure. One or more exciters were used with a sinusoidal signal which were tuned to one of the eigenvalues. If the sensors were accelerometers, the 90 degree out of phase response of the transducers was measured, or if they were velocity sensors, the in phase response was measured as an estimate of the eigenvector at the point on the structure where the sensor was mounted. The transducers had to be roved over the structure for each frequency. This was a very time-consuming procedure.

In the early '60s, a technological breakthrough occurred with the development of the tracking filter. With this development, frequency response functions could be measured with swept sine testing in a timely fashion (timely for that period of time; it took 15 minutes or more to measure a single frequency response function). The frequency response functions included all of the significant eigenvalues. This development led to the idea of using a parameter estimation process to determine the eigenvalues and eigenvectors from the measured frequency response functions. Analytically, the frequency response functions could be formulated in terms of modal parameters. This led to a great deal of research on how to estimate modal parameters from frequency response functions. The formulation for solving for the modal parameters generated a set of nonlinear equations which were very difficult to solve. Remember that in the early '60s, computers had limited capabilities and were expensive to use.

The second technology breakthrough occurred in the middle '60s with the advent of the Fast Fourier Transform (FFT). Up till the time this breakthrough occurred, computing the Fourier coefficients using a Discrete Fourier Transform (DFT) took approximately 15 minutes on a mainframe computer of that time. Computer time cost about $1000 per hour at that time, so it was costly. After the implementation of the FFT, this time dropped to several seconds. Using broadband excitation signals, frequency response measurements could be made in seconds instead of minutes.

Impact testing was developed in the late '60s based on the FFT. Broadband random, pseudo random, and fast sine sweeps were all being investigated as possible excitation signals in the period of the late '60s and early '70s. This was a time of extensive research on signal processing techniques using the FFT, excitation methods, and parameter estimation methods for extracting modal parameters.

The Fourier Analyzer systems which were developed at this time were limited to a maximum of 4 channels and 10 bit ADCs at a cost of approximately $100,000. The computers which were built into these analyzers were limited to between 16K and 64K bytes of memory. FFT times, programmed in microcode on these machines, took approximately 1 second for 1024 FFT block sizes. In comparison, today using a $2000 PC and a FFT algorithm programmed in a high level language takes a fraction of a millisecond to compute.

Modal testing using the FFT machines gradually replaced the swept sine testing during the early to mid '70s. Modal tests were performed by either impact testing or roving a single triaxial accelerometer over the structure. Frequency response measurements were made between the exciter location and accelerometer locations. These frequency response functions were curve fit using parameter estimation algorithms.

The cost of the Fourier Analyzer dropped and more channels were added during the late '70s. During the late '70s, Multiple-Input-Multiple-Output (MIMO) methods were developed. It became obvious that it was important to have consistent sets of data. Multiple measurement parameter estimation algorithms were being developed. These algorithms were sensitive to inconsistent data. This led to the idea that all the data should be taken simultaneously in order to get the most consistent data. The problem was that the cost of these systems would be too expensive. At the time, it appeared that the cost of the data acquisition could not be significantly reduced, but the cost of the response sensors could. In the early '80s, this led to the development of the Structcell^{TM} system, which was a very inexpensive transducer system. The characteristics of the sensor were compromised in order to get more consistent data (errors of 20 to 30 percent were common in modal parameters determined from inconsistent data sets). The Strutcells' sensitivities could fluctuate by 5 percent or more. The net result was an improvement in the modal parameter estimates. At the time, this was a "chicken or egg" problem. In order to get an inexpensive system with many channels, the cost of the transducer system was compromised. This led to the development of a multiple channel data acquisition system. Once the multiple channel data acquisition system had been purchased, then the transducer system could be updated. It ultimately led to the development of inexpensive data acquisition systems.

The test setup for a large scale modal test today is to premount several hundred response sensors (a large test may have over a thousand) and to excite the system with a small number of exciters using uncorrelated broadband input signals. A randomizing input signal tends to result in data sets which can be better processed using the current MIMO parameter estimation algorithms.

For these large tests, cable management is very important in order to keep track of the transducers and to minimize book-keeping errors.

Operating Inputs (Ambient Testing)

There is an increased interest in using ambient inputs to estimate modal parameters of a structure. There was an initial interest in this technology in the early '70s, primarily in the civil engineering area. There was a great deal of work on the development of using Auto Regressive Moving Average (ARMA) algorithms to process this ambient data. One of the major assumptions is that the inputs are white random noise, or at least have a smooth input spectrum. There should be no peaks or valleys in the input spectrum because these peaks can be identified as resonances of the system. In general, these applications were of minimum interest to the modal analysis community as a whole in the early '70s. There was also some interest in the aerospace industry for processing flight data in the mid '70s using ambient inputs.

In the '90s, a tremendous interest has developed in using modal parameters for the monitoring of infrastructures. In the '90s, there have been an increasing number of researchers looking at the measurement and parameter estimation aspects of this problem.

This area of research is also driving one of the next advances in data acquisition systems. Since these systems are large and located in remote sites, transducers which have digital outputs and can be located and synchronized over a network are being developed. This is going to significantly impact the data acquisition aspects of ordinary modal testing in the near future.

Free Decay Testing

Free Decay Testing is a special case of ambient testing. For this case, the system is given an initial condition (displacement, velocity, etc.) and the free decay response of the system is measured. It should be noted that the unit impulse response of the system is a free decay. Therefore, the data can be processed using any of the parameter estimation algorithms used for processing frequency response or unit impulse measurements. This is a technique which can be used to test large or very small structures. It is a good field testing method. The data has to be measured simultaneously, or has to be measured relative to a reference transducer which is included in every free decay data set. Again, the advent of inexpensive multiple channels makes this an attractive method.

SENSORS-CALIBRATION-CABLE MANAGEMENT

Over the past 30 years there has been significant advances in sensors, signal conditioning and cable management systems. The most popular sensing systems used over the past 30 years have used piezoelectric sensing elements. Both charge-type and those with integrated electronics have been used. The integrated electronics have dominated since the mid '70s. The sensitivity of these sensors has increased, the size has been reduced, the dynamic range has significantly improved, and the cost has been reduced. In the '70s, data acquisition systems were limited to just a few channels, and the volume of sensors supplied to the modal analysis community was small. In the '80s, multiple channel data acquisition systems became popular, and less expensive sensing systems were developed. The standard large channel system used two to four load cells and 128 to 256 response channels (accelerometers). Improved cabling systems were developed so that large numbers of channels could easily be managed.

Sensors

The most popular response sensor has been the accelerometer, but microphones, strain gages, and scanning laser systems are also used. For smaller light-weight systems, the laser system is often the sensor of choice. There is currently research into making the laser more adaptable for making multiple axis measurements at a point, including rotations. However, the laser is still limited to line-of-sight, which limits its applications.

Ambient testing of large structures is becoming more popular, and smart sensors and digital sensors which can be distributed along a network are being developed. These transducers can be accessed over the Internet, making remote sensing a possibility. For large structures, the wiring cost can be significantly reduced. For example, for a large bridge, the wiring cost can easily exceed the cost of the sensors and signal conditioning. For this case, a single network cable or a wireless network can be used which significantly reduces the wiring cost.

Calibration

For large channel count systems, calibration has become a more important part of the pretest activity. Every sensor should be calibrated before and after every modal test. This can be a formidable task, since there are large numbers of sensors. For these cases, array calibrators are useful. There are commercial array calibrators, which can calibrate up to 120 channels simultaneously.

For the newly developed smart sensors, the calibration is stored in the sensor. If the sensor is moved during the test to a different acquisition channel, the calibration moves with sensor.

There are small hand held end to end calibrators which can be used with the modal test system to check the calibration or to calibrate a sensor during the test. Remember, the test data is only as good as the calibration.

Cable Management

For large channel count tests, the cable management system becomes a very important part of the acquisition system. The book-keeping can become one of the more difficult parts of a test. Cable management systems were developed in the mid '80s where keeping track of the cables was not necessary. The cable was arbitrarily connected from a location on the test structure to the data acquisition system. When the transducer was installed, it would self-identify. Bar-coded labels were used to read the serial number of the sensor and the location of the sensor on the structure. This tremendously reduced the test setup time and minimized much of the book-keeping.

The smart transducers which are currently being developed will take this a step further. Digital network transducers will make this possible over the Internet.

DATA ACQUISITION

Over the years, there has been a continuing and accelerating level of development of data acquisition systems. The data acquisition systems became important with the advent of the FFT algorithm in the mid '60s. The initial ADCs used with the FFT system were 10 bits resolution and sampled at 50K samples per second. These ADCs were $3000-$5000 per channel. A low pass filter was required for each acquisition channel to eliminate aliasing problems. The original filters had adjustable cutoff frequencies and typically had a 48 dB per octave roll-off after the cut-off frequency.

The cost of the data acquisition systems changed slowly and the resolution had increased to 12 bits by the late '70s. The big change occurred when the anti-aliasing filter was incorporated into the data acquisition channel. A fixed frequency anti-aliasing filter (or possibly several cut-off fixed frequencies) was used and the data was digitally filtered and resampled to get the right effective sampling rate. This also allowed the acquisition channel to perform a zoom transform. By the early to mid '80s, the cost of an acquisition channel was a little over $1000.

The big change in the resolution occurred in the mid '80s, with the commercialization of digital music (CD players). The digital music was 16 bits and 16 DACs and ADCs were developed for this market. Since the CD players were a consumer product, the cost of these ADCs dropped significantly. These were delta sigma ADCs which provided basic anti-aliasing protection at the chip level. A small analog anti-aliasing filter would normally be included in the designed ADC to guarantee anti-aliasing protection. 16 bits is approximately 96 dB of dynamic range in the ADC. This reduced the requirements on the transducer signal conditioning. The number of input gain settings could be reduced, which in turn reduced the cost of the front end. However, the cost of the acquisition channel was only slightly reduced, but the resolution had increased and the signal condition requirements were reduced.

Over the period of time from the mid '60s, the transducers also evolved significantly. During the '60s, the accelerometers were either force balance (servo accelerometers) or charge type. The servos were very good accelerometers but were fragile, large and expensive. The charge transducers were smaller but required a charge amplifier, and the cables had to be protected to guarantee a good impedance balance. If the cables became dirty, this could significantly affect the low end frequency response. Also, for the smaller transducers, the capacitance of the cable was important. Simply flexing the cable could change the capacitance and change the sensitivity of the sensor. Low noise cables were required which were expensive and fragile. A third type of transducer was becoming popular in the mid '60s. These were piezo-electric sensors with built in electronics. This transducer operated over a two wire system and required inexpensive conditioning. The major limitation was that these transducers had a relatively high noise floor, and as a result, limited dynamic range. By the early '80s, these transducers were the most popular transducer for modal testing. Most of the commercial data acquisition systems had built-in conditioning for these sensors. Over the years, the noise floor of the built-in electronics improved dramatically, thereby reducing one of the big limitations of these sensors. These transducers also have a very low output impedance, which means that unshielded cabling can be used, thereby reducing cabling cost.

In the mid '80s, a very inexpensive transducer system was developed. This transducer used a sensing element which was essentially a microphone element. This transducer could use very inexpensive cabling. The cabling system was made modular so that patch panels and ribbon cables could be used in an integrated cable management system. The signal conditioning amplifiers for these sensors included a self-identifying feature. This meant that the system could determine which transducer was plugged into which data acquisition channel. Bar-coded labels were put on the sensors so that the calibration factors could be accessed. The points on the structure were also bar-coded, so that the location and direction of the sensor could easily be input into the modal analysis software. This cable management capability was very important for instrumenting structures with many sensors. The initial sensors were very inexpensive. The sensitivity of these sensors would vary approximately five percent due to environmental conditions. Even with this variation, improved modal data was obtained due to improved data consistency. In the late '80s and early '90s, this system was updated with inexpensive piezo-electric shear element transducers which matched the characteristics of existing modal sensors. The sensitivities of these sensors are stable.

Currently, there are efforts to build smart sensors which have digital data stored in the sensor. This data includes the serial number, calibration data, location, and direction of the sensor. This is part of the IEEE 1451 standard, which will specify the communications protocol for digital sensors. This includes sensors with built-in ADCs and IP addresses so that they can communicate over a network. The network interface, or NCAP, can also handle the synchronization of these transducers.

SIGNAL PROCESSING

Signal processing and parameter estimation are two areas where the most effort has been spent in the past thirty years in developing the experimental modal analysis area. A compete review of this work would be extensive. There are hundreds of references in the literature and there are a number of good text books on signal processing. In this review, only the high points will be addressed. This review is the equivalent of one day short course relative to a complete university course in signal processing. In order to have a good understanding of the signal processing aspects of modal analysis, it is important understand the properties of the Fourier transform and the properties of the Laplace transform. These transformations explain the time-frequency-Laplace domain relationships. In terms of sampled data, the z transform is also important.

The input and output data is digitized and processed in most cases using the Fast Fourier Transform (FFT) algorithm to compute power spectra, cross spectra, frequency response functions, coherence functions, etc. This processed data is used to estimate modal parameters using various curve fitting or system identification processes.

The FFT is essentially a Fourier series approximation of the Fourier transform integral equation. As a result of this approximation, there are serious potential errors in the computation of frequency response functions using the FFT algorithm. A Fourier series assumes that the time function is a periodic function. If the time function is periodic and matches the period of the Fourier series, the results are a good approximation of the Fourier transform. It is also a good approximation if the time function is a completely observed transient. For this case, the Fourier series coefficients fall on the envelope of the Fourier transform. If the time signal is nonperiodic and is not completely observed in the window of the FFT, then serious errors can exist. These errors are referred to as 'leakage errors'. In general, there is no way to completely eliminate this leakage error. The effects of this error can be minimized using various types of sampling windows. The type of window used depends on the characteristics of the time signal and, to a lesser extent, on what the data is being used for. There are many good references on window functions and their effects on controlling leakage errors. All books on FFTs discuss windows and leakage in great detail. Leakage is a major error and has caused much grief in processing measurement data.

If the measurement data are sampled, there are likely to be errors associated with sampling. The maximum frequency that can be computed from sampled data is determined by the sampling frequency. This maximum frequency is equal to one half the sampling frequency. The maximum frequency is independent of the algorithm used to process the time data. If the data contains information greater than one half the sampling frequency, this information will fold back into the computed frequency range and cause an aliasing error. To eliminate this error, the time data should be passed through a low-pass filter to remove this high frequency content before sampling. All commercial Fourier Analyzer systems have anti-aliasing filters to remove this high frequency data. However, this can be a problem if a system is built up from components. For example, when using a PC and a plug-in data acquisition board, if the data acquisition board does not have anti- aliasing protection, then filters will have to be included.

The data used in most parameter estimation algorithms is either frequency response functions or unit impulse functions. The frequency response functions are simply the Fourier transforms of the unit impulse functions, and vice versa. In other words, they contain the same information about the system, and which one is used depends on the parameter estimation algorithm that is used to extract the modal parameters.

The frequency response function of a system is the Fourier transform of its output divided by the Fourier transform of its input. In terms of sampled data, the frequency responses are estimated by taking the FFT of the input and outputs and processing this data to get an estimate. The FFT operates on a rather small block of time data. The FFT of the input and output blocks of data contain leakage errors if the data are not completely observed (periodic functions which are periodic in the window or completely observed transients). This leakage error, as mentioned above, can be a serious error. In order to reduce these errors, different input signals can be used. In fact, there has been considerable research on the types of input signals and the types of windows which should be used to measure frequency response signals. The choice of the signal type and windows depends on the kind of measurement noise that is present in the data. The noise can be broken up into three types of noise:

a. Non-Coherent Noise - The noise is due to electrical noise on the transducer signals, due to unmeasured excitation sources, etc., which are non-coherent with respect to the measured input signals or to some other signal which is used in the averaging process. Zero mean non-coherent noise can be eliminated by averaging with respect to a reference signal. This reference signal could be the input signal in terms of a spectrum averaging process, or could be a synchronization or trigger signal in terms of cyclic averaging or a random decrement process.

b. Signal Processing Noise - The signal processing itself may generate noise. For example, "leakage" is a classic source of noise when using FFTs for computing frequency domain measurements. This type of noise is reduced or eliminated by using completely observed time signals (periodic or transient), by using various types of windows, or by increasing the frequency resolution.

c. Nonlinear Noise - If the system is nonlinear, then free decay, frequency response, or unit impulse function measurements may be distorted, which consequentially causes problems when estimating modal parameters. Nonlinear distortion noise can be eliminated by linearizing the test structure before testing, or by randomizing the input signals to the structure. This will cause the nonlinear distortion noise to become non-coherent with respect to the input signal. The nonlinear noise can then be averaged out of the data in the same manner as ordinary non-coherent noise.

In the early to mid '70s, there was a significant amount of research done on excitation techniques. To summarize this research in a sentence or two, if the system is lightly damped and leakage is a problem, then a periodic or a completely observed transient signal should be used. If there is nonlinear noise, then a randomizing signal should be used. Straight averaging then takes care of the non-coherent noise. For heavily damped systems, pure random is normally used, and for lightly damped systems, burst random or periodic random signals are used. Most systems have some degree of nonlinearities, and random signals are best suited for these cases. For field testing, the transient methods are convenient (impact testing, step relaxation). In recent years there has been renewed interest in excitation methods that use cyclic averaging and other techniques for reducing leakage.

Frequency response estimation algorithms are another area where there has been a significant amount of research. A least squares technique is used to estimate the frequency response from the measured input output data. In general, a great deal of time data is used in the estimation of frequency response functions. This amount of data is necessary to average out the non-coherent and nonlinear noise. Like all pseudo inverse solutions, there is no unique answer. The answer depends on the error object function and any weighting functions applied to the data. There are several frequency response estimators:

1. H_1 which assumes all the noise is on the output.
2. H_2 which assumes all the noise is on the input.
3. H_v which assumes the noise is on both the input and output, in a manner where the noise vector is perpendicular to the solution vector.

At a particular frequency, if all the noise is on the output, then H_1 will estimate the true frequency response of the system in the limit of the averaging process. In other words, if the objective function is satisfied, then the estimator will give the right answer if enough averages are taken to average the noise to zero.

It is important to make good measurements, since the parameter estimation does not compensate for bad measurements.

PARAMETER ESTIMATION

Modal parameters can be estimated from a variety of different measurements. These measurements can include free decays, forced responses, frequency responses, and unit impulse responses. These measurements can be processed one at a time, or in partial or complete sets simultaneously. The measurements can be generated with no measured inputs, a single measured input, or multiple measured inputs. The data can be measured individually or simultaneously. In other words, there is tremendous variation in the types of measurements and in the types of constraints that can be placed on the testing procedures.

There has been at least a hundred different modal parameter estimation algorithms developed in the past 30 years. The ones covered in this review are the ones that have become commercially available. Many users have developed their own algorithms, or have some sort of special adaptation of a standard algorithm.

Modal parameter estimation algorithms became important in the mid '60s when it became convenient to measure frequency response functions. The frequency response functions can be expressed in terms of modal parameters as follows:

$$H_{pq}(\omega) = \sum_{r=0}^{N} \frac{A_{pq_r}}{j\omega + \lambda_r} \qquad where \quad A_{pq_r} \approx \Psi_{p_r} \Psi_{q_r} \qquad 5$$

In this equation, the eigenvalues (λ) and residues (A) (function of eigenvectors(Ψ)) are the modal parameters. The frequency response is measured and is known, and the modal parameters are unknown. The modal parameters are estimated from this expression by a parameter estimation process.

The frequency response function is measured between a given input point and a given output point as function of frequency. The temporal information (frequency) is a function of the eigenvalue, and the spatial information is a function of the eigenvector. In order to measure the eigenvector, a number of frequency responses must be measured as a function of response point or input point in order to sample the spatial information.

Unfortunately, a set of nonlinear equations must be solved in order to estimate the modal parameters from the measured frequency response functions. In the mid '60s, this was a difficult problem since these nonlinear equations are very ill-conditioned. It took very good starting values to have the solution converge. Approximate solutions based on Single-Degree-of Freedom (SDOF) algorithms were used to obtain the starting values. These SDOF algorithms proved to be useful tools for getting quick estimates of the modal parameters. As a result, these algorithms are useful for troubleshooting applications and are still used. Quadrature response method, circle fitting, and SDOF polynomial method are the more useful of these algorithms.

In the late '60s and early '70s, an algorithm which was being used to analyze noise signatures from submarines turned out to be the first really successful parameter estimation algorithm for fitting measured frequency response functions. The algorithm was the Complex Exponential Algorithm (CEA). The complex exponential algorithm has historically been derived from the Prony algorithm, which was developed over one hundred years ago. The method simply takes a time history, which is composed of a summation of damped exponentials, and estimates the amplitude, frequency and damping of the individual terms in the summation. Of course, this corresponds to computing the eigenvalues and residues for a measured initial condition or impulse response of a system. The Prony algorithm is in reality a very clever algebraic manipulation for solving a nonlinear set of equations for the eigenvalues. When this algorithm was first used in modal analysis to process unit impulse functions, it was a revelation. Over the next several years, this method was extended to handle multiple impulse measurements in a least squares sense. This method was referred to as the Least Squares Complex Exponential (LSCE) algorithm. At the same time that the least squares complex exponential algorithm was being developed, the Ibrahim Time Domain (ITD) method was also being developed. Both of these methods were designed to take multiple impulse or initial condition measurements from a single initial condition or a single reference input and compute a set of eigenvalues and eigenvectors. The mathematical derivations were completely different, and as a result these methods were thought of as being completely different methods. In reality, both methods are similar and can be derived from a common starting point which will be discussed later in this review. All of these techniques fit unit impulse response functions which are computed by taking the inverse Fourier transform of the frequency response functions.

In the early '80s when MIMO testing became popular, the LSCE method was extended to handle multiple references. This was a major step forward since the multiple references allowed repeated roots to be estimated, and in general did a better job of uncoupling closely coupled modes. Again, the basic philosophy was one of fitting the response of the system. This was the Polyreference Time Domain (PTD) method. Again, algebraic manipulation was used in deriving this method. It was very difficult to see any common threads between the ITD method and CEA methods (CEA, LSCE and PTD). The next major development occurred in the mid '80s, and this was the development of the Eigenvalue Realization Algorithm (ERA). This algorithm was developed from a viewpoint used by control engineers and represented a breakthrough view point. Instead of fitting a model to the solution, the concept of fitting a model to determine the underlying system equations was developed (generalized equations of motion). This was the approach used by the engineers looking at large civil engineering structures in the early '70s, who were trying to determine modal parameters of civil engineering from unmeasured ambient inputs. Unfortunately, the modal analysis community did not pick up on this method as a general parameter estimation procedure in the early '70s. The approach used an Auto-Regressive Moving Average (ARMA) modeling method. The ARMA consisted of two types of terms, auto-regressive terms which were simply regression coefficients, and moving average terms which in general handled the unmeasured inputs and noise. The AR terms are linear but the MA terms are nonlinear. This complicated the solution of the set of equations. In the ERA case, the responses were assumed to be free decays and the MA could be neglected. Therefore, a simple set of linear equations could be solved to determine the AR terms.

Historically, ARMA models were developed as a set of finite difference equations and were specific to the time domain. However, with the general use of both the

Fourier transform and the Laplace transform, the frequency and Laplace domain information are used in an interchangeable manner. The time, frequency and the Laplace variables are simply transforms of each other and correspond to the temporal information. The definition of an ARMA model has been expanded from the finite difference formulation to include the time, frequency and Laplace domains. One of the problems is that the definition of the ARMA model has been expanded by many researchers in different areas of science, to the point where there is general confusion over what is an ARMA model. In the context of the applications described in this paper, only the matrix polynomial formulation of the ARMA model will be used and expressed as a function of the time, frequency or the Laplace variable with only the AR (ARX model) terms being used. This method will simply be referred to as the Unified Matrix Polynomial Approach (UMPA). The basic form of the UMPA model is given in the discrete time and frequency domains below:

Time Domain

$$\sum_{i=0}^{N} [A_i]\{x_{m+i}\} = \sum_{j=0}^{M} [B_j]\{f_{m+j}\}$$

6

Frequency Domain

$$\sum_{i=0}^{N} (j\omega_k)^i [A_i]\{X(\omega_k)\} = \sum_{j=0}^{M} (j\omega_k)^j [B_j]\{F(\omega_k)\}$$

7

where N is the order of the UMPA model. The number of eigenvalues are determined by the order of the model times the size of the size of the A matrices. The eigenvalues are computed from a companion matrix constructed from the A coefficients. The $\{X\}$ vector is the measured responses and the $\{F\}$ vector is the measured forcing functions.

The A and B coefficients are both auto-regressive terms, and can be determined by solving a set of linear equations from the measured response and input forces. Once the A's and B's are known, then an eigenvalue problem can be formulated to determine the eigenvalues. For the ITD and ERA methods, which are formulated as low order models, the eigenvector is also obtained from the eigenvalue solution. For PTD, which uses a high order model, the eigenvector is only estimated at the reference points (exciter locations). The complete eigenvector is estimated in a second phase.

In the time domain, if free decays (unit impulse functions) are used, then there are no external forces acting on the system during the free decay, and the A terms are the only important terms for determining the responses of the system. The CEA, LSCE, PTD, ITD and ERA methods can all be derived using the UMPA model formulation based on a model with just A terms. These methods are all related, and the relationships are made clear. These methods are also the time domain algorithms which have been commercialized and are currently the most popular time domain methods.

There are frequency domain equivalents to all of the above techniques, and they can easily be derived. For the frequency domain case, the B terms have to be included. In the frequency domain, the force always acts at the frequency of the response. The Polyreference Frequency Domain (PFD) algorithm is a commercial example. The PFD algorithm is actually a frequency domain implementation of a low order model closer to ERA than to PTD.

The differences between all of the above-mentioned methods are simply the order of the polynomial and the number of references. CEA is a high order model with a single reference. PTD is a high order model with the number of references equal to the number of inputs. ITD is a low order model with a single reference, and ERA is a low order model with multiple references. PFD is a low order frequency domain model with multiple references. All of these techniques are commercially available.

The procedure for using these parameter estimation algorithms is as follows:

1.   Measure Input-Output Responses
   a. Can be time histories, Fourier spectra
   b. Averaged data: power spectra, cross spectra, correlation functions, etc.
   c. Frequency Response Functions, Unit Impulse Responses, etc.
   d. Random decrement measurements
2. Determine type of UMPA model to use (domain, model order-matrix size)
3. Determine [A]'s and [B]'s of the UMPA model using a pseudo-inverse procedure.
   a. Condensation procedures
      i.   Least Squares
      ii.   Total Least Squares
      iii. Singular Value Decomposition (SVD)
      iv. Coherent Averaging
4. Solve for eigenvalues from UMPA model
5. Solve for the eigenvectors

Important

Determination of the model order-matrix size is important, as indicated above. As a result, there have been a number of techniques which have been developed to aid in the determination of the model order-matrix size. For example, the rank of the matrices used in computing the pseudo-inverse can be used as an indicator for determining the model order. Singular Value Decomposition (SVD) is another technique used to evaluate the rank of the pseudo-inverse problem. For the multiple reference cases, various mode indicator functions can help (Multiple Variant Mode Indicator function, Complex Mode Indicator Function(CMIF), to name several). These techniques establish a starting point for the model order. In general, with the time domain techniques, the model order-matrix size should be over specified by a factor of 1.5 to 2. Over specifying allows the technique to have other degrees-of-freedom which are useful in reducing the influence of measurement noise, distortion errors, data inconsistencies, etc. The over specifying gives better answers for the system eigenvalues, but generates computational or noise eigenvalues which are difficult to sort out.

There are a number of techniques which are useful in sorting out these computational eigenvalues (modes). Stability diagrams, modal indicator functions, Modal Assurance Criteria (MAC) tests, and jack knife procedures, are some of the more important techniques. The commercial software vendors have tried to make this sorting process as simple as possible. However, sorting out computational modes is still the biggest problem with using the multiple reference time domain methods. It requires experience and judgement to use the sorting tools effectively. Novice users have great difficulty using these tools on difficult data.

There is currently renewed research activity on other parameter estimation strategies which do not have the computational mode problems. As mentioned in the introduction, technology often folds back on itself. Methods which were used earlier are being rediscovered and modified to take advantage of newer computational and measurement technologies. One of the more successful new parameter estimation methods is the CMIF parameter estimation method, which is based on a SDOF method which has been used since the late '50s. The CMIF method can process multiple reference frequency response measurements using the peak picking or quadrature response technique for measuring the eigenvector. If there are n references, then n quadrature estimates of the eigenvector can be obtained for a given frequency. The n estimates of the eigenvector can be loaded as the columns of a matrix. The matrix size is the number of response points by the number of references. A SVD can be performed on the matrix. If the quadrature eigenvector is a good approximation of the true eigenvector, there should only be one significant Singular Value (SV). The SVD vector associated with the largest SV is the best least squares estimate of the dominant vector that spans the space of the n quadrature estimates. This SVD vector is a good estimate of one of the system eigenvectors if the quadrature vectors are picked at one of the natural frequencies of the system.

The CMIF algorithm is implemented by performing a SVD on the frequency response matrix at each spectral line. The SVs of this SVD operation are plotted as a function of frequency. See Figure 2 for a graphical description of the CMIF

142

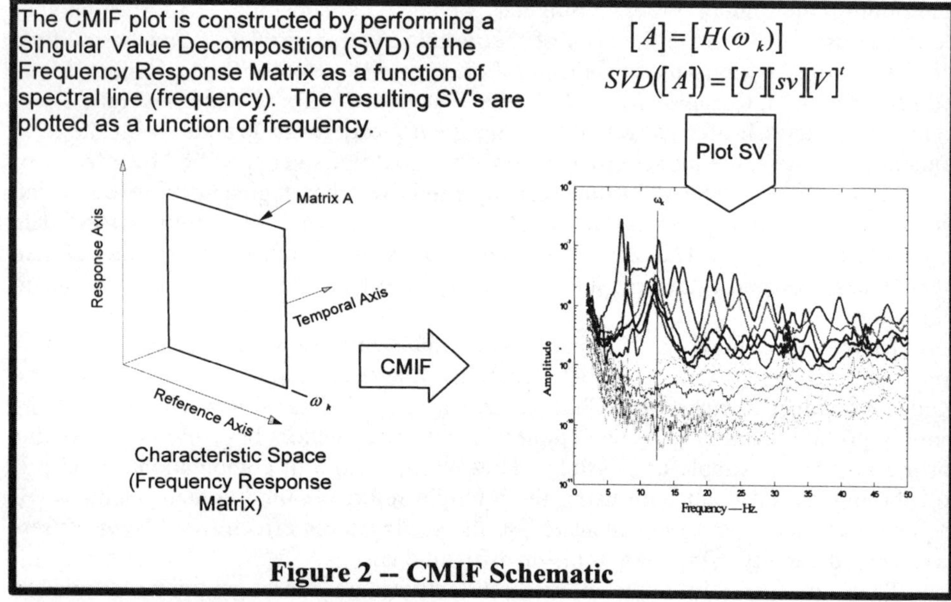

$$[A] = [H(\omega_k)]$$
$$SVD([A]) = [U][sv][V]^t$$

Figure 2 -- CMIF Schematic

Algorithm. The peaks in the largest SV curve correspond to the eigenvalues of the system. In fact, the plot of the largest SV curve corresponds to information similar to the information contained in a power spectrum plot generated from all the measured frequency response functions. Repeated or closely coupled eigenvalues are indicated when two or more of the SV curves reach peaks at the same frequency. The data processing is independent of frequency or temporal information. This model corresponds to a zero order UMPA model. This type of model is a *spatial domain* model. It does not estimate the eigenvalues directly. The eigenvalues are estimated in a second phase. The eigenvectors are used as modal filters to filter the frequency response data and generate a series of Enhanced Frequency Response Functions (eFRFs). These eFRFs can be fit with a simple SDOF polynomial method to get the eigenvalues and modal scale factors.

CMIF is a relatively simple multiple reference algorithm which works well for cases where many references are available. The CMIF does an excellent job of uncoupling closely coupled modes, but at some point it starts to collapse. In the early '70s when the quadrature method was no longer useful, a more sophisticated SDOF polynomial curve fitting method was used. This method assumed a single mode but included higher order B coefficients to handle important out of band eigenvalues. This same approach has been used to extend the CMIF method. It is called the Enhanced Mode Indicator Function (EMIF). It uses a simplified version of the PFD algorithm. Instead of adding extra eigenvalue terms to handle noise and distortion, extra B terms are added. CMIF is used to estimate the number of eigenvalues in a small frequency band and the PTD method, set for a fixed number of eigenvalues with extra B terms, is used to estimate the modal parameters for the frequency band in question. This method eliminates the difficult step of sorting out computational modes. The EMIF method is meant to supplement the CMIF method. CMIF is the method of choice and EMIF is only used in the difficult areas.

Circle fitting, another of the early SDOF methods, is currently being used to understand and characterize modes that exhibit non-linear behavior. This is an important application when solving real problems.

As mentioned previously, there is increasing interest in extracting modal parameters from ambient inputs. Currently, there is considerable research in developing and refining modal parameter estimation algorithms for ambient testing. All of the standard methods will work with ambient response data if assumptions about the input forces can be made. When exciters are used to excite the structure, the resulting measurements are relatively good. Adding extra terms to the AR model to handle the noise is a reasonable strategy to get a linear solution. For ambient testing where there is significantly more noise, then including the Moving Average (MA) terms is useful. When the MA terms are included, the solution is nonlinear and an iterative solution may be necessary.

There have been hundreds of papers written on modal parameter estimation over the past 30 years. Instead of referring to individual papers, the best reference is the IMAC proceedings where most of the work has been documented for the past 15 years. The first 5 or 6 proceedings of IMAC included a comprehensive biography of modal analysis up to the time of those proceedings. The IMAC proceeding are available on CDs from SEM (Society for Experimental Mechanics, Inc.). These CDs make it relatively easy to search for developments in the modal analysis area.

MODAL MODEL ASSIMILATION AND VALIDATION

A typical modal test consists of a number of individual tests for different exciter configurations and test setups. A modal model has to be assembled from all of these tests.   From one exciter configuration, a fraction of the modal parameters can be determined. Other exciter configurations supply additional information. More than one estimate of each modal parameter is normally available from a complete modal test. Averaging and sorting the modal database generated during the testing is required to obtain the best modal model. Once the modal model is generated, then it can be validated using a number of techniques:

1)   Synthesis of arbitrary FRF
2)   Reciprocity-Curve fitting
3)   Perturbation
        a) Mass Additive
        b) Constraints
4)   Forced Response

144

5)   Orthogonality
6)   Model Correlation

The first four items use the modal model to predict measurements or system performance which can be experimentally checked using data taken during the modal test. Frequency response measurements can be synthesized from the modal model and checked against measurements made during the testing programs. These could be measurements used in extracting the modal parameters, or measurements made specifically to validate the model.

Most of the parameter estimations algorithms do not enforce reciprocity, so this can be used to check the model. If the modal model demonstrates reciprocity, then input and output predictions should obey reciprocity.

The modal model should be able to predict simple modifications to the structure, such as simple mass modifications. Masses can be added at important connection points during the test and frequency response measurements made on the modified structure and checked against results predicted by the modal model.

One of the more severe tests is to predict responses due to inputs to the model, and to compare these against the measured responses. Inputs signals can be input at selected points and the response time histories can be predicted at other points. The predicted time histories can be compared against the measured time histories.

The last two cases are checks against FEM models. The first is to check orthogonality with respect to the analytical mass matrix, to determine if the measured eigenvectors satisfy orthogonality. The second is to correlate the model with FEM results. Of course, these last two methods are making assumptions about the validity of the analytical model. These tests primarily provide a check on the analytical model.

CONCLUDING REMARKS

As mentioned throughout this review, this is only a brief overview of the high points of experimental modal testing. No specific references are cited in the review, since it covers the complete area of modal testing, and there are hundreds of reference articles. A complete overview of the complete area of modal analysis can be found in the IMAC proceedings.

Vibration-Based Damage Detection

Charles R. Farrar, Scott W. Doebling, Technical Staff Members
Thomas A. Duffey, Consultant
Engineering Analysis Group
MS P-946
Los Alamos National Laboratory
Los Alamos, New Mexico, 87544 USA

ABSTRACT

Many aerospace, civil, and mechanical systems continue to be used despite aging and the associated potential for damage accumulation. Therefore, the ability to monitor the structural health of these systems is becoming increasingly important. A wide variety of highly effective local non-destructive evaluation tools are available. However, damage identification based upon changes in vibration characteristics is one of the few methods that monitor changes in the structure on a global basis. A summary of developments in vibration-based damage detection that have taken place over the last thirty years is first presented. Recent research has begun to recognize that the vibration-based damage detection problem is fundamentally one of statistical pattern recognition and this paradigm is described in detail. This paradigm is composed of four parts: 1.) Operational evaluation; 2.) Data acquisition and cleansing; 3.) Feature selection and data compression, and 4.) Statistical model development. Throughout this summary of the statistical pattern recognition paradigm, applications of this technology to rotating machinery and civil engineering infrastructure will be used to demonstrate various aspects regarding the implementation of the health monitoring process. The rotating machinery applications represent a mature technology where vibration-based damage detection has made the transition from a research topic to industry practice. The civil engineering applications represent a field that is still primarily the focus of research efforts. This summary identifies technical challenges that must be addressed if vibration-based structural health monitoring is to gain wider application. Finally, a discussion of future directions for this technology is presented.

1.   INTRODUCTION

In the most general terms, damage can be defined as changes introduced into a system that adversely affect its current or future performance. Implicit in this definition is the concept that damage is not meaningful without a comparison between two different states of the system, one of which is assumed to represent the initial, and often undamaged, state. This discussion is focused on the study of damage identification in structural and mechanical systems. Therefore, the definition of damage will be limited to changes to the material and/or geometric properties of these systems, including changes to the boundary conditions and system connectivity, which adversely affect the current or future performance of these systems.

The interest in the ability to monitor a structure and detect damage at the earliest possible stage is pervasive throughout the civil, mechanical and aerospace engineering communities. Current damage-detection methods are either visual or localized experimental methods such as acoustic or ultrasonic methods, magnetic field methods, radiograph, eddy-current methods and thermal field methods (Doherty, 1987). All of these experimental techniques require that the vicinity of the damage is known *a priori* and that the portion of the structure being inspected is readily accessible. Subjected to these limitations, such experimental methods can detect damage on or near the surface of the structure. The need for quantitative global damage detection methods that can be applied to complex structures has led to the development of, and continued research into, methods that examine changes in the vibration characteristics of the structure.

The basic premise of vibration-based damage detection is that damage will alter the stiffness, mass or energy dissipation properties of a system, which, in turn, alter the measured dynamic response of the system. Although the basis for vibration-based damage detection appears intuitive, its actual application poses many significant technical challenges. The most fundamental challenge is the fact that damage is typically a local phenomenon and may not significantly influence the lower-frequency global response of a structure that is normally measured during vibration tests. Stated another way, this fundamental challenge is similar to that found in many engineering fields where there is a need to capture the system response on widely varying length scales and such system modeling has proven difficult. Another fundamental challenge is that in many situations vibration-based damage detection must be performed in an *unsupervised learning* mode. Here, the term *unsupervised learning* implies that data from damaged systems are not available. These challenges are supplemented by many practical issues associated with making accurate and repeatable vibration measurements at a limited number of locations on complex structures often operating in adverse environments.

In an effort to emphasize the extent of the research efforts to date in vibration-based damage detection, a brief summary of applications that have driven developments in this field over the last thirty years is first presented. Recent research has begun to recognize that the vibration-based damage detection problem is fundamentally one of statistical pattern recognition and this paradigm is described in detail. Current damage detection methods are then summarized in the context of this paradigm. This summary is supplemented with examples of structural health monitoring applied to rotating machinery and bridge structures. The rotating machinery application represents a mature technology where vibration-based damage detection has made the transition from a research topic to industry practice. The bridge applications represent a field that is still primarily the focus of individual research efforts with only recent and limited marketing of this technology on a commercial basis.   Concluding comments will compare and contrast the rotating machinery applications with the bridge applications. These comments will focus on the future research directions for this technology and some of the economic considerations that are driving this research.

2.   HISTORICAL PERSECTIVE

It is the authors' speculation that damage or fault detection, as determined by changes in the dynamic properties or response of systems, has been practiced in a qualitative manner, using acoustic techniques, since modern man has used tools. More recently, this subject has received considerable attention in the technical literature and a brief summary of the developments in this technology over the last thirty years is presented below. Specific references are not cited; instead the reader is referred to (Doebling, et al. 1996, 1998) for a recent review of literature on this subject.

The development of vibration-based damage detection technology has been closely coupled with the evolution, miniaturization and cost reductions of Fast Fourier Transform (FFT) analyzers and digital computing hardware.   To date, the most successful application of vibration-based damage detection technology has been for monitoring rotating machinery.   The rotating machinery application has taken an almost exclusive non-model based approach to damage detection.   The detection process is based on pattern recognition applied to time histories or spectra generally measured at a single point on the housing of the machinery during normal operating conditions.   Often this pattern recognition is performed only in a qualitative manner. Databases have been developed that allow specific types of damage to be identified from particular features of the vibration signature. For these systems the approximate damage location is generally known, making a single channel FFT analyzer sufficient for most periodic monitoring activities. Typical damage that can be identified includes loose or damaged bearings, misalign shafts, and chipped gear teeth.   Today, commercial software integrated with measurement hardware is marketed to help the user systematically apply this technology to the operating equipment.

During the 1970s and 1980s the oil industry made considerable efforts to develop vibration-based damage detection methods for offshore platforms.   This damage detection problem is fundamentally different from that of rotating machinery because the damage location is unknown and because the majority of the structure is not readily accessible for measurement. To circumvent these difficulties, a common methodology adopted by this industry was to simulate candidate damage scenarios with numerical models, examine the changes in resonant frequencies that were produced by these simulated changes, and correlate these changes with those measured on a platform. A number of very practical problems were encountered including measurement difficulties caused by platform machine noise, instrumentation difficulties in hostile environments, changing mass caused by marine growth and varying fluid storage levels, temporal variability of foundation conditions, and the inability of wave motion to excite higher vibration modes. These issues prevented adaptation of this technology, and efforts at further developing this technology for offshore platforms were largely abandoned in the early 1980s.

The aerospace community began to study the use of vibration-based damage detection during the late 1970s and early 1980s in conjunction with the development of the space shuttle. This work has continued with current applications being investigated for the National Aeronautics and Space Administration's space station and reusable launch vehicle. The Shuttle Modal Inspection System (SMIS) was developed to identify fatigue damage in components such as control surfaces, fuselage panels and lifting surfaces. These areas were covered with a thermal protection system making them inaccessible and, hence, impractical for conventional local non-destructive examination methods. This system has been successful in locating damaged components that are covered by the thermal protection system. All orbiter vehicles have been periodically subjected to SMIS testing since 1987. Space station applications have primarily driven the development of experimental/analytical damage detection methods. These approaches are based on correlating analytical models of the undamaged structure with measured modal properties from both the undamaged and damaged structure. Changes in stiffness indices as assessed from the two model updates are used to locate and quantify the damage. Since the mid-1990s, studies of damage detection for composite materials have been motivated by the development of a composite fuel tank for a reusable launch vehicle.

The civil engineering community has studied vibration based damage assessment of bridge structures since the early 1980s. Modal properties and quantities derived from these properties such as mode-shape curvature and dynamic flexibility matrix indices have been the primary features used to identify damage in bridge structures. Environmental and operating condition variability present significant challenges to the bridge monitoring application. The physical size of the structure also presents many practical challenges for vibration-based damage assessment. Regulatory requirements in eastern Asian countries, which mandate the companies that construct the bridges periodically certify their structural health, are driving current research and commercial development of vibration-based bridge monitoring systems.

In summary, the review of the technical literature presented by (Doebling et al. 1996, 1998) shows an increasing number of research studies related to vibration-based damage detection. These studies identify many technical challenges to the adaptation of vibration-based damage detection that are common to all applications of this technology. These challenges include better utilization of the nonlinear response characteristics of the damaged system, development of methods to optimally define the number and location of the sensors, identification of the features sensitive to small damage levels, the ability to discriminate changes in features caused by damage from those caused by changing environmental and/or test conditions, the development of statistical methods to discriminate features from undamaged and damaged structures, and performance of comparative studies of different damage detection methods applied to common data sets. These topics are currently the focus of various research efforts by many industries including defense, civil infrastructure, automotive, and semiconductor manufacturing where multi-disciplinary approaches are being used to advance the current capabilities of vibration-based damage detection.

3. VIBRATION-BASED DAMAGE DETECTION AND STRUCTURAL HEALTH MONITORING

The process of implementing a damage detection strategy is referred to as *structural health monitoring*. This process involves the definition of potential damage scenarios for the system, the observation of the system over a period of time using periodically spaced measurements, the extraction of features from these measurements, and the analysis of these features to determine the current state of health of the system. The output of this process is periodically updated information regarding the ability of the system to continue to perform its desired function in light of the inevitable aging and degradation resulting from the operational environments. Figure 1 shows a chart summarizing the structural health-monitoring process. The topics summarized in this figure are discussed below. Examples of structural health monitoring applied to rotating machinery and to bridges are used to further illustrate the topics summarized in Fig. 1.

3. 1. Operational Evaluation

Operational evaluation answers three questions in the implementation of a structural health monitoring system:

1. How is damage defined for the system being investigated and, for multiple damage possibilities, which are of the most concern?

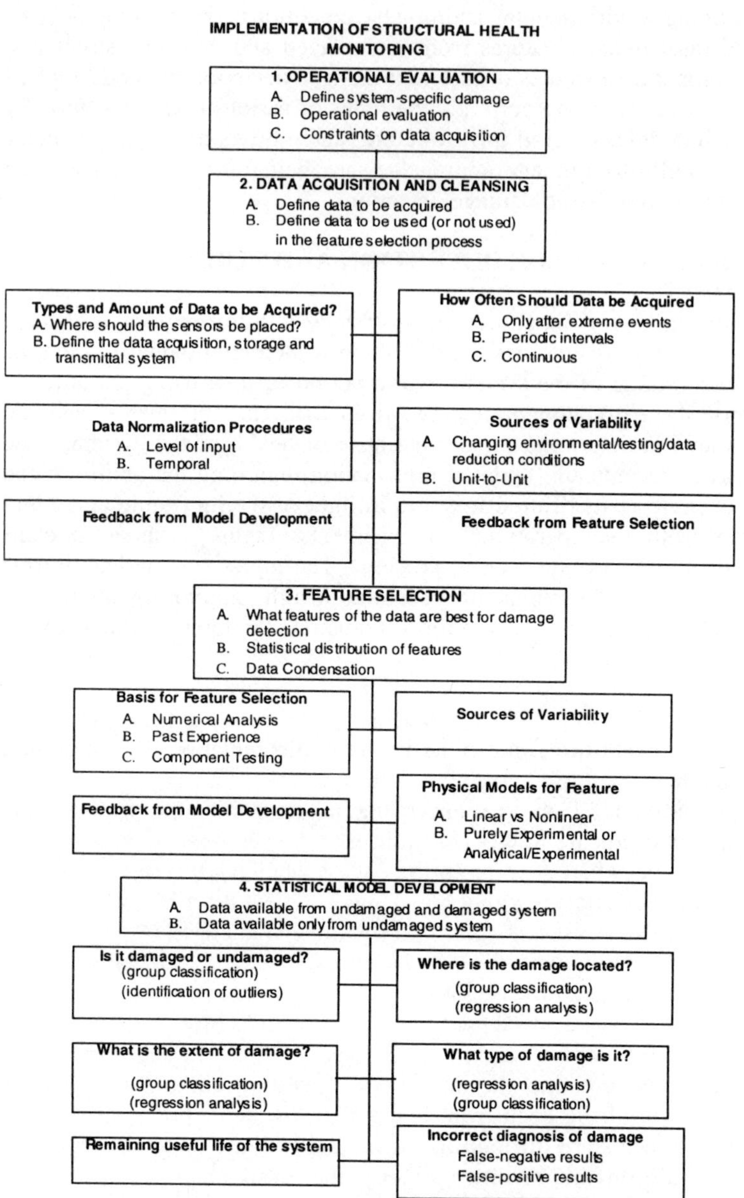

Figure 1. Flow chart for implementing a structural health monitoring program.

2.   What are the conditions, both operational and environmental, under which the system to be monitored functions?

3.   What are the limitations on acquiring data in the operational environment?

Operational evaluation begins to set the limitations on what will be monitored and how the monitoring will be accomplished. This evaluation starts to tailor the health monitoring process to features that are unique to the system being monitored and tries to take advantage of unique features of the postulated damage that is to be detected.

3.1.1. Operational evaluation for rotating machinery

The definition of damage is often very straightforward for rotating machinery. Many times there are a limited number of damage scenarios and the possible damage locations are known *a priori*. The primary operational limitation on acquiring data is that the machine will typically be in operation and performing its normal function or will be in a transient start-up or shutdown mode. In its *in situ* environment many other machines will most likely produce additional vibration sources that must be accounted for in the damage detection process. Limitations to acquiring vibration data can vary widely. For many applications the limitations will be based on administrative criteria such as the availability of personnel to make the necessary measurements. In other applications the machine may be located in hazardous environments that limit access time.

3.1.2. Operational evaluation for bridges

The definition of damage for bridges can vary widely as many bridges are one-of-a-kind structures. In many cases it is difficult to specifically define the damage to be monitored because there has not been enough experience with failures of the particular structure being studied or similar structures. For large bridges it is imperative that the damage definition be restricted so as to make the monitoring possible with a cost-effective number of sensors. The need to perform the monitoring in a manner that does not impede traffic flow dictates that the monitoring occurs when the bridge is subjected to changing traffic patterns coupled with typical environmental variability. These operational and environmental constraints pose a formidable challenge to bridge health monitoring.   Traffic flow can vary on a 24-hr cycle, and a weekly cycle. Environmental variability can occur over a 24-hr cycle as shown in Fig. 2 (Farrar, et a.l., 1997), over a seasonal cycles as discussed in (Askegaard and Mossing, 1988), and intermittent cycles caused by varying rainfall conditions. Portions of the bridge may not be accessible for instrumentation during normal operating conditions. Traffic usually prevents the topside of the deck from being instrumented. For large bridges many of the structural elements are difficult to instrument because it is impractical to access them in a safe and economic manner. Finally, the sensors and associated data transmission hardware are subjected to harsh environments that make equipment reliability a serious issue (Nigbor, and Diehl, 1997).

152

3. 2. Data Acquisition and Cleansing

The data acquisition portion of the structural health monitoring process involves selecting the types of sensors to be used, selecting the location where the sensors should be placed,

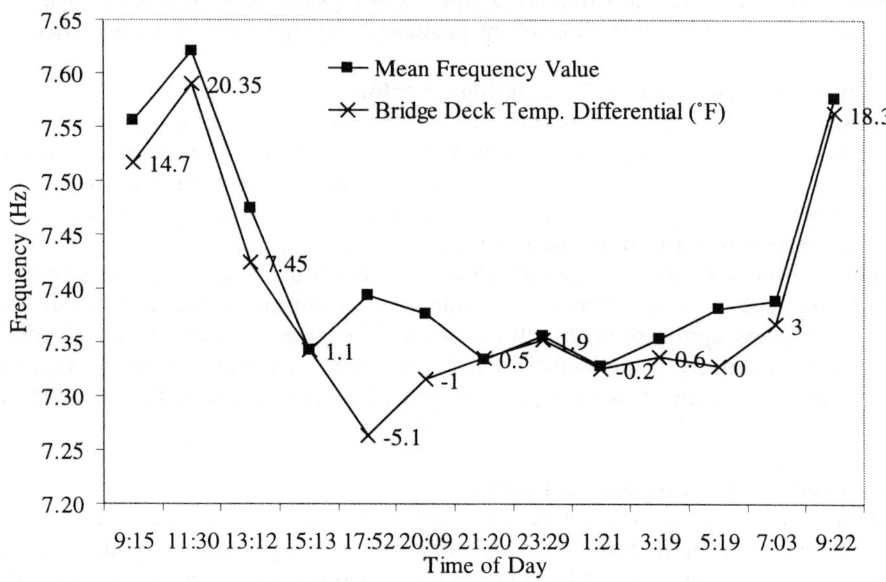

Figure 2. The change in the first mode resonant frequency over a 24-hr period measured on the Alamosa Canyon Bridge in southern New Mexico.

determining the number of sensors to be used, and defining the data acquisition/storage/transmittal hardware. This process is application specific. Economic considerations play a major role in these decisions. Another consideration is how often the data should be collected. In some cases it is adequate to collect data immediately before and at periodic intervals after a severe event. However, if fatigue crack growth is the failure mode of concern, it is necessary to collect data almost continuously at relatively short time intervals.

Because data can be measured under varying conditions, the ability to normalize the data becomes very important to the damage detection process. One of the most common procedures is to normalize the measured responses by the measured inputs. When environmental or operating condition variability is an issue, the need can arise to normalize the data in some temporal fashion to facilitate the comparison of data measured at similar times of an environmental or operational cycle. Sources of variability in the data acquisition process and with the system being monitored need to be identified and minimized to the extent possible. In general, all sources of variability cannot be eliminated. Therefore, it is necessary to make the appropriate measurements such that these sources can be statistically quantified.

Data cleansing is the process of selectively choosing data to accept for, or reject from, the feature selection process. The data cleansing process is usually based on knowledge gained by individuals directly involved with the data acquisition. Finally, it is noted that the data acquisition and cleansing portion of a structural health-monitoring process should not be static. Insight gained from the feature selection process and the statistical model development process provides information regarding changes that can improve the data acquisition process.

3.2.1. General data acquisition issues common to rotating machinery and bridges

Data acquisition issues for rotating machinery and bridge applications include the sensor type and the number of sensors, sensor location, sensor mounting, environmental effects on the sensors, signal recording, and record duration. Averaging, windowing and similar data processing parameters must also be determined. Another issue is determining the steps that should be taken to make the data acquisition as repeatable as possible.

3.2.2. Data acquisition for rotating machinery

Accelerometers are the primary vibration transducer used for damage detection and condition monitoring of rotating machinery. Piezoelectric accelerometers have a broad operating frequency range and are well suited to monitoring of roller bearings and gear trains. Accelerometers are typically used in conjunction with single-channel signal analyzers so that the machinery vibration output signal can be viewed in the frequency domain as well as a function of time, i.e., amplitude-frequency, amplitude-time, and waterfall plots. Velocity transducers and non-contact displacement transducers are also widely used. Non-contact (Eddy current) displacement transducers find application for monitoring shaft motion and position relative to fluid-film bearings. A set of two transducers, mounted at right angles, is often used to determine the orbit of the shaft in its bearing. Data acquisition recording equipment and transducers used to monitor rotating machinery are discussed in detail in (Mitchell, 1992; Crawford, 1992; Eisenmann and Eisenmann, 1997; Hewlett-Packard Application Note 243-1, 1997; Taylor, 1994; and Wouk, 1991). The selection and placement of appropriate transducers depends upon the type of machinery and its construction. A detailed discussion of sensor placement for rotating machinery applications is given in (Eisenmann and Eisenmann, 1997).

3.2.3. Data acquisition for bridges

As with the rotating machinery the primary sensors used for bridge health monitoring are piezoelectric accelerometers. Force-balance accelerometers and electric resistance and vibrating-wire stain gages are also widely used. More recently, fiber–optic sensors utilizing Bragg grating (Todd, et al., 1999) have been studied as a means of increasing channel counts for bridge monitoring in a cost-effective manner. In addition to the motion measuring devices, anemometers and temperature sensors such as thermocouples are used to characterize the environmental variability. Data transmission and recording are done in manners similar to most mechanical vibration applications as summarized in (McConnell, 1995). Recent developments in wireless data acquisition systems have shown promise for large civil engineering structures (Straser, 1998). When ambient excitation sources are used, all channels of data are typically recorded simultaneously. The damage to be monitored and the number of available sensors typically dictate sensor placement. Most bridge health monitoring studies reported to date utilize between 15 and 50 sensors. At the extreme is the Tsing Ma Bridge in Hong Kong that has been instrumented with 600 sensors at an approximate cost of US$16 million. Data acquisition intervals are not well defined for bridges as most studies have been conducted in a research mode where long term monitoring has not been the focus of the study.

3. 3. Feature Selection

The data features used to distinguish the damaged structures from undamaged ones receives the most attention in the technical literature. Inherent in the feature selection process is the condensation of the data. The operational implementation and diagnostic measurement technologies needed to perform structural health monitoring typically produce a large amount of data. Condensation of the data is advantageous and necessary, particularly if comparisons of many data sets over the lifetime of the structure are envisioned. Also, because data may be acquired from a structure over an extended period of time and in an operational environment, robust data reduction techniques must retain sensitivity of the chosen features to the structural changes of interest in the presence of environmental variability. The best features for damage detection are typically application specific. Numerous features are often identified for a structure and assembled into a feature vector. In general, a low dimensional feature vector is desirable. It is also desirable to obtain many samples of the feature vectors. There are no restrictions on the types or combinations of data contained in the feature vector. As an example, a feature vector may contain the first three resonant frequencies of the system, the time when the measurements were made, and a temperature reading from the system. A variety of methods are employed to identify features for damage detection. Past experience with measured data from a system, particularly if damaging events have been previously observed for that system, is often the basis for feature selection. Numerical simulation of the damaged system's response to simulated inputs is another means of identifying features. The application of engineered flaws, similar to ones expected in actual operating conditions, to laboratory specimens can identify parameters that are sensitive to the expected damage. Damage accumulation testing, during which significant structural components of the system under study are subjected to a realistic accumulation of damage, can also be used to identify appropriate features. Fitting linear or nonlinear, physical-based or non-physical-based models of the structural response to measured data can also help identify damage-sensitive features. Common features used in vibration-based damage detection studies are briefly summarized below. A more detailed summary can be found in (Doebling, et al., 1996, 1998).

Basic modal properties, mode shape curvature changes, dynamic flexibility matrices and stiffness indices from updated finite element models are features most commonly used for damage detection in bridges. Features derived from the time-histories, the spectral pattern or from nonlinear response of the system are primarily used for rotating machinery damage detection studies.

3.3.1. Basic modal properties

The most common features that are used in damage detection, and that represent a significant amount of data condensation from the actual measured quantities, are resonant frequencies and mode-shape vectors. These features are identified from measured response time-histories, most often absolute acceleration, or spectra of these time-histories. The technology required to accurately make these measurements is summarized in (McConnell, 1995). Often these spectra are normalized by spectra of the measured force input to form frequency response functions. Well-developed experimental modal analysis procedures are applied to these functions or to the measured-response spectra to estimate the system's modal properties (Ewins, 1995, and Maia and Silva, 1997).

The amount of literature that uses resonant frequency shifts as a data feature for damage detection is quite large. The observation that changes in structural properties cause changes in vibration frequencies was a primary impetus for developing vibration-based damage identification technology. In general, changes in frequencies cannot provide spatial information about structural changes. For applications to large civil engineering structures the low sensitivity of frequency shifts to damage requires either very precise measurements of frequency change or large levels of damage. An exception occurs at higher modal frequencies, where the modes are associated with local responses. However, the practical limitations involved with the excitation and identification of the resonant frequencies associated with these local modes, caused in part by high modal density and low participation factors, can make them difficult to identify.

Figure 3 shows incremental damage made by torch cuts in the main girder of a bridge as part of a controlled damage detection study (Farrar, et al., 1994). Also shown in Fig. 3 is the first mode resonant frequency of the undamaged structure (dam 0) and changes in the first mode frequencies for each subsequent damage case. Although difficult to see, error bars corresponding to the 95 percent confidence intervals on the identified resonant frequencies are also shown (Doebling and Farrar, 1998). The increases in frequency associated with the first two damage cases are the result of systematic errors in the testing procedure. In terms of absolute value, these changes are more significant than those produced by the third damage case. This study indicates some of the difficulties in using resonant frequencies as a damage indicator for large civil engineering structures.

Damage detection methods using mode shape vectors as a feature generally analyze differences between the measured modal vectors before and after damage. Mode shape vectors are spatially distributed quantities; therefore, they provide information that can be used to locate damage.

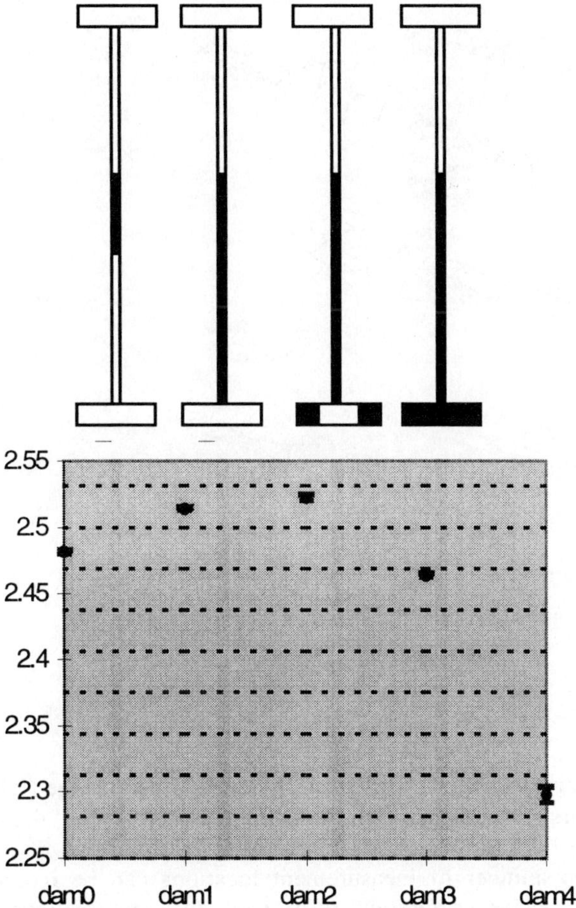

Figure 3. Bridge damage cases and corresponding changes in the first mode resonant frequency.

158

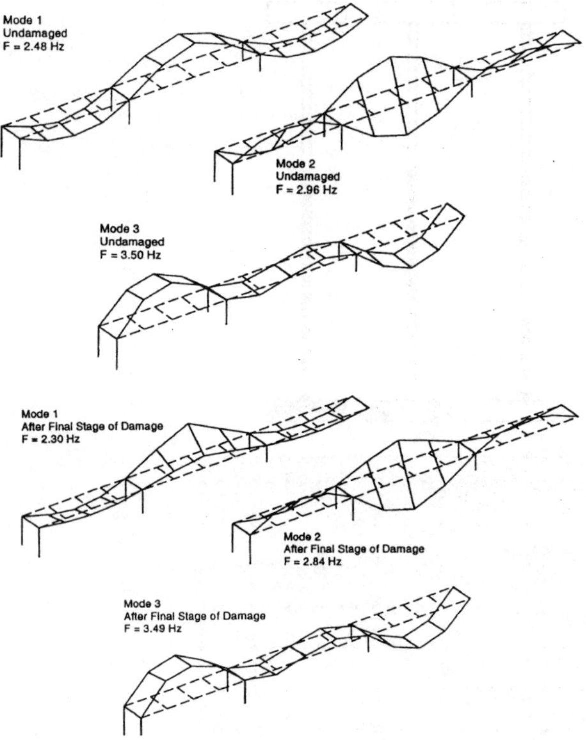

Figure 4. First three modes identified on the undamaged bridge (left) compared with the first three modes identified after the final level of damage (right).

However, a large number of measurement locations can be required to accurately characterize mode shape vectors and to provide sufficient resolution for determining the damage location.

Figure 4 shows the first three modes shapes and corresponding resonant frequencies measured on the undamaged bridge and similar modes and resonant frequencies after the introduction of the final level. After the final damage was introduced, the location of the damage is clearly indicated when Mode 1 in Fig. 4 (final stage of damage) is compared with the corresponding damaged mode in Fig. 4 (undamaged). Intermediate damage levels produced mode-shape changes that were not statistically significant from the undamaged case (Doebling and Farrar, 1998). It should also be noted that when damage is located at a node for a particular mode, as it is for Mode 3, even the most severe damage case produces no significant change in that mode shape or corresponding resonant frequency. Again, this example illustrates the relative insensitivity of the lower-frequency global modes and resonant frequencies of a large civil engineering structure to local damage.

3.3.2. Mode shape curvature changes

An alternative to using mode shapes to obtain spatially distributed features sensitive to damage is to use mode shape derivatives, such as curvature. Mode shape curvature can be computed by numerically differentiating the identified mode shape vectors twice to obtain an estimate of the curvature. These methods are motivated by the fact that the second derivative of the mode shape is much more sensitive to small perturbations in the system than is the mode shape itself. Also, for beam- and plate-like structures changes in curvature can be related to changes in strain energy, which has been shown to be a sensitive indicator of damage. A comparison of the relative statistical uncertainty associated with estimates of mode shape curvature, mode shape vectors and resonant frequencies showed that the largest variability is associated with estimates of mode shape curvature followed by estimates of the mode shape vector. Resonant frequencies could be estimated with least uncertainty (Doebling, et al., 1997).

3.3.3. Dynamically measured flexibility

Changes in the dynamically measured flexibility matrix indices have also been used as damage sensitive features. The dynamically measured flexibility matrix is estimated from the mass-normalized measured mode shapes and measured eigenvalue matrix (diagonal matrix of squared modal frequencies). The formulation of the flexibility matrix is approximate because in most cases all of the structure's modes are not measured. Typically, damage is detected using flexibility matrices by comparing the flexibility matrix indices computed using the modes of the damaged structure to the flexibility matrix indices computed using the modes of the undamaged structure. Because of the inverse relationship to the square of the modal frequencies, the measured flexibility matrix is most sensitive to changes in the lower-frequency modes of the structure.

3.3.4. Updating structural model parameters

Another class of damage identification methods is based on features related to changes in mass, stiffness and damping matrix indices that have been correlated so that the numerical model predicts, as closely as possible, the identified dynamic properties (resonant frequencies, modal damping and mode shape vectors) of the undamaged and damaged structures. These methods solve for the updated matrices (or perturbations to the nominal model that produce the updated matrices) by forming a constrained optimization problem based on the structural equations of motion, the nominal model, and the identified modal properties (Friswell and Mottershead, 1995). Comparisons of the matrix indices that have been correlated with modal properties identified from the damaged structure to the original correlated matrix indices provide an indication of

160

damage that can be used to quantify the location and extent of damage. Degree of freedom mismatch between the numerical model and the measurement locations can be a severe limitation on performing the required matrix updates.

3.3.5.   Time-history and spectral pattern methods

Non-model based approaches that examine changes in the features derived directly from measured time histories or their corresponding spectra have been used extensively by the rotating machinery industry. There exist numerous detailed charts of anticipated characteristic faults of a variety of machines and machine elements and corresponding features in the measured time histories or spectra (Mitchell, 1992; Crawford, 1992). These features have been widely used to successfully detect the presence, location and type of fault, and the degree of damage. Commercially available software specifically designed for the isolation of faults based on vibration signatures is readily available.

Qualitative features include, for example, the presence of peaks in acceleration spectra at certain multiples of shaft rotational frequency and their growth or change with time. The important qualitative features are quite distinct to the type of machine element, the specific fault, and in some cases to the level of damage. Therefore, it may be possible to locate the defective machine element, isolate the specific fault in the element, and determine the level of damage based purely on these qualitative features. As an example, Figure 5 shows that changes in a power spectral density function can be used to monitor the deterioration in a vacuum blower's bearings.

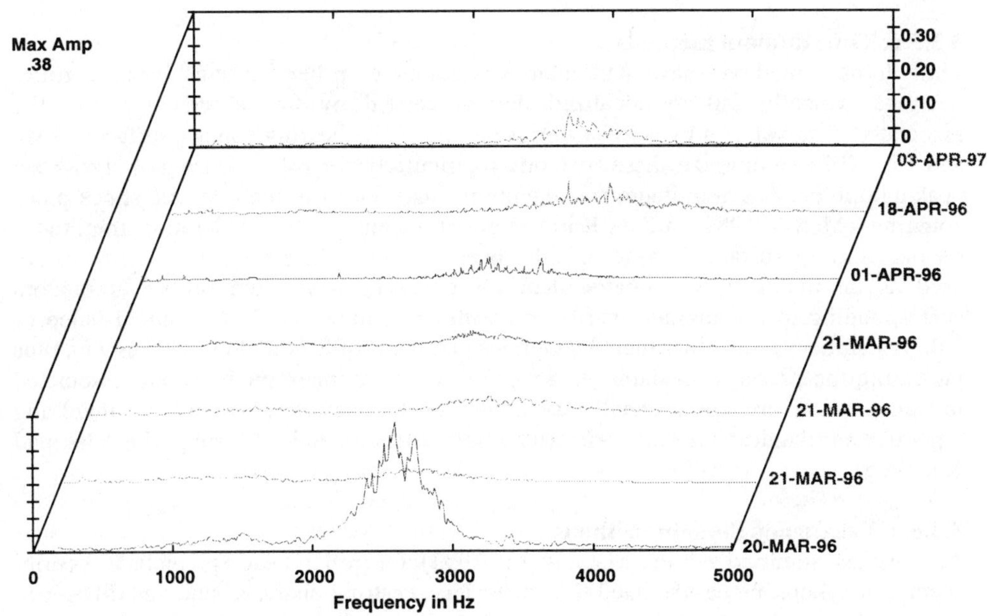

Fig. 5. Power spectral density measured on a vacuum blower. The March 20th measurement was made when maintenance personnel felt that the equipment was running at a hotter temperature than normal. Bearings were replaced on March 21st. Subsequent plots show spectra corresponding to the degrading bearings. (Courtesy of Intel Corp.)

Early studies using qualitative features were based on the concept that detection of each fault is fundamentally different. Recent progress has, however, been reported on generalized failure prediction indices capable of monitoring the condition of a wide variety of manufacturing equipment (Roth and Pandit, 1999).

Quantitative features fall into the following categories: time-domain methods, transformed-domain methods, and time-frequency methods. Included in transformed-domain methods are the traditional frequency-domain methods as well as the cepstrum (the inverse Fourier transform of the logarithm of the Fourier spectra magnitude squared) techniques. Briefly, frequency domain methods characterize features in machine vibrations over a given time window. Time domain and time-frequency methods have application to non-stationary faults, i.e., those associated with machines that exhibit different phenomena in different phases of the machine cycle.

3.3.5.1. Time domain methods

Time-domain methods have particular application to roller bearings because roller bearings typically fail by localized defects caused by fatigue cracking and the associated removal of a piece of material on one of the bearing contact surfaces. (Ma and Li, 1993) summarize these methods (particularly for roller bearing analysis) as: peak amplitude, rms amplitude, crest factor analysis, kurtosis analysis, and shock pulse counting. (Martin, 1989) utilizes Kurtosis measurements of the acceleration amplitudes for detection of surface damage in roller bearings. If surface roughness attributes are used as an indicator of damage, then for a good surface, the profile is random corresponding to a Gaussian distribution with an infinite-sample theoretical value of 3.0. A Kurtosis value other than 3.0 denotes that the profile is no longer Gaussian, thus indicating the presence of damage. Proprietary time-domain methods and associated instrumentation are commercially available for the detection of defects involving repetitive mechanical impacts, primarily associated with roller bearings (Le Bleu and Xu, 1995)

3.3.5.2. Frequency domain methods

Approaches summarized in (Ma and Li, 1993) for roller bearings include Fourier spectra of synchronized-averaged time histories, cepstrum analysis, sum and difference frequency analysis, the high frequency resonance technique, and short-time signal processing. Quantitative evaluation of gear faults using cepstrum peaks as a harmonic indicator is proposed in (Tang, et al., 1991). Thresholds distinguishing normal, moderate and serious wear in gears are determined quantitatively. Other cepstral approaches for spectral-based fault detection applied to helicopter gearboxes are presented in (Kemerait, 1987).

3.3.5.3. Time-frequency methods

Time-frequency methods have their application in the investigation of rotating machinery faults exhibiting non-stationary vibration effects. Non-stationary effects are associated with machinery in which the dynamic response differs in the various phases associated with a machine cycle. Examples include reciprocating machines, localized faults in gears, and cam mechanisms. The wavelet transform is discussed in (Chui, 1992) and is applied to fault detection and diagnosis of cam mechanisms in (Dalpiaz and Rivola, 1997) and to a helicopter gearbox in (Wang and McFadden, 1996). An application to fault detection utilizing three widely differing methods falling in the above categories (Fourier transform, power cepstrum, and wavelet transform) as applied to two meshing spur gears with an induced local fault on one gear is shown in (Petrilli, et al., 1995). A comparative study of various quantitative features that fall into the time-domain and frequency-domain categories is presented in (Elbestawi and Tait, 1986).

3.3.6. Nonlinear Methods

Identification of the basic modal properties, mode-shape curvature changes and dynamic flexibility is based on the assumption that a linear model represents the structural response before and after damage. However, in many cases the damage will cause the structure to exhibit nonlinear response. Therefore, the identification of features indicative of nonlinear response can be a very effective means of identifying damage in a structure that originally exhibited linear response. The specific features that indicate a system is responding in a nonlinear manner vary widely. Figure 6 shows the Wigner-Ville transform applied to vibration response data from both uncracked and cracked cantilever beams (Prime and Shevitz, 1996). The time-frequency plots in Fig. 6 show the generation of resonant frequency harmonics in the freely vibrating, cracked cantilever beam as well as the change in stiffness state as the crack opens and closes. For an extreme event such as an earthquake, the normalized Arias intensity provides an estimate of the structure's kinetic energy as a function of time and has been successfully used to identify the onset of nonlinear building response subjected to damaging earthquake excitations (Straser, 1998). Deviations from a Gaussian probability distribution function of acceleration response amplitudes for a system subjected to a Gaussian input have been used successfully to identify that loose parts are present in a system. Temporal variation in resonant frequencies identified using canonical variate analysis is another method to identify the onset of damage (Hunter, 1999). In general, features based on the nonlinear response of a system have only been used to identify that damage has occurred. Few methods have been described that locate the source of the nonlinearity. Because all systems exhibit some degree of nonlinearity, it is a challenge to establish a threshold for which changes in the nonlinear response features are indicative of damage. The statistical model

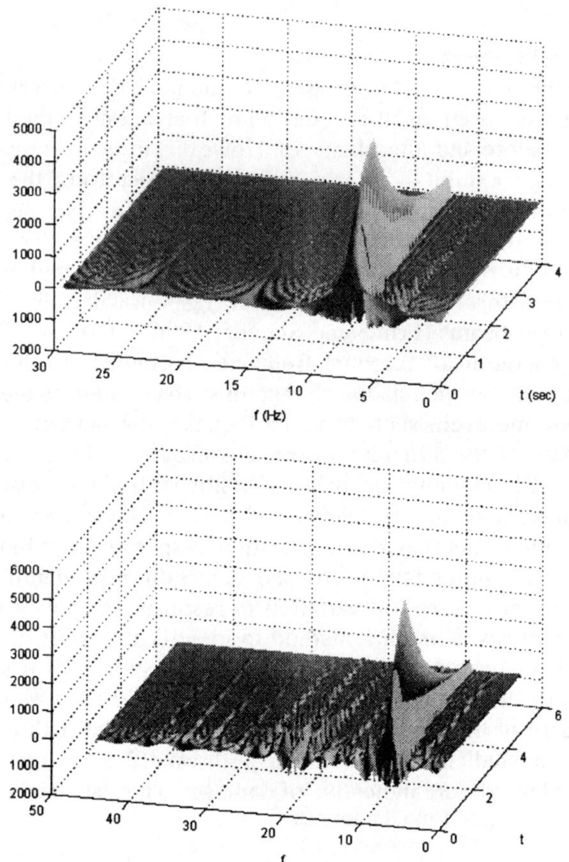

Figure 6. Wigner-Ville transforms of the free-vibration acceleration-time histories measured on an uncracked cantilever beam (left) and a cracked cantilever beam (right).

building portion of the structural health monitoring process is essential for establishing such thresholds. Note that the previously discussed features based on time-history and spectral pattern changes are often the result of nonlinear response caused by the damage.

3. 4. Statistical Model Development

The portion of the structural health monitoring process that has received the least attention in the technical literature is the development of statistical models to enhance the damage detection. Almost none of the hundreds of studies summarized in (Doebling, et al, 1996, 1998) make use of any statistical methods to assess if the changes in the selected features used to identify damaged systems are statistically significant. (Doebling and Farrar, 1998) present one of the few applications of statistical analysis to a bridge structure damage detection study. However, there are many reported studies for rotating machinery damage detection applications where statistical models have been used to enhance the damage detection process (e.g. Petrilli, et al., 1995; Chin and Danai, 1991; and Stevenson, et al., 1991).

Statistical model development is concerned with the implementation of the algorithms that operate on the extracted features to quantify the damage state of the structure. The algorithms used in statistical model development usually fall into three categories. When data are available from both the undamaged and damaged structures, the statistical pattern recognition algorithms fall into the general classification referred to as *supervised learning*. *Group classification* and *regression analysis* are supervised learning algorithms. *Unsupervised learning* refers to algorithms that are applied to data not containing examples from the damaged structure. All of the algorithms produce statistical distributions of the measured or derived features to enhance the damage detection process.

The damage state of a system can be described as a five-step process along the lines of the process discussed in (Rytter, 1993) to answers the following questions: 1. Is there damage in the system (existence)?; 2. Where is the damage in the system (location)?; 3. What kind of damage is present (type)?; 4. How severe is the damage (extent)?; and 5. How much useful life remains (prediction)? Answers to these questions in the order presented represents increasing knowledge of the damage state. The statistical models are used to answer these questions in an unambiguous and quantifiable manner. Experimental structural dynamics techniques can be used to address the first two questions. To identify the type of damage, data from structures with the specific types of damage must be available for correlation with the measured features. Analytical models are usually needed to answer the fourth and fifth questions unless examples of data are available from the system (or a similar system) when it exhibits varying damage levels.

Finally, an important part of the statistical model development process is the testing of these models on actual data to establish the sensitivity of the selected features to damage and to study the possibility of false indications of damage. False indications of damage fall into two categories: 1.) False-positive damage indication (indication of damage when none is present), and 2). False-negative damage indications (no indication of damage when damage is present). Although the second category is detrimental to the damage detection process and can have serious implications, false-positive readings also erode confidence in the damage detection process.

3.4.1 Supervised learning: group classification

Group classification attempts to place the features into respective "undamaged" or "damaged" categories in a statistically quantifiable manner. Informally, skilled individuals can use their experience with previous undamaged and damaged systems and the changes in the features associated with previously observed damage cases to deduce the presence, type and level of damage. This is an example of informal *supervised learning*. For example, it is possible to examine acceleration signals in the frequency or time domain and deduce in some cases, from the presence and location of peaks, the type, location, and extent of damage of a rotating machinery component. As previously cited, extensive tables are commercially available to facilitate this process.

More formal methods founded in machine learning have been applied to damage detection. These methods place features into either an undamaged category or one or more damaged categories (Chin and Danai, 1991). The classification techniques fall into three general categories: Bayesian Classification, K^{th}-nearest neighbor rules, and artificial neural network classifiers (Lin and Wang, 1993).

A specific example of a Bayesian group-classification statistical model applied to damage detection process is shown in Fig. 7. Here, a linear discriminant operator, "Fisher's Discriminant" (Bishop, 1995), is applied to the problem of identifying structural deterioration in progressively damaged concrete columns (Farrar, et al., 1999).

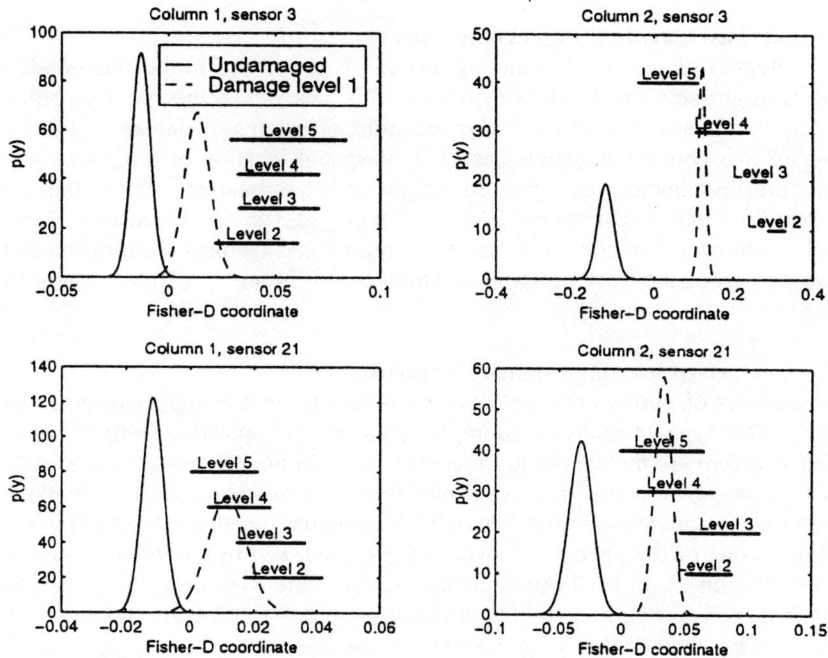

Figure 7. Distribution of LPC-generated feature vectors projected onto the Fisher coordinate. Solid horizontal lines represent the widths of the distributions for higher damage levels.

Accelerometer time histories are recorded from sensors attached to the columns while the columns are excited with an electro-dynamic shaker. Linear prediction coding (LPC) coefficients convert the accelerometer time-series data into multi-dimensional samples representing the dynamic response of the system during a brief segment of the time series. Fisher's discriminant is then used to find the linear projection of the LPC data distributions that best separates data from undamaged and damaged systems. Fisher's discriminant is defined such that the separation of features from the undamaged columns and the columns at incipient damage is maximized. The same projection is then used with subsequent damage cases. For the cases shown, the method captures a clear distinction between undamaged and damaged vibration features. Data from subsequent tests can then be classified using Bayesian methods as belonging to either the undamaged or damaged group.

3.4.2. Supervised learning: regression analysis

Another category of statistical modeling that can be employed in the damage detection process is regression analysis. Typically, this analysis refers to the process of correlating data features with particular locations or extents of damage. The features are mapped to a continuous parameter, such as spatial location or a remaining-useful-life temporal parameter, as opposed to group classification where the features correspond to discrete categories such as "damaged" or "undamaged". Regression analysis for damage detection applications requires that features from the undamaged structure and from the varying damage cases or levels are available to generate the mapping.

3.4.3. Unsupervised learning: density estimation

Finally, analysis of outliers is employed when data are not available from a damaged structure. This type of analysis attempts to answer the question: when data from a damaged structure are not available for comparison, do the observed features indicate a significant change from the previously observed features that cannot be explained by extrapolation of the feature distribution? Multivariate probability density function estimation is one of the primary statistical tools employed in this type of analysis. A particular difficulty with performing an analysis of outliers is that as the feature vectors increase in dimension, large amounts of data are needed to properly define the density function. The trigger level indicators of anomalous response that have been incorporated in commercial bridge monitoring systems can be thought of as providing indicators of outliers (Nignor and Diehl, 1997). However, to date, these indicators have not been based on a statistical analysis of the measured data.

4. COMPARISON OF BRIDGE AND ROTATING MACHINERY APPLICATIONS

Throughout this summary, applications of vibration-based damage detection to rotating machinery and bridge structures have been used to emphasize various issues and current limitations of vibration-based damage detection schemes. These bridge and rotating machinery applications represent the extremes in successful applications of the damage detection technology. In general, the application of vibration-based damage detection to rotating machinery has made the transition from a research topic to successful implementation by practicing engineers. In contrast, vibration-based damage detection in larger structures, such as bridges, has been studied for many years, but this application has, in most cases, not progressed beyond the research phase. In conclusion, a summary directly comparing these two applications is presented. This comparison further emphasizes some of the research directions that must be followed if vibration-based damage detection for large structural systems is to gain the same acceptance that it has in the rotating machinery industry.

1. Motivation: Damage detection in bridges has been primarily motivated by the prevention of loss of life; damage detection in rotating machinery is motivated largely by economic considerations often related to minimizing production downtime. Clearly, there are exceptions where bridges are being monitored to facilitate timely and cost-effective maintenance and where failure of rotating machinery can have life-safety implications, as for example, fracture of jet engine turbine blades.

2. Availability: Highway bridges are generally one-of-a-kind items with little or no data available from the damaged structure. Rotating machines are often available in large inventories with data available from both undamaged and damaged systems. It is much easier to build databases of damage-sensitive features from these inventories and, hence, supervised machine learning can be much more readily accomplished for rotating machinery.

3. Definition of Damage: For rotating machinery there are a finite number of well-defined damage scenarios and the possible locations of that damage are limited to a fairly small spatial region. Many bridge damage detection studies do not specifically define either the damage type or location.

4. Operational Evaluation: In practical health-monitoring applications, measured vibration inputs are not applied to either class of system. Rotating machinery typically exhibits response to a harmonic-like input, while traffic tends to produce inputs that are typically assumed to be random in nature.

5. Data Acquisition: Because the approximate location of the damage is generally known, vibration test equipment for rotating machinery can consist of only a single sensor and a single-channel FFT analyzer. Monitoring of bridges is normally performed with few channels distributed over a relatively large spatial region. For damage ID on a highway bridge, 30-50 data acquisition channels represent a sparsely instrumented bridge. A permanent *in situ* data acquisition system for bridge structures can be represent a significant capital outlay and further funds would be needed to maintain such a system over extended periods of time.

6. Feature Selection: A well developed database of features corresponding to various types of damage has been developed by the rotating machinery community. Many of these features are qualitative in nature and have been developed by comparing vibration signatures from undamaged systems to signatures from systems with known types, locations and levels of damage. Many of the features observed in the vibration signatures of rotating machinery result from nonlinear behavior exhibited by the damaged system. Features used to identify damage in bridge structures are most often derived from linear modal properties such as resonant frequencies and mode shapes. These features are identified before and after damage and require a distributed system of sensors. Few studies report the development of damage-sensitive features for bridge structures based on nonlinear response characteristics.

7. Statistical Model Building: The rotating machinery literature reports many more studies that investigate the application of statistical pattern classifiers to the damage detection process than have been reported for civil engineering infrastructure applications. Rotating machinery is often sited in a relatively protected environment and operates under relatively consistent conditions. The primary sources of extraneous vibration inputs are other rotating machinery in the vicinity. Changes in damage-sensitive features caused by environmental and operational variability are significant and must be accounted for in bridge applications through statistical pattern classifiers. However, the literature shows little application of this technology to bridge damage detection studies.

Clearly, the application of vibration-based damage detection to rotating machinery is a much more mature technology than that associated with large engineering infrastructure. This comparison shows that a pressing need for the large system applications is to define a limited number of damage scenarios to be monitored. Such a limitation will reduce the requirement for an expensive and difficult to maintain distributed sensing system. Advances in low-cost, wireless instrumentation and data acquisition systems can make a major contribution to the large structures applications. Further developments of sensitive analog-to-digital converter technology will reduce the noise floor in measurements allowing for better assessment of the high-frequency structural response. It is postulated that this high frequency response is more sensitive to local damage. Also, to account for variability in ambient loading conditions and environmental variability, it is imperative that the statistical pattern classifier technology must be adopted and further developed by researchers in this field. Without this technology it will be difficult to determine if changes in the identified features are caused by damage or are caused by varying operational/environmental conditions. Identifying new damage-sensitive features, particularly those that are based on nonlinear, time-varying response, should always be a focus of research efforts. Finally, there is a pressing need to make measurements on large one-of-a-kind structures such as bridges and buildings. Experience gained from analyzing data from *in situ* structures will be instrumental in developing new damage sensitive features as well as defining new and improved hardware for the vibration measurements.

5. CONCLUDING COMMENTS

The development of robust vibration-based damage detection technology has many elements that make it a potential "Grand Challenge" for the structural dynamics community. First, almost every industry wants to detect damage in its structural and mechanical infrastructure at the earliest possible time. Industries' desire to perform such monitoring is based on the tremendous economic and life-safety benefits that this technology has the potential to offer. The semiconductor manufacturing industry is adopting vibration–based damage detection to help minimize the redundancy in rotating machinery required to prevent inadvertent downtime in their fabrication plants. Such downtime can cost these companies on the order of US$10 million per hour. Driven by life-safety concerns resulting from recent structural failures, regulatory agencies in Asia are requiring the builders of large infrastructure to instrument the structures and periodically certify their health.

Significant future developments of this technology will, in all likelihood, come by way of multi-disciplinary research efforts encompassing fields such as structural dynamics, signal processing, motion and environmental sensing hardware, computational hardware, data telemetry, smart materials, and statistical pattern recognition, as well as other fields yet to be defined. Finally, the problem of global structural health monitoring is significantly complex and diverse that it will not be solved in the immediate future. Like so many other technology fields, advancements in vibration-based structural health monitoring will most likely come in small increments requiring diligent, focused and coordinated research efforts over long periods of time.

Finally, a web site that is dedicated to vibration-based damage detection and that contains many of the papers and reports referenced in this study, including links to other damage detection web sites, can be accessed at http://ext.lanl.gov/projects/damage_id/.

6.   REFERENCES

1.   Askegaard, V. and P. Mossing (1988) "Long Term Observation of RC-bridge Using Changes in Natural Frequencies," *Nordic Concrete Research*, **7**, pp. 20-27.
2.   Bishop, C. M. (1995) *Neural Networks for Pattern Recognition*, Oxford University Press, Oxford, UK.
3.   Chin, H. and K. Danai (1991) "A Method of Fault Signature Extraction for Improved Diagnosis", *Journal of Dynamic Systems, Measurement, and Control*, **113**, pp. 634-638.
4.   Crawford, A. R. (1992) *The Simplified Handbook of Vibration Analysis*, Computational Systems, Inc., Knoxville.
5.   Chui, C. K. (1992) *Wavelet Analysis and its Applications Vol I: An Introduction to Wavelets*, Academic Press.
6.   Dalpiaz, G. and A. Rivola (1997) "Condition Monitoring and Diagnostics in Automatic Machines: Comparison of Vibration Analysis Techniques", *Mechanical Systems and Signal Processing*, **11**, No.1, pp. 53-73.
7.   Doebling, S. W. and C. R. Farrar (1998) "Statistical Damage Identification Techniques Applied to the I-40 Bridge Over the Rio Grande", *Proc. of the 16th International Modal Analysis Conference*, Santa Barbara, CA, pp. 1717-1724.
8.   Doebling, S. W., C. R. Farrar and R. Goodman (1997) "Effects of Measurement Statistics on the Detection of Damage in the Alamosa Canyon Bridge,"

Proceedings 15th International Modal Analysis Conference, Orlando, FL, pp. 919-929.

9. Doebling, S. W., C. R. Farrar, M B. Prime, and D W. Shevitz, (1996) "Damage Identification and Health Monitoring of Structural and Mechanical Systems From Changes in their Vibration Characteristics: A literature Review, Los Alamos National Laboratory report LA-13070-MS.

10. Doebling, S. W., C. R. Farrar, M. B. Prime and D. W. Shevitz, (1998) "A Review of Damage Identification Methods that Examine Changes in Dynamic Properties," *Shock and Vibration Digest* **30** (2), pp. 91-105.

11. Doherty, J. E. (1987) "Nondestructive Evaluation," Chapter 12 in *Handbook on Experimental Mechanics*, A. S. Kobayashi Edt., Society for Experimental Mechanics, Inc.

12. Eisenmann, R. C., Sr. and R.C. Eisenmann, Jr. (1997) *Machinery Malfunction Diagnosis and Correction: Vibration Analysis and Troubleshooting for the Process Industries*, Hewlett-Packard Professional books, Prentice-Hall, Upper Saddle River, NJ.

13. Elbestawi, M. A. and H. J. Tait (1986) "A Comparative Study of Vibration Monitoring Techniques for Rolling Element Bearings", *Proceedings of the 4th International Modal Analysis Conference*, Los Angeles, CA, pp. 1510-1517.

14. Ewins, D. J. (1995) *Modal Testing: Theory and Practice*, Research Studies Press Ltd..

15. Farrar, C. R., W. E. Baker, T.M. Bell, K.M. Cone, T. W. Darling, T. A. Duffey, A. Eklund, and A. Migliori (1994) "Dynamic Characterization and Damage Detection in the I-40 Bridge Over the Rio Grande," Los Alamos National Laboratory report LA-12767-MS.

16. Farrar, C. R., S. W. Doebling, P. J. Cornwell, and E. G. Straser (1997) "Variability of Modal Parameters Measured on the Alamosa Canyon Bridge," *Proceedings 15th International Modal Analysis Conference*, Orlando, FL, pp. 257-263.

17. Farrar, C. R., D. A. Nix, T. A. Duffey, P. J. Cornwell, and G. C. Pardoen (1999) "Damage Identification with Linear Discriminant Operators", *Proceedings of the 17th International Modal Analysis Conference*, Orlando, FL, pp. 599- 607.

18. Friswell, M. I. and J. E. Mottershead (1995) *Finite Element Modal Updating in Structural Dynamics*, Kluwer Academic Publishers, Dordrecht, The Netherlands.

19. Hewlett Packard Company Application Note 243-1 (1997) *Effective Machinery Measurements using Dynamic Signal Analyzers*.

20. Hunter, N. F. (1999) "Bilinear System Characterization from Nonlinear Time Series Analysis," in *Proceedings of the 17th International Modal Analysis Conference*, Orlando, FL, pp. 1488-1494.

21. Kemerait, R. C. (1987) "A New Cepstral Approach for Prognostic Maintenance of Cyclic Machinery", *Proceedings of IEEE Southeastcon '87*, Tampa, FL, pp. 256-262.

22. Le Bleu, J., Jr. and M. Xu (1995) "Vibration Monitoring of Sealess Pumps Using Spike Energy", *Sound and Vibration*, pp. 10-16.

23. Lin, C.-C. and H.-P. Wang (1993) "Classification of Autoregressive Spectral Estimated Signal Patterns Using an Adaptive Resonance Theory Neural Network," *Computers in Industry*, **22**, pp. 143-157.

24. Ma J. and C. J. Li (1993) "Detection of Localized Defects in Rolling Element Bearings Via Composite Hypothesis Test", *Symposium on Mechatronics*, DSC-Vol. 50/PED-Vol. 63, American Society of Mechanical Engineers, pp. 295-300.

25. Maia, N. M. M. and J. M. M. Silva, Edts., *Theoretical and Experimental Modal Analysis*, Research Studies Press Ltd.

26. Martin, H. R. (1989) "Statistical Moment Analysis as a Means of Surface Damage Detection", *Proceedings of the 7th International Modal Analysis Conference*, Las Vegas, NV, pp. 1016-1021.

27. McConnell, K. G. (1995) *Vibration Testing Theory and Practice*, John Wiley and Sons, Inc., NY.

28. Mitchell, J. S. (1992) *Introduction to Machinery Analysis and Monitoring*, PenWel Books, Tulsa.

29. Morgan, D. P. and C. L. Scofield (1992) *Neural Networks and Speech Pattern Processing*, Kluwer Academic Publishers, Boston, MA

30. Nigbor, R. L. and J. G. Diehl (1997) "Two Year's Experience Using OASIS Real-Time Remote Condition Monitoring System on Two Bridges," *Structural Health Monitoring Current Status and Perspectives*, F. K. Chang Edt., pp. 410-417.

31. Petrilli, O., B. Paya, I. I. Esat, and M. N. M. Badi (1995) "Neural Network Based Fault Detection Using Different Signal Processing Techniques as Pre-Processor", *Structural Dynamics and Vibration*, American Society of Mechanical Engineers, NY, pp. 97-101.

32. Prime, M. B. and D. W. Shevitz (1996) "Linear and Nonlinear Methods for Detecting Cracks in Beams," *Proceedings of the 14th International Modal Analysis Conference*, Dearborn, MI, pp. 1437-1443.

33. Rabiner, L.P. and R.W. Shafer (1978) *Digital Processing of Speech Signals*, Prentice-Hall, Inc., Englewood Cliffs, NJ.

34. Roth, J. T. and S. M. Pandit (1999) "Condition Monitoring and Failure Prediction for Various Rotating Equipment Components", *Proceedings of the 17th International Modal Analysis Conference*, Orlando, FL, pp. 1674-1680.

35. Rytter, A. (1993) "Vibration based inspection of civil engineering structures," Ph. D. Dissertation, Dept. of Building Technology and Structural Eng., Aalborg Univ., Denmark.

36. Stevenson, W. J., D. L. Brown, R. W. Rost, and T. A. Grogan (1991) "The Use of Neural Nets in Signature Analysis - Rotor Imbalance", *Proceedings of the 9th International Modal Analysis Conference*, Florence, Italy, pp. 1283-1288.

37. Straser, E. G. (1998) "A Modular, Wireless Damage Monitoring System For Structures," Ph. D. Dissertation, Dept. of Civil Eng., Stanford Univ., Palo Alto, CA.

38. Tang, H., J-Z. Cha, Y. Wang, and C. Zhang (1991) "The Principle of Cepstrum and its Application in Quantitative Fault Diagnostics of Gears", DE-Vol. 38, *Modal Analysis, Modeling, Diagnostics, and Control - Analytical and Experimental*, American Society of Mechanical Engineers, pp. 141-144.

39. Taylor, J. I. (1994) *Back to the Basics of Rotating Machinery Vibration Analysis*, Vibration Consultants, Inc., Tampa Bay, FL.

40. Todd, M. D., C. C. Chang, G. A. Johnson, S. T. Vohra, J. W. Pate and R. L. Idriss (1999) "Bridge Monitoring Using a 64-Channel Fiber Bragg Grating System," *Proceedings of the 17th International Modal Analysis Conference*, Orlando, FL, pp. 1719-1725.

41. Wang, W. J. and P. D. McFadden (1996) "Application of Wavelets to Gearbox Vibration Signals for Fault Detection", *Journal of Sound and Vibration*, **192**, No. 5, pp. 927-939.

42. Wouk, V. (1991) *Machinery Vibration Measurement and Analysis*, McGraw-Hill, New York.

Vibration Testing: Reviewing the State of the Art- Prospects and Challanges

Norman F. Hunter, Technical Staff Member
Mechanical Testing Group
MS-C931
Los Alamos National Laboratory
Los Alamos, New Mexico, 87545 USA

1.0   INTRODUCTION

Formal Shock and vibration testing gained popularity during and following World War II. These tests were designed to ensure that the typical test unit survived the stresses induced by anticipated field environments. The test system consisted of items that were, in service, exposed to significant shock or vibration levels. Typical systems included gun mounts, warheads, shipping containers, and automotive or aircraft components. Field environments expose test items to a wide range of environments, either simultaneously or in succession, including elevated or reduced temperatures, impacts, static loads, or vibrations. This contribution emphasizes vibration testing.

From a relatively simple overtest whose only goal was ensuring unit survival, vibration and shock tests progressed to complex, heavily instrumented tests recording data from force transducers, accelerometers, and strain gauges. Tests are designed to ensure unit survival, to evaluate the structural effects of the environment, to estimate bounds on the survivable input and to verify analytical models. This expanded test philosophy leads to lighter, more efficiently designed systems. Estimation of bounds on survivability and verification of analytical models strongly shaped the process of environmental testing during the last decade.

This paper presents a brief history of vibration testing, outlines the current process of shock and vibration testing, suggest some improvements, and, in a broader sense, challenges, for future testing. Several excellent histories of Shock and Vibration testing are documented in the literature (Pusy, 1996). Section 2 of this contribution draws heavily on Pusy and Smallwood (Pusy, p. 187 ff.) to outline the history of vibration testing. Steps in the vibration testing process are outlined in Section 3. Section 4 summarizes issues involved in current testing, with particular emphasis on trends and challenges. The critical factors of cost, data integrity, and systematic coordination of testing and analysis are reviewed in some detail.

2.0   A BRIEF HISTORY OF VIBRATION TESTING

Shock and vibration testing has been practiced since ancient times, albeit in an intuitive and strongly practically oriented manner. Shipbuilders, for example, considered the repetitive impacts of wind and wave when designing wooden vessels. Primitive archers intuitively designed their arrows to withstand the shock of encountering bone and sinew. J.E. Gordon (Gordon, 1978), in his excellent book, "Structures, or Why Things Don't Fall Down" documents fatigue. failures soon after the industrial revolution and notes that fatigue is the result of "the cumulative effect of fluctuating loads". The steady improvement of intuitive methods led to the more formalized testing, and to the current emphasis on mathematical models. Currently our knowledge of failure mechanisms caused by the "cumulative effects of fluctuating loads" leaves much to be desired, especially if the material in question is a nonmetal.

Formalized shock and vibration tests began during and following World War II. Vibration testing arose because of the perverse nature of physical systems to fail in unexpected ways. The failure mode was often quite unrelated to behavior under static loads. It was evident that simple static load models, while often very useful in the design of civil engineering structures, were insufficient for systems subjected to repetitive loads. The spectacular torsional, vibration failure of the Tacoma Narrows bridge in 1940 (Gordon, 1978) is a classic example.

Shock loads are distinct from vibration loads. Vibration implies the repetitive application of forces for many natural periods of a structure. In contrast, most shocks are transient events, lasting for relatively short time periods (milliseconds to minutes). Shocks last at most for few natural periods of a structure. It is fair to say that one structure's shock is another structure's vibration. In most cases, shock and vibration are distinct entities. Shocks sometimes induce very high values of acceleration over small physical regions of a structure, values that would certainly lead to failure if the duration or physical extent were increased. Shock testing emphasizes high amplitude transient events in contrast to vibration, where the emphasis is on long lasting, lower amplitude repetitive events.

The history of shock and vibration testing is the story of steadily increasing sophistication in test hardware, test control, and in the interpretation of test results.

Vibration Machines

Early vibration machines utilized rotating eccentric weights, but during the mid 1950s electrodynamic vibration machines, conceptually similar to those used today, made their appearance (Pusy and Smallwood, 1996). The electronic vacuum tube amplifiers of the 1950s and 60s were followed by solid state amplifiers in the 1970s.

Electrodynamic vibration machines consist of a moving part (the armature) which oscillates relative to a fixed stator (or field coil) (Mcconnel, 1995). Alternating electric currents coupled either mechanically using wires, or inductively, to the moving armature induce the oscillatory forces driving the armature motion. A constant electric current in the stator produces a steady state magnetic field against which the alternating field of the armature reacts. The electrodynamic vibration machine is fundamentally a large loudspeaker without the cone. Some small vibration machines used in modal vibration testing and generating tens of pounds of force use permanent magnets rather than fixed stator coils and thus more closely resemble loudspeakers.

The second class of vibration machines are electrohydraulic. Electrohydraulic vibration machines move a piston using fluid pressure. Electrohydraulic machines are displacement driven devices, since a given oil volume corresponds to a given piston displacement. Electrohydraulic machines readily produce low frequency vibrations, but the oil flow rate through the valve(s) feeding the piston limits the high frequency performance, as does the high but finite compliance of the oil fluid column. Electrohydraulic machines are readily designed to produce large forces. The force rating of an electrohydraulic actuator is based on the product of the piston area (usually some number of square inches) times the fluid pressure (usually 3,000 psi.) For typical forces, say 10,000 lbs, electrohydraulic vibration machines reproduce frequencies from zero to perhaps 500 Hz.

Electrodynamic vibration machines generally reproduce higher frequencies well. Low frequencies that require relatively large displacements are difficult to generate on electrodynamic vibration machines, due to the difficulty of producing a strong, uniform magnetic field over distances ranging from inches to feet. The contrasting natural advantages of each method led to an emphasis on electrohydraulic machines for earthquake simulation and other low frequency applications and electrodynamic machines for simulation of missile launch vibrations and aircraft vibration. Truck transportation environments can be produced by either type of machine.

Since 1960 steady progress has been made in the design and application of electrodynamic and electrohydraulic vibration testing machines. High current, solid state electronics replaced the high voltage vacuum tube power amplifiers. Improved materials led to stiffer, more durable armatures and test fixtures. These are significant mechanical changes, but the biggest changes have occurred in testing philosophy, data acquisition, and data processing.

Intuitive views of vibratory motion originally led to an emphasis on sinusoidal vibration testing. It is natural to excite vibratory motion with a sine wave of slowly varying frequency. The amplitude of the input sine wave corresponds to the input vibration level, and the amplitude of the response sine wave corresponds to the response acceleration level. Analog electronics readily estimates these amplitudes. A sinusoidal input to a linear system naturally excites a single frequency sinusoidal response. The relative simplicity of the electronics, coupled with the conceptual simplicity, led to considerable sinusoidal testing in the 1950s and 1960s. Sinusoidal testing was straightforward. Harmonics at multiples of the excitation frequency corresponded to nonlinear behavior in the test item. Nonlinear behavior, as evidenced by considerable harmonic generation, was a normal feature of system tests during the era of sinusoidal testing.

Random testing gained popularity with the advent of filter banks that allowed adjustment of the RMS value in each frequency range. Typically, using the analog controllers the test frequency range from 5-2000 Hz. was controlled in 80 bandwidths each 25 Hz wide. The level in each band was adjusted using a slide wire potentiometer. Random tests were often conducted following a swept sinusoidal test.

Random testing is usually more realistic than sinusoidal testing. Multiple frequencies are induced simultaneously on the test item. This is a much more realistic simulation of inputs that occur as a result of environments like truck transport, or missile launch. At the same time random testing requires more sophisticated analysis. This analysis became available with the analog filter banks and later, in a much more elaborate and convenient form, with digital computers.

Data Acquisition and Data Processing

Modern data acquisition methods began with the use of accelerometers in the 1950s. Accelerometer based measurements have endured for nearly 50 years, supported by ever more sophisticated electronics, data acquisition, and data processing. The ready availability of digital data acquisition and processing revolutionized data processing during the 1970s and early 1980s. Accelerometers now include integral charge amplifiers so most of the signal conditioning is done inside the accelerometer case. Time series data are readily gathered from up to a hundred transducers. The fast fourier transform (FFT) implements digital filter algorithms to replace the older analog filter banks. Frequency domain averaging is implemented to produce power spectra and transfer functions. The transfer functions are then further processed to obtain modal frequencies and mode shapes. Compared to the situation a decade ago, enormous volumes of data are processed into modal models. Modal models are very compact, effective descriptions of linear systems. Their main limitation is an overemphasis on linear behavior as a descriptor of vibration response. With the advent of powerful data acquisition and processing algorithms, nonlinear data descriptors will supplement the traditional linear models.

3.1   Process Outline

The goals of vibration testing have expanded to include:

1.   Ensuring unit survival in the expected field environment.

2.   Qualifying an analytical model of the test unit in such a manner that the analytical model can both interpolate response to environments reasonably related to the test environment, and extrapolate response to environments outside of the test envelope. For example, knowing the response of a test item to 1.0 and 3.0 G's of random vibration, we interpolate the response to a 2 G input and extrapolate the response to a 5.0 G input. Interpolation is reasonably straightforward. Extrapolation is much more difficult and demands a very good understanding of

system dynamics and system failure mechanisms. To some extent, extrapolation demands further testing to confirm that an unexpected failure will not occur.

The process of vibration testing is outlined in Figure 1. Links between testing and analysis are emphasized. Each step in this process is critical, from the test definition through fixture design, preliminary testing, and test conduct, to reconciliation of the testing and analysis. Each step of this process is discussed in detail below.

180

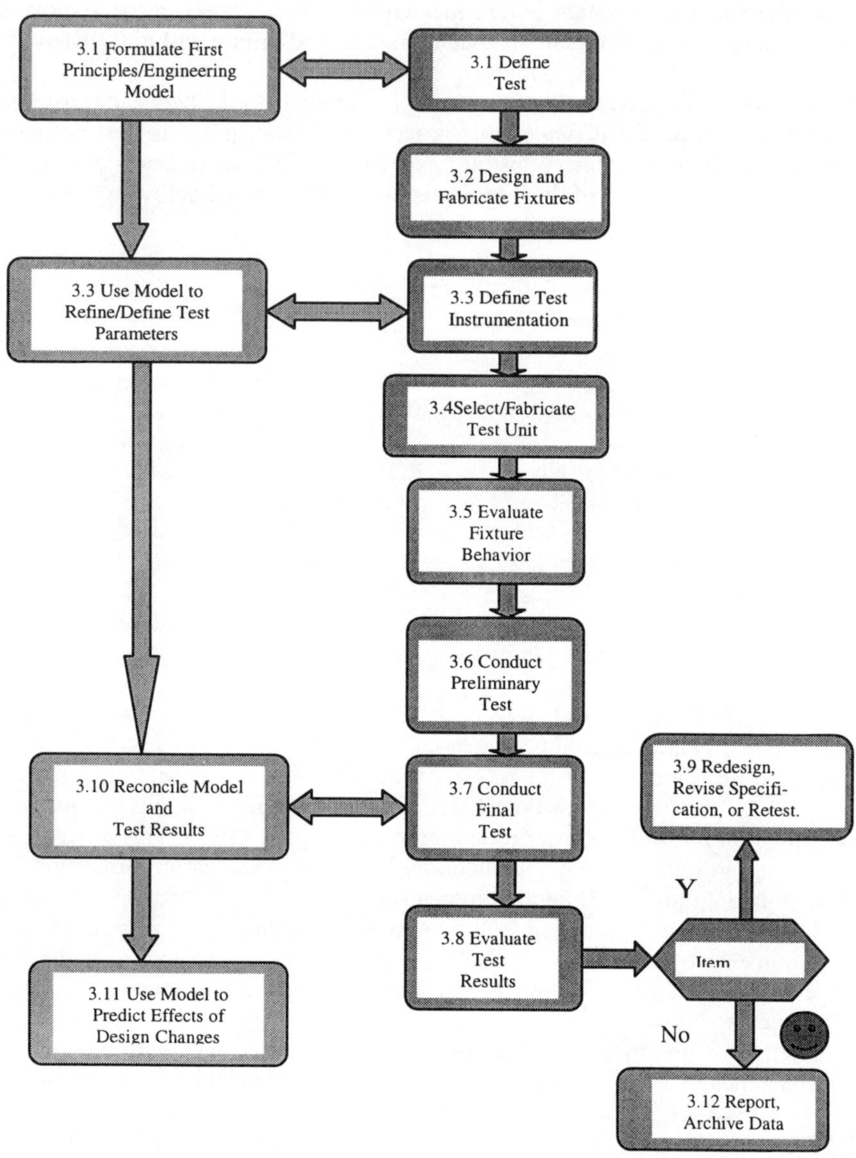

3.1 Define the Test Environment --- Formulate First Principles Engineering Model.

One major goal of vibration testing is the simulation of field vibrations in a controlled laboratory environment. Sophisticated engineering judgements are required to select the features of the field environment which most affect the test item. Some features, for example small changes in temperature, or vibrations above 3,000 Hz., are omitted from the laboratory test on the grounds that they don't play a significant part in test item survival. The laboratory environment does reproduce field force or acceleration levels, frequencies, and, to some extent, test item boundary conditions. Whether justified or not, different environments are usually tested separately. This separation is based on the practical difficulty of simultaneously generating diverse environments. The effects of quasi-static accelerations and vibration are tested separately, for example, because producing both in a laboratory test involves mounting a vibration machine on a centrifuge. Consequently laboratory tests simulate each environment in isolation. A conservative attitude, insight, and analysis attempt to compensate for interactions of the various environments. Ultimately, field testing exercises each environment simultaneously.

The sources of field environments are selected field measurements, engineering judgement, and analytical models. Many environmental requirements, specifically including vibration, are embodied in Military Standards (Mil Standard 810E) or in Stockpile to Target Sequences (STS documents). Vibration environments are traditionally separated into random, sinusoidal, and shock. With sophisticated digital control systems combined swept sine on random and swept random on random environments are produced in the laboratory as improved simulations of certain field environments.

Random test levels are defined by the input power spectrum, sinusoidal tests by frequency bounds, sweep rate and peak g level, and shock tests by a combination of the pulse form and the shock spectrum. Test levels are based on "enveloping" the field measurements. In the case of the Power Spectrum enveloping smooths the measured field spectrum while simultaneously rounding toward higher values. This procedure, is inherently conservative, since it rounds doubtful or noisy values upward. Problems occur if the rounding process is applied to deterministic values, for example, when low response levels are rounded upward in the vicinity of notches in the response transfer function. Since test values are rounded upward, testing at too low a level is not typically a problem. Testing at too high a level can, and often does, occur, and is referred to as overtesting.

Overtesting is an artifact of the enveloping process combined with the ability of vibration machines to generate high forces in any given frequency range. Consider a test item with a lightly damped, internal, 100 Hz. resonance. This resonance absorbs energy from the structure at 100 Hz. and, in the field, tends to lower the acceleration at the test item mounting point, producing a notch in the power spectrum. The frequency of this notch varies somewhat between different field tests as a consequence of the variation between different, nominally identical structures. Enveloping the measured field accelerations smooths over the accleration at the narrow notch frequencies and consequently overestimates the interface accelerations in the notch frequency range. The vibration machine readily generates the forces necessary to meet the unrealistic power spectrum even though the internal resonances absorb considerable energy at the notch frequency. These forces may damage the test item, when, in fact, no damage would occur in the field. The tendency to overtest when using enveloped acceleraton specta has been recognized for many years and several procedures used to minimize the degree of overtest. These procedures include acceleration limiting, force control, and simulation of mechanical impedance at the fixture-test item interface.

Limiting simply bounds spectral response of many (or all) response accelerometers to values not significantly exceeding those observed in field tests. Internal test item resonances increase local acceleration response values, limiting reduces the forces input to the test item at just these frequencies. It is a relatively simple procedure and quite effective.

Force control is based on measurement of field forces at interfaces and control of the force input at the test item interface in the laboratory. It is more precise than is limiting, but is harder to implement. Force transducers are structural links, require modifications to existing structures to utilize, and are consequently rarely used in field tests. Also, enveloping field forces leads to some problems analogous to those encountered in enveloping field accelerations. In fact, at interfaces, both acceleration and force must be specified for a fully realistic test.

The mechanical impedance approach solves the overtesting problem, which results because the source, the vibration machine, is unaffected by the test item impedance. A combination of field impedance measurements and use of force transducers in the laboratory allows intelligent simulation of the true interface characteristics (Scharton,1995).

An analytical model should be constructed prior to test definition to help define the test parameters. Excellent communication between the analyst and the test engineer is a critical component of this process. Prior to testing, the analytical model exercises the test structure in the field environment, modifies the boundary conditions, and exercises the structure in the laboratory, assessing the effects of the changed boundary conditions and qualifying the fixture design.

Definition of the test vibration test environment and subsequent testing usually deals with tests that excite vibration along each axis of the test item sequentially. This is somewhat akin to the process of separating the effects of different environments and testing each in sequence. It is intuitively evident that simultaneous excitation of several axes is more realistic than traditional single axis testing, since the field environment excites all axes simultaneously.

Simultaneous excitation of several axes of a test item is a challenging problem, but not outside the bounds of possibility. Controlled simultaneous excitation of multiple axes of vibration requires sophisticated mechanical coupling between several independently driven vibration machines. Further, the excitation signal for each vibration machine must, in general, take into account acceleration and/or force coupling between the vibration machines. The basic paper in the field is one by Smallwood, who formulated an algorithm for multi-shaker vibration control.

Analytical model building has been revolutionized during the past 30 years. Dynamic models have progressed from hand calculated single-degree-of-freedom models used to represent complex systems some 30 years ago to sophisticated finite element computer models. With continuing improvements in computer hardware, finite element models have grown in complexity, both in the number of elements in a simulation (from tens to hundreds to hundreds of thousands), and in the type of elements simulated. (Beam elements, gap elements, nonlinear elements). These models achieve accuracy's undreamed of a few decades ago. At the same time the models and their associated tests have a long way to go – many mechanical phenomena are poorly understood from a first principles standpoint, including damping and the behavior of sliding and bolted joints. In some cases it is fair to say that we are just learning the necessary measurements needed to understand the vibration behavior of mechanical joints. In any case, in the author's opinion, there is insufficient communication between analytical model builders and experimentalists. Model formmulation and test planning should be concurrent with lots of information interchange.

3.2   Design and Fabricate Fixtures
Analyze Fixtures in light of the Environment
The ideal vibration fixture transmits only the desired forces to the test item, while simultaneously simulating the field boundary conditons. Practical vibration fixtures are quite limited when compared to the ideal. Typical vibration fixtures are rigid and massive. This produces excellent transmission of forces to the test item, but generally does a poor job of simulating the more flexible boundaries present in the field. Fixtures are usually monolithic metal structures. Such structures are by nature lightly damped. The combined characteristics of light damping and massive rigid design mean that it is usually possible for powerful vibration machines to transmit sufficient force to the test item to achieve desired acceleration levels. It is equally easy to over test so the limiting, and impedance control techniques discussed in 3.1 are of paramount importance. Rigid fixtures, in other words, shift the burden of proper control from mechanical construction to electronic control algorithms.

A hierarchy of improvements in fixture design can be envisioned. First, existing designs can be analytically enhanced, given estimates of modal frequencies available from current finite element codes. Second, an analytical model which treated the vibration machine and the test item as known fixed, entities, could potentially optimize the test fixture using such cost functions as known item field boundary conditions and cross motion requirements. Mechanically, composite fixtures could be developed with improved (higher) damping characteristics.

184

3.3 Define Test Instrumentation
Coordinate Instrumentation Locations with Analytical Model

For several generations test instrumentation has consisted of four types of transducers: accelerometers, strain gauges, force transducers, and displacement gauges. Piezoelectric accelerometers are far and away the dominant transducer. With increasing frequency, velocity field measurements are made using laser Doppler velocometers. These full field measurements are very useful for panel-like or 2 dimensional surfaces. Scanning vibrometers provide excellent spacial resolution for modal tests when sinusoidal excitation is used. Point velocity measurements are also useful during full level vibration tests. These laser devices are very useful, but have not supplanted accelerometers for several reasons, including the necessity of visual access to the surface and the relatively high cost of a scanning laser system.

In many cases intuition serves as the basis for selecting both the number of transducers and the response locations. A better method combines intuition with information from the finite element model. For example, transducer locations may be selected for model validation, or for observations at locations of maximum response.

One fundamental difference between the analytical model and the test (or modal model) is the number of degrees of freedom. The modal model has 20-200 degrees of freedom while the finite element model has tens of thousands to millions of degrees of freedom. Measured test responses don't have thousands of degrees of freedom, even with many measurement locations. A hundred degrees of freedom may be detected, but not a thousand. Reconciliation of the test model and analytical model requires two things: condensation of the analytical model into the critical degrees of freedom (or at least those degrees of freedom observed in the test) combined with the good, well selected test measurements. There may be concerns with fixturing and with test to test repeatability, but four factors are especially critical in model validation: First, unit to unit variability is generally an order of magnitude larger than test to test variability, even with units nominally identical in construction. This unit to unit variability is currently very difficult to quantify without conducting multiple tests. Second, acceleration measurements are fundamentally measures of the derivatives of the state quantities velocity and displacement. Acceleration measurements are incapable of quantifying state initial conditions. The generally limited low frequency response of many accelerometer models means that displacement shifts, for example in joints, are undetectable. Third, rotational accelerations are difficult to measure. Fourth, analytical and test parameters must be coupled through the measurements. Somehow the situation where a model parameter is sensitive to a measurement that we cannot accurately obtain must be avoided. For example, if a 0.1% change in a measurement changes a model parameter by 35% then the model parameter cannot be validated by the existing test measurements.

3.4 Select/Fabricate the Test Unit

The test unit is a sample from a random test unit population just as the test environment is a sample from a random population of possible environments As a first approximation, test units may be assumed identical, but realistically test units vary. In many cases the variance is unknown, and in most cases the variance occurs in a high dimensional space. More precisely, if a test of a given item typically shows 20 degrees of freedom (10 resonances), then the difference between a large set of test items occurs in at least a 20 dimensional space. Sometimes multiple units are available, but usually a single unit is tested. From a particular population of test items and a particular set of environments whose probability density functions are illustrated in Figure 2, the best combination is the test item and environment that maximizes the information obtained from the test.

Population of Environments   Population of Test Items.

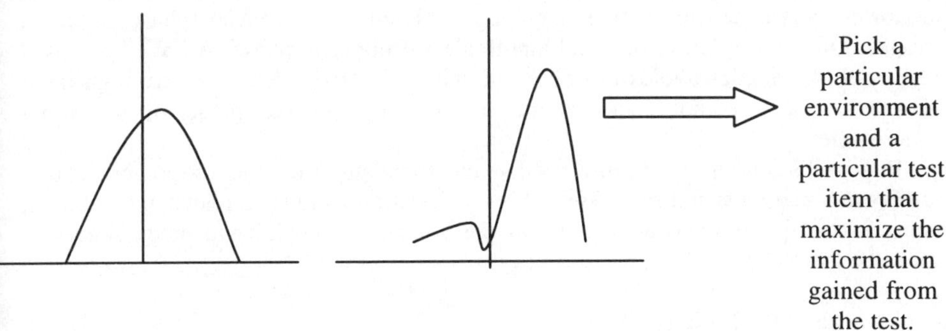

Environment PDF   Test Item Population.

Maximizing information implies selection of a test item with a particular set of characteristics. Items of information include unit survival, boundaries to failure, and, components most likely to fail.   The test should clarify critical parameters. Unfortunately, just as the test item population exists in a space of many dimensions, so does the information gained from the test. Current test planning attempts to maximize the information obtained from a test, though often in the absence of a good analytical model.

3.5. Evaluate the fixture - Reconcile the Fixture and the Analytical Fixture Model

Most vibration tests (excluding modal tests) take place in three stages. The first stage is fixture evaluation. The second stage uses a mockup test unit. The third stage tests the actual unit. The first of these steps, fixture evaluation, exercises the bare, or unloaded fixture under test conditions. The fixture transmits forces from the vibration machine to the test unit. The fixture also serves as a boundary condition at the fixture-test unit interface. Both the test and field interface forces are usually nonuniform at frequencies above the first interface resonance. The pattern of the field interface forces typically does not match the force pattern in the laboratory environment. Vibration control systems are used to duplicate the average forces or accelerations observed in the field. Non phase sensitive spectral averages, which combine spectral magnitudes, are superior to phase sensitive time averages. At some frequencies phase sensitive averages allow cancellation of equal amplitudes of opposite phase. An average based on spectral magnitudes avoids this problem, which otherwise leads to very high input forces when a test unit rocks about an axis through the plane of the input accelerometers.

Reconciliation of the fixture model and the data from the fixture evaluation qualifies the analytical fixture model. This qualification includes a realistic, test based, fixture damping value and adjustment of the model frequencies and mode shapes to match the test values.

3.6   Conduct Preliminary Test
Reconcile Model and Test Results

A preliminary vibration test is conducted at levels less than the desired test level. Traditionally low level tests are conducted from −18 Db (1/8) level to −3 Db (0.707) level. Decibel notation is used on many vibration control systems.

Low level testing exercises the fixture-test item interface. Vibration controllers operate by exercising the test system with low level vibration (usually random), determining the system transfer function, modifying the drive spectrum by the inverse of the transfer function, and repeating the process. Once control at a given level is achieved, the test level is increased. This process exposes the test item to a series of environments, whose total duration may exceed that of the field environments. The slow increase in test level allows the control system to adjust to a transfer function that changes with test level as a result of system nonlinearities. Usually more severe nonlinear behavior occurs at the higher test levels. A major difference between system level testing and modal testing is the effect of nonlinear behavior. At modal test levels (pounds to tens of pounds of force and fractions of a g to several g accelerations) most test items are nearly linear. System level tests may require thousands of pounds of force and tens of g responses. Buckling, variable stiffnesses, and mechanical joints are usually nonlinear at these levels. The modal model is a useful baseline for the system level test, but cannot predict higher level responses in detail. The first full level system mode is often 10% lower in frequency than the corresponding modal test frequency. Higher frequency modes are poorly correlated with the modal modal, until, at frequencies above 1000 Hz., modes occuring in the system level test that are completely absent in the modal model. Harmonic generation may dominate the response at higher frequencies in a system level test, even though, at modal test levels, harmonic generation is insignificant.

Some nonlinearity is handled by the vibration controller, in the sense that the control input does not exceed the desired power spectrum. The drive signal, based on the inverse of the best transfer function estimate, adjusts for changes in resonant frequencies with increasing test level. Harmonic generation is another problem. Harmonics mean that the drive at a frequency f_0 produces a response at some harmonic nf_0. Linear control systems assume that the response at nf_0 is caused by the drive signal at nf_0. As the test level increases, the harmonic response at nf_0 increases. Eventually the nf_0 response exceeds the desired response level. Even with zero drive at nf_0 an overtest results. This commonly occurs at twice or three times the frequency of some low frequency system mode. The harmonic response is also not truely random since it is related to the response at f_0. Usually an overtest due to harmonic response is accepted and documented as a limitation of the laboratory environment. Note that a fully qualified analytical model produces similiar nonlinear behavior to that observed in the test.

3.7   Conduct Final Test
Reconcile Model and Test Results
The test is conducted to levels based on the field environment. Test duration may be realistic or based on some compromise. For some simulations, like missile flight, the field environment duration is equal to that of the test. For much longer duration tests, like truck transport, test duration is less than the duration of the field environment. A compromise is reached based on increasing test amplitude and decreasing test duration. Any such test depends on engineering judgement, perhaps rationalized by some fatigue related criteria (Miner's Rule).

3.8   Evaluate Test Results
Validate Analytical Model
Even today, in an era of model validation, a major criteria of test success is unit survival. Component failure may be a realistic reflection of a field failure or an artifact of the testing process. In practice, failure of a test item in a vibration environment leads to serious investigations of the possibility that some spectral component was overemphasized, and that the test was an overtest. Overtests, unfortunately, are fairly common. Two factors contribute to the prelevance of overtesting. First, test

specificaitons are usually conservative. Second, force limiting, acceleration limiting, and control of mechancial impedence are not always used, resulting in very large force inputs at some frequencies.

Evaluate results in terms of new signal processing methods.
The behavior of the test unit during the full level test is the final validiation of the test fixture and the analytical model. Reconciliation of the test item response and the corresponding analytical model response is the final qualification of the analytical model. Model qualification at full level is usually more difficult than modal model qualification, because the increased test levels excite nontrivial nonlinear behavior.

A major question regarding test-model correlation deals with the probabilistic nature of the test item properties. The probabilistic nature of the test item selection and the model selection should be considered. In most cases a direct comparison of the response time series from the test and analytical model is inappropriate. Time series values are generally very sensitive to small perturbations. Some less sensitive measure is required for the initial comparison. This measure emphasizes important data features and condenses the time series values into a readily understandable format. Condensation reduces the effect of nondeterministic (from the measaurement standpoint) events by averaging them out of the estimation process as noise.

The Power Spectrum is one convenient and commonly used condensation for test-model comparison. The modal model, composed of mode shapes, frequencies, and damping values, is the most popular. The modal model relates directly to the eigenvalues and eigenvectors of the mathematical model and the power spectrum to the averaged magnitude of the response Fourier spectrum.

Test-model comparisons look pretty good by the measures defined above. The broader and less precise the measure used for comparison (fewer degrees of freedom), the better is the test-model comparison. For example, matching the first modal frequencies is relatively straightforward. Matching the first six modal frequencies is usually much more difficult. Some suggestions for better test-model correlation include:

1. Better definition of the test item assembly and its specific place in the random population of potential test items.
2. Improved accuracy of test measurements in concert with better measurement locations.
3. Improved understanding of the fundamental physics underlying the vibration behavior of structures, especially mechanical joints and damping.

Briefly summarized, better test-model correlation requires better measurements, both in the characterization of the test assembly and during the testing process, and improved understanding of physical processes.

Is there any way around this path of ever increasing detail outlined above. Perhaps, in some sense, such a way exists based on knowledge of the essential features affecting the modes evident in test responses.

3.8-3.12 Evaluate Test Results

Despite the increased precision, faith in models, and multitude of measurements, the fundamental fact of unit survival is important. In the author's experience the results of vibration tests fall into three categories:

1. The test unit survives with no obvious damage.
2. The test unit is damaged, but the damage is attributed to an overtest. The test specification is revised (usually based on either better field data or on a mechanical impedance related function) and the unit passes the revised test.
3. The unit fails, as do attempts to rationalize changes in the specification, and the unit is redesigned.

These cases, curiously enough, are in order of probability. It seems that most items survive vibration, but a few fail. Of those few that fail, revision of the specification eliminates most failures. Finally a few items still fail, and these are redesigned.

3.11 Final Model Qualification

Final qualification of the analytical model depends on successful prediction of an environment outside the bounds of the data used to reconcile the model to the test data. In the most straightforward qualification, some of the test data is used to qualify the model and the model is used to predict the remainder. This is, for example, a common procedure in statistical model validation. A more difficult challenge is the prediction of unit responses outside of the envelope of the test. This is much more challenging and likely requires full awareness of the probabilistic nature of the test item and of the environment. It is the author's firmly felt opinion that, in general, a model used in this manner should be validated using several tests beyond those used for model reconciliation.

CHALLENGES

The next few decades will hopefully see many advances in environmental testing. Some of these have been mentioned in Section 3 above. These are summarized in Chart form below:

The Major Challenges

The Process	The Challenge
3.1 Develop Test Specifications	1. Accurately Quantify The Random Process Generating the Environment. 2. Quantify the Degree of Conservatism in the Testing Process.
3.2 Design and Fabricate Fixtures	1. Compile the Basic Principles of Fixture Design. 2. Analyze Fixtures before we Build Them. 3. Build Fixtures to simulate Real World Boundary Conditions.
3.3 Define Test Instrumentation	More Accurate Measurement, more and better chosen measurement locations. (A new form of measurement sensor?).
3.4 Fabricate Test Unit	Using both our Past Experience and Analytical Models, Know Where to Use Real Parts and Where to Use Mockups. Precisely quantify assembly parameters and where the unit falls is the possible test unit population.
3.5 Evaluate/Qualify Fixture	Analysis Concurrent with test fixture evaluation to Suggest Realizable fixture Improvements.
3.6 Conduct Preliminary Test	Minimize Pre Test Time and Effort.
3.7 Conduct Test	1. Better Control and Simulation – Multi-Shaker and Multi-Axis Testing. 2. Realistic Simulation of Real World Boundary Conditions. 3. Simulation/Control of Non Gaussian and Non Linear Behavior.
3.10 Reconcile Testing and Analytical Model	Know when to modify parameters and when to change the model. Better understanding of structural physics, especially joints and damping.
3.11 In Event of Failure, Redesign or Change the Test Specifications	Fix the Problem, not the blame.
3.12 Report and Archive Test Data	How Many Reports and Data from tests 5 or 10 years ago are readily Available? How many lessons have to be relearned?

4.0 CONCLUSION – A NEW PARADIGM: TESTING FOR MODEL VALIDATION

Vibration testing has advanced significantly over recent decades. Comparatively speaking, enormous volumes of acceleration data are now available for virtually every vibration test. These data are readily compressed into frequency response functions and modal models. The modal models are compared with corresponding analytical models. Model updating techniques are used to adjust analytical model parameters to minimize the differences between model and test frequencies and mode shapes. These are very positive developments.

Model and testing limitations lie in areas other than the direct recording and translation of acceleration data to modal models. Two major limitiations of models are the inability to model damping and the limited degree to which nonlinear behavior is incorporated into model construction and model validation. Much more thorough understanding of damping mechanisms in real structures is required. Currently we know enough to bound the range of damping values for some typical structures. A much more thorough understanding of mechanical joints is required to accurately model structural joints. Some studies now underway offer a beginning to such understanding.

Except for some special cases, nonlinear behavior in vibration testing is not considered in most models. The linear modal model is assumed to be a sufficiently accurate approximation to the test item. To a first approximation this is usually true. As more sophisticated and accurate measurements become available, modeling nonlinear behavior becomes possible and probably essential. When the results of an analytical model are extrapolated to regions outside the region of the test data.

In summary, we have, as a testing and analysis community, come a long way in the last four decades. Many challenges remain, pre-eminent among these are realistic models of damping and nonlinearity, improved test measurements, consideration of the probabilistic nature of the test item, and clear definition of the relationships between tests and models.

REFERENCES

Gordon, J.E., Structures, or Why Things Don't Fall Down, DaCapo, 1986.

McConnell, Kenneth, 1995, Vibration Testing: Theory and Practice.

Scharton, 1995, Vibration Test Force Limits Derived From Frequency Shift Method, Journal of Spacecraft and Rockets, Vol. 32, No. 2, March-April, 1995.

Smallwood, David, 1982, Random Vibration Control System for Testing a Single Test Item with Multiple Inputs. Advances in Dynamic Analysis and Testing. SAE Aerospace Congress and Exposition, 1982, Anaheim, Calif, USASAE Spec Publ; 1982; p.55-61.

Smallwood, David, Shock and Vibration at the National Laboratories, in 50 Years of Shock and Vibration Technology, Pusy, 1996.

Updating of Analytical Models - Review of Numerical Procedures and Application Aspects

Michael Link

Light Weight Structures and Structural Mechanics Laboratory
University of Kassel, Germany

ABSTRACT

In this article basic procedures for computational updating of analytical model parameters are presented. The procedures reviewed include

- the numerical estimation techniques for solving the updating equations and
- the type   of the   residuals formed by the test/analysis differences to be minimised including force and response equation errors, eigenfrequency (resonance and antiresonance), mode shape and frequency response errors.

All the procedures allow to handle incomplete test vectors, where the number of measured degrees of freedom (DOF) is much less than the DOF no. of the computational model which is a prerequisite for computational updating of large order finite element models. Application aspects are also addressed including

- the influence of different parametrisations defining the type and the location of the erroneous model parameters,
- the requirements to be posed on the initial analysis model and
- the assessment of the final model quality.

The application aspects are illustrated using an industrial application example.

194

1. INTRODUCTION

The validation of analytical models in practise is mainly based on comparing experimental modal analysis results with the analytical predictions. Despite the high sophistication of analytical (Finite Element) modelling practical applications often reveal considerable discrepancies between analytical and test results. In recent times some effort has therefore been spent in the development of mathematical procedures for updating analytical mass and stiffness matrices using dynamic test data. The books of Natke [1] and Friswell/ Mottershead [2] represent comprehensive monographs containing the most relevant techniques. The present article contains a slightly extended version of ref.[49a].

The requirement for updating design parameters selected by the analyst like local mass and stiffness parameters results in iterative methods using non-linear optimisation in conjunction with least square procedures. The success of these methods is governed not only by the skill of the analyst to assume an appropriate initial analysis model but also the source and the location of the erroneous parameters to be corrected. The mathematical procedures presented below assume that these requirements are met.

In practical applications the source and location of the errors can be manifold resulting in non-unique updated matrices all of them fulfilling the mathematical minimisation criteria. For example, using a physical design parameter like a bending stiffness to update a discretisation error caused by a coarse finite element mesh would not be consistent with the real error source and would therefore destroy the physical significance of the design parameters. The updated model plays the role of a substitute model which at least has to fulfil the requirement of reproducing the test data used for updating. Generally we call all updating approaches where the real error source and location is not consistent with the assumed error source and location an <u>inconsistent</u> updating approach. The practical applicability of any localisation and updating procedure requires its ability and robustness with respect to handle

- incomplete test data where the no. of measurement DOF's is less than the no. analytical DOF,
- local and global physical modelling errors related to parameters like the stiffness or the mass of a single or a group of finite elements,
- inconsistent assumptions with respect to location and type of modelling error (model structure error) and
- measured data polluted with random noise and unavoidable (small) systematic errors.

2. MATHEMATICAL BACKGROUND OF MODEL PARAMETER ESTIMATION

Parameter estimation techniques aim at fitting the parameters of a given initial analytical model in such a way that the model behaviour corresponds as close as possible to the measured behaviour. The resulting parameters represent estimated values rather than true values since the test data are unavoidably polluted by unknown random and systematic errors. Also the mathematical structure of the initial analysis is not unique depending on the idealisations made by the analyst for the real structure. The method of extended weighted least squares is summarised in the following since it represents the most important estimation technique.

The first step in parameter estimation is the definition of a residual containing the difference between analytical and measured structural behaviour, for example the difference between analytical and measured eigenfrequencies. The weighted least squares technique requires to define a weighting matrix $\mathbf{W}_v$ accounting for the importance of each individual term in the residual vector ε :

$$\varepsilon_w = \mathbf{W}_v \varepsilon = \mathbf{W}_v (\mathbf{v}_M - \mathbf{v}(\mathbf{p}_i)) \tag{1}$$

$\mathbf{v}_M$ represents the measured and $\mathbf{v}(\mathbf{p})$ the corresponding analytical vector which is a function of the parameters $\mathbf{p}_i$ ($i = 1,...n_p$ = number of correction parameters). The weighted squared sum of the residual vector yields the objective function

$$J = \varepsilon_w^T \varepsilon_w = \varepsilon^T \mathbf{W} \varepsilon \ \rightarrow \ \min, \qquad \mathbf{W} = \mathbf{W}_v^T \mathbf{W}_v \tag{2}$$

whose minimisation yields the unknown parameters. In general the model vector $\mathbf{v}$ represents a non- linear function of the parameters resulting in a non- linear minimisation problem. One of the techniques to solve this non- linear optimisation problem is to expand the model vector into a Taylor series truncated after the linear term according to

$$\mathbf{v}(\mathbf{p}) = \mathbf{v}_a + \mathbf{G}\,\Delta\mathbf{p} \tag{3}$$

where $\mathbf{v}_a = \mathbf{v}\big|_{\mathbf{p}=\mathbf{p}_a}$ represents the model vector at the linearisation point $\mathbf{p} = \mathbf{p}_a$.

$\mathbf{G} = \dfrac{\partial \mathbf{v}}{\partial \mathbf{p}}\bigg|_{\mathbf{p}=\mathbf{p}_a}$ represents the   sensitivity matrix (order (m,n_p) with $m :=$ no. of

measurements and $n_p :=$ no. of parameters) and $\Delta\mathbf{p} = \mathbf{p} - \mathbf{p}_a$ represents the vector of the parameter changes.   Eq.(3) introduced into eq.(2) yields the linear residual

$$\varepsilon_w = \mathbf{W}_v \varepsilon = \mathbf{W}_v (\mathbf{v}_M - \mathbf{v}_a - \mathbf{G}\Delta\mathbf{p}) = \mathbf{W}_v (\mathbf{r}_a - \mathbf{G}\Delta\mathbf{p}) \tag{4}$$

where $\mathbf{r}_a = \mathbf{v}_M - \mathbf{v}_a$ contains the residual at the linearisation point. Of course, this formulation includes the special case when the model vector is a linear function of the parameters resulting in a constant sensitivity matrix. The stepwise calculated minimum of the objective function with respect to the parameter changes is obtained from the derivative of the objection function $\partial J / \partial \Delta\mathbf{p} = 0$ yielding the linear system of equations

196

$$W_v G \Delta p = W_v r_a \tag{5}$$

with the solution

$$\Delta p = \underbrace{(G^T W G)^{-1} G^T W}_{Z^T} r_a = Z^T r_a . \tag{6}$$

The condition of the sensitivity matrix G plays an important role for the accuracy and the uniqueness of the solution. It is clear that in the case when less measurements than parameters ($m < n_p$) are available eq.(4) leads to an underdetermined system whose solution is not unique. Even if a minimum norm or a minimum parameter change solution is selected the resulting parameters will in general not retain their physical meaning. In parameter updating the number of measurements should always be made larger than the number of parameters ($m > n_p$) which yields overdetermined equation systems. If in practical applications it is not possible to increase the number of measurements it is recommended to reduce the number of parameters by applying parameter localisation techniques like those described in [3]- [5] in order to retain only the most erroneous and sensitive parameters.

Instead of decomposing $(G^T W G)$ in eq.(6) the numerical solution of the overdetermined system is preferably done via QR or singular value decomposition in order to check the condition of G . The singular value decomposition (SVD) of the matrix G is defined by (see e.g.[6])

$$G_{(m,n_p)} = U_{(m,m)} \Sigma_{(m,n_p)} V^T_{(n_p,n_p)} \tag{7}$$

U and V represent orthogonal matrices with the properties $U^T U = V^T V = V V^T = I_{n_p}$ and $\Sigma = \mathrm{diag}\,(s_1 \ldots s_r \ldots s_{n_p})$. The singular values s_r are the roots of the eigenvalues of $G^T G$.

$$\text{If rank}(G) = R: \qquad \Sigma = \mathrm{diag}\,(s_1 \ldots s_r \ldots s_R, s_{R+1} = 0 \ldots s_{n_p} = 0) \tag{8}$$

The pseudo inverse is calculated from:

$$G^+ = V \Sigma^+ U^T \tag{9}$$

with $\qquad \Sigma^+ = \mathrm{diag}\,(1/s_1 \ldots 1/s_R, 1/s_{R+1} = 0 \ldots 1/s_{n_p} = 0)$

In practical cases a clear rank defect represented by zero singular values and caused by linear dependent parameter sensitivities can often not be detected, for example due to the influence of measurement noise. In this case the singular values can be truncated below a certain threshold of the ration s_R / s_1 . In any case it is recommended to consider a priori only parameters exhibiting linear independent sensitivities. With these definitions the solution to Eq.(4) can be expressed by

$$\Delta \mathbf{p} = \mathbf{G}_v^+ \mathbf{r}_{av} \tag{10}$$

with $\mathbf{G}_v = \mathbf{W}_v \mathbf{G}$ and $\mathbf{r}_{av} = \mathbf{W}_v \mathbf{r}_a$.

The statistical properties of the solution are calculated from the mean values and the covariance matrix of the estimate. After substituting the unknown true vector $\Delta \mathbf{p}^0$ into the eq.(4) we obtain

$$\varepsilon^0 = \mathbf{v}_M - \mathbf{v}_a - \mathbf{G}\Delta \mathbf{p}^0 = \mathbf{r}_a - \mathbf{G}\Delta \mathbf{p}^0$$

which is a measure for the for the random measurement error. With $\mathbf{r}_a = \mathbf{G}\Delta \mathbf{p}^0 + \varepsilon^0$ an estimate vector is calculated from

$$\Delta \hat{\mathbf{p}} = \mathbf{Z}^T \mathbf{r}_a = \mathbf{Z}^T (\mathbf{G}\Delta \mathbf{p}^0 + \varepsilon^0) = \Delta \mathbf{p}^0 + \mathbf{Z}^T \varepsilon^0 \tag{11}$$

The mean values of this estimate is calculated by the expectation operation

$$E(\Delta \hat{\mathbf{p}}) = \Delta \mathbf{p}^0 + E(\mathbf{Z}^T \varepsilon^0) \tag{12}$$

If the matrix $\mathbf{Z}$ defined in eq.(6) is statistically independent of ε^0 and if it is assumed that the mean of the measurements error vector equals zero , i.e. $E(\varepsilon^0) =$ **0,** then $E(\mathbf{Z}^T \varepsilon^0)$ also equals zero and

$$E(\Delta \hat{\mathbf{p}}) = \Delta \mathbf{p}^0 \tag{13}$$

The mean values of the estimate in this case are equal to the true values. Such an estimate is called an unbiased estimate. Often the above assumptions are not valid in particular when the sensitivity matrix is corrupted by measurement errors or when the model vector used to calculate $\mathbf{Z}$ is a non- linear function of the parameters, i.e. the estimate is biased. Procedures like the instrumental variable technique [7] directed to reduce the bias are useful in model updating, however, updating methods should be preferred   where the sensitivity matrix is not directly corrupted by measurement errors.

The covariance matrix $\mathrm{cov}\,(\Delta \hat{\mathbf{p}})$ represents a measure for the deviation of the estimate depending on the covariance matrix of the measurement vector $\mathrm{cov}\,(\varepsilon^0) = E(\varepsilon^0 \varepsilon^{0T})$. With the assumption of an unbiased estimate

$$\mathrm{cov}\,(\Delta \hat{\mathbf{p}}) = E[(\Delta \hat{\mathbf{p}} - \Delta \mathbf{p}^0)(\Delta \hat{\mathbf{p}} - \Delta \mathbf{p}^0)^T] = \mathbf{Z}^T E(\varepsilon^0 \varepsilon^{0T})\mathbf{Z} . \tag{14}$$

198

Substituting the inverse of the measurement covariance matrix into eq.(14) as a weighting matrix , i.e. $\mathbf{W} = [E(\boldsymbol{\varepsilon}^0 \boldsymbol{\varepsilon}^{0T})]^{-1}$, yields the covariance matrix of the estimate in the form

$$\text{cov}\,(\Delta\hat{\mathbf{p}}) = (\mathbf{G}^T \mathbf{W} \mathbf{G})^{-1} \tag{15}$$

This result shows the importance of the numerical condition of the sensitivity matrix $\mathbf{G}$ as well as the influence of the measurement errors on the error of the parameter estimate. In the special case $E(\boldsymbol{\varepsilon}^0 \boldsymbol{\varepsilon}^{0T}) = \sigma_v^2 \mathbf{I}$, i.e. when it is assumed that all measured quantities have the same standard deviation σ_v the covariance matrix of the parameter estimate is

$$\text{cov}\,(\Delta\hat{\mathbf{p}}) = \sigma_v^2 (\mathbf{G}^T \mathbf{G})^{-1} . \tag{16}$$

In [7] it is shown that the covariance matrix of the estimate is always a minimum when the above assumption for the weighting matrix is used, i.e. the covariance matrix is always smaller than the covariance matrix resulting from the standard unweighted LS with $\mathbf{W} = \mathbf{I}$. Such estimation procedures are called Markov estimates in estimation theory, they yield the best linear unbiased estimate.

The classical weighted LS method described above can be extended in cases where it difficult to obtain a convergent solution because of an ill- conditioned sensitivity matrix. The objective function (2) is extended by the requirement that the parameter changes $\Delta\mathbf{p}$ shall be kept minimal

$$J(p) = \boldsymbol{\varepsilon}^T \mathbf{W} \boldsymbol{\varepsilon} + \Delta\mathbf{p}^T \mathbf{W}_p \Delta\mathbf{p} \;\rightarrow\; \min \tag{17}$$

When the parameters are unbounded the minimisation (17), now with respect to the parameter changes $\Delta\mathbf{p}$, yields the following linear problem to be solved within each iteration step which represents the linearisation point a:

$$(\mathbf{G}^T \mathbf{W} \mathbf{G} + \mathbf{W}_p) \Delta\mathbf{p} = \mathbf{G}^T \mathbf{W}\,\mathbf{r}_a \tag{18}$$

Of course, any other mathematical minimisation technique could also be applied, in particular when the parameters shall be constrained by upper and lower bounds. In case of $\mathbf{W}_p = \mathbf{0}$ the solution of eq. (17) represents the standard weighted least squares solution, otherwise the solution is affected by the choice of the weighting matrix $\mathbf{W}_p$. In [8] we proposed to relate this matrix to the inverse of the squared sensitivity matrix acc. to

$$\mathbf{W}_p = w_p \mathbf{B}$$

$$\text{where} \quad \mathbf{B} = \frac{\text{mean}(\mathbf{g})}{\text{mean}(\mathbf{g}^{-1})} \mathbf{g}^{-1} \quad \text{and} \quad \mathbf{g} = \text{diag}\,(\mathbf{G}^T \mathbf{W} \mathbf{G}) \tag{19}$$

This definition allows to constrain $\Delta\mathbf{p}$ according to the sensitivity of the parameters. In consequence the parameters p_k remain unchanged if their sensitivity approaches zero ($W_{pk} \rightarrow 0$). The weighting factor w_p allows to scale $\mathbf{W}_p$ with respect to $\mathbf{B}$. $\mathbf{B} = \mathbf{I}$ represents the classical Tikonov regularisation [9] used to solve ill conditioned systems of equations. The question remains who to choose the regularisation parameter w_p. Several proposals for the selection of $\mathbf{W}_p$ have been proposed, e.g. in refs.[10] - [12]. Hansen [13] proposed to balance the norms $n_\varepsilon = \varepsilon^T \mathbf{W} \varepsilon$ and $n_p = w_p \Delta\mathbf{p}^T \mathbf{B} \Delta\mathbf{p}$ of the two terms in the extended objective function (17) with respect to minimising $n_\varepsilon + n_p$ as a function of w_p. This minimum is localised as the corner of the so-called L-curve obtained from plotting n_ε versus n_p. In [14] model updating applications of this and other regularisation methods were investigated. MATLAB algorithms for the different procedures have been developed and described by Hansen [15].

3.   DEFINITION OF UPDATING PARAMETERS

Starting point are the assumptions to be made for the type and the location of the erroneous parameters to be updated which are contained in the finite element system matrices of the elastodynamic equation motion:

$$(-\omega^2 \mathbf{M} + \mathbf{K} + j\omega\, \mathbf{D})\mathbf{y} = \hat{\mathbf{K}}\,\mathbf{y} = \mathbf{f} \tag{20a}$$

where the system matrices $\mathbf{M}$, $\mathbf{K}$ and $\mathbf{D}$ of order (n,n) represent the mass, the stiffness and the damping matrix, ω the excitation frequency, $\mathbf{f}$ the excitation force vector and $\mathbf{y}(j\omega)$ the complex frequency response vector ($j = \sqrt{-1}$). The special case of the undamped eigenequation is also considered in this article:

$$(-\omega_0^2 \mathbf{M} + \mathbf{K})\mathbf{y}_o = \mathbf{0} \tag{20b}$$

with ω_o and $\mathbf{y}_o$ denoting the undamped eigenfrequency and eigenvector, respectively.

In the most popular approach first introduced by Natke [16] the system matrices are updated by substructure matrices according to

$$\mathbf{K} = \mathbf{K}_A + \sum \alpha_i \mathbf{K}_i \tag{21a}$$

$$\mathbf{M} = \mathbf{M}_A + \sum \beta_j \mathbf{M}_j \tag{21b}$$

$$\mathbf{D} = \mathbf{D}_A + \sum \gamma_k \mathbf{D}_k \tag{21c}$$

where

200

$[\alpha_i \ \beta_j \ \gamma_k] = [p_s]$:= unknown correction (design) parameters,

$i = 1, 2, ..., I;$ $\quad j = 1, 2, ..., J;$ $\quad k = 1, 2, ..., K;$ $\quad s = 1, 2, ..., S$

$S = I + J + K$:= no. of correction parameters,

$\mathbf{K}_A, \mathbf{M}_A, \mathbf{D}_A$:= (n, n) analytical (initial) stiffness, mass and damping matrix,

$\mathbf{K}_i, \mathbf{M}_j, \mathbf{D}_k$:= assumed correction substructure matrices (elements or

element groups) defining source and location of modelling error.

The correction sub-matrices defined above can be considered as the first derivative of the updated matrices with respect to a physical or geometrical model parameter :

$$\mathbf{K}_i = \alpha_A \, \partial \mathbf{K}/\partial \alpha_i \ , \quad \mathbf{M}_j = \beta_A \, \partial \mathbf{M}/\partial \beta_j \quad \text{and} \quad \mathbf{D}_k = \gamma_A \, \partial \mathbf{D}/\partial \gamma_k \tag{22}$$

These derivatives are constant like in the case of a beam element with α representing Young's modulus where the stiffness matrix is a linear function of the modulus. The derivatives must not be constant like in the case when α represents the shear modulus of a Timoshenko beam where the stiffness matrix terms are non- linear functions of the shear modulus. α_A , β_A and γ_A denote the initial parameters used to make the parameter changes dimensionless. Other parametrisations related to generalised elements or substructures have been proposed in [23] and [37].

The success of parameter updating is governed not only by the skill of the analyst to assume an appropriate initial analysis model but also to assume the right source and location of the erroneous parameters to be corrected. In practical applications the source and location of the errors can be manifold resulting in non-unique updated matrices all of them fulfilling the mathematical minimisation criteria.

4.   DEFINITION OF TEST/ ANALYSIS RESIDUALS

Another important assumption the analyst has to make is the choice of the residuals formed by the differences of the predicted analytical and the measured behaviour. In the present investigation we have considered the following residuals which have most often been applied in the past:

- eigenvalues and anitresonances,
- mode shapes,
- (weighted) input forces and
- frequency response functions (FRF).

In [17] the authors present a comprehensive selection of these and other residuals with special consideration of statistically based weighting and the statistical properties of the parameter estimates. In the following we concentrate on those residuals which allow to update individually selected structural design parameters. Other residuals exist (e.g. [26],[46], [47]) allowing to update directly all the terms of the system matrices under physical constraints like conservation of matrix symmetry and connectivity. These methods are not included here since past experience with such methods was restricted to small order systems.

The linearised undamped **eigenvalue residuals** are defined by the differences between measured (index M) and analytical undamped eigenvalues at the linearisation point a:

$$\varepsilon_\lambda = \lambda_M - \lambda = \mathbf{r}_{\lambda a} - \mathbf{G}_\lambda \Delta \mathbf{p} \qquad (\lambda = \omega_0^{\,2}) \tag{23a}$$

where

$$\mathbf{r}_\lambda = \lambda_M - \lambda_a \tag{23b}$$

$:=$ residual vector containing test/ analysis differences of eigenvalues and

$$\mathbf{G}_\lambda = \left[...\partial\lambda / \partial p_s ... \right] = \left[... \partial\lambda/\partial\alpha_i ... \partial\lambda/\partial\beta_j ... \right]_{\alpha=\alpha_a, \beta=\beta_a} \tag{23c}$$

$:=$ sensitivity matrix at point a.

If the undamped problem is considered $\mathbf{G}_\lambda$ can be calculated by differentiation of the undamped eigenvalue equation and by substituting the parametrisation of eqs. (21a-b) which results in

$$\partial\lambda/\partial\alpha_i = \mathbf{y}_o^T \mathbf{K}_i \mathbf{y}_o \quad \text{and} \quad \partial\lambda/\partial\beta_j = -\lambda\, \mathbf{y}_o^T \mathbf{M}_j \mathbf{y}_o \tag{23d-e}$$

for the sensitivities with respect to the i-th stiffness and the j-th mass parameter. ($\mathbf{y}_o$ = real mode shape normalised to unit modal mass).

In [41] and [42] the authors proposed to expand the residuals by including the **antiresonances** which appear as zeros in the (undamped) frequency response functions thus enlarging the measured information. The zeros are calculated from the eigenvalue problem $(-\omega_0^2 \mathbf{M} + \mathbf{K})_{ij} \mathbf{z} = \mathbf{0}$ where the indices i and j denote the system with row i and column j deleted. For i = j (driving point FRF) the system can physically be interpreted as being grounded at DOF i = j which means that the above sensitivity expressions of eqs.(23) can directly be applied for the grounded system matrices. It can be shown that in this case the antiresonances are located between the resonances (interlacing property) . In the general case i $\neq$ j the reduced matrices become non- symmetric resulting in negative and/or complex antiresonances in conjunction with the loss of the interlacing properties and the loss of physical interpretation. In [43] the author found that the antiresonance sensitivities can be expressed by a sum of eigenvalue and mode shape sensitivities. In cases where the influence of out-of-frequency range modes is not negligible the antiresonances thus contain additional independent information compared to the

202

conventional eigenvalue and mode shape residuals taken from a limited frequency range. However, it should be noted that even in the case where the influence of the out of range modes is negligible the addition of even redundant information is favourable since it allows to average out unavoidable measurement errors in the parameter estimation process. The study on updating a model using only eigenvalues and antiresonances (i.e. no mode shape residuals) was conducted in [44] in which case the influence of the mode shape sensitivities is only implicitly contained in the antiresonance information. Special attention was directed to problems when $i \neq j$ antiresonances are used in addition to the $i = j$ antiresonances.

The linearised real **mode shape residuals** are obtained from the differences of the measured modes at the reduced set of $n_M < n$ measured DOF's denoted by the index c:

$$\varepsilon_y = y_{oM} - y_{oc} = r_{ya} - G_y \Delta p \tag{24a}$$

where
$$r_{ya} = y_{oM} - y_{oca} \tag{24b}$$

$\quad :=$ residual vector with test/ analysis differences of eigenvectors at point a and

$$y_{oca} = (y_{ocl} \cdots y_{ocr} \cdots y_{ocR})_a \quad := \text{real model modes,}$$
$$(r = 1, 2 \dots R := \text{no. of measured modes})$$

$y_{oM} \quad :=$ corresponding measured modes,

$$G_y = [\dots \partial y_{oc} / \partial p_s \dots] = \left[\dots \partial y_{oc} / \partial \alpha_i \dots \partial y_{oc} / \partial \beta_j \dots \right]_{\alpha=\alpha_a, \beta=\beta_a} \tag{24c}$$

$\quad :=$ sensitivity matrix at linearisation point a.

The calculation of the mode shape sensitivity matrix involves a major numerical effort. The modal method of Fox and Kapoor [18] is widely used due to its simplicity of implementation. It is based on expanding the gradients by a weighted sum of the eigenvectors

$$\partial y / \partial p_s = \sum_r y_{or} c_r \quad (r = 1, \dots R \leq n = \text{model order}) \tag{25a}$$

which yields after substitution into the derivative of the eigenequation the gradients with respect to the stiffness parameters:

$$\frac{\partial y_o}{\partial \alpha_i} = -\sum_{s=1}^{R} y_{os} y_{os}^T K_i y_o / (\lambda_s - \lambda) \quad \text{for } \lambda_s \neq \lambda, y_{os} \neq y_o$$

$$\frac{\partial y_o}{\partial \alpha_i} = 0 \qquad \text{for } \lambda_{os} = \lambda, y_{os} = y_o \tag{25b}$$

The gradients with respect to the mass parameters follow from:

$$\frac{\partial \mathbf{y}_o}{\partial \beta_j} = -\sum_{s=1}^{R} \mathbf{y}_{os}\mathbf{y}_{os}^T \mathbf{M}_j \mathbf{y}_o / (\lambda_s - \lambda) \quad \text{für } \lambda_s \neq \lambda, \lambda_s \neq \lambda$$

$$\frac{\partial \mathbf{y}_o}{\partial \beta_j} = -0.5 \mathbf{y}_{os}\mathbf{y}_{os}^T \mathbf{M}_j \mathbf{y}_{os} \quad \text{für } \lambda_s = \lambda, \lambda_s = \lambda$$

(25c)

This expansion is exact if R= n modes are used. For R < n the expansion represents an approximation depending on the number of modal terms. Corrections to this approach have been investigated by several authors [19]. Eqs.(25b,c) also show that the convergence of the expansion will decrease for neighboured eigenvalues $\lambda_s \approx \lambda$. Lallement [20] proposed a procedure to overcome this difficulty. An investigation of several procedures developed in the past was given by Balmes [21] with respect to using other reduced projection bases than the modal basis in eq.(25a).

Since the sensitivity matrices are derived from the (updated) analytical model they do not contain measurement errors which is an essential prerequisite for an unbiased estimate. However it should be kept in mind that due to the iterative process the sensitivity matrices depend on the parameters from the previous iteration step calculated from the noise polluted residuals in eqs.(23b) and (24b). Another advantage stems from the fact that the mode shape residuals and the sensitivities need only to be calculated for the measured DOF's, i.e. the analytical model must neither be condensed nor must the measured mode shapes be expanded to the unmeasured DOF's. It must be noted that the modal residual have to be formed between paired mode shapes. Most often the correct mode shape correlation is checked using the modal assurance criterion MAC= $(\mathbf{y}_M^T \mathbf{y}_A)^2 / (\mathbf{y}_M^T \mathbf{y}_M \mathbf{y}_A^T \mathbf{y}_A)$ which approaches one if the measured mode $\mathbf{y}_M$ and the analytical mode $\mathbf{y}_A$ are fully correlated. Mode pairs with MAC values smaller than a certain threshold (for example MAC < 0.7) should not be included in the residuals.

Using the deviations between the analytical and experimental mode shapes in the residual vector suffers from the disadvantage that the updated model is fitted to the test data in a least square's sense as close as possible although the test data are uncertain . In [22] and [23] we presented a technique of relaxing this requirement by applying a model based smoothing procedure to the experimental modes. This derivation is repeated here for completeness. The result is that the modal data of the updated model are not fitted as exactly as possible to the original experimental data but more to the smoothed data which allows to cancel not only random but also systematic measurement errors . The idea behind this approach is to bring simultaneously together both sides, the test data and the analytical parameters. Of course, both types of modifications must be kept within realistic bounds.

The smoothed (n,R_M) modal matrix $\Phi = [\varphi_1 \ldots \varphi_R]$ is expressed by a linear combination of the expanded measured modal matrix $\mathbf{Y}_M$ via an unknown transformation matrix $\mathbf{Q}$

$$\Phi = Y_M \, Q \tag{26}$$

where Φ and Y_M have order (n, R_M) and Q has order (R_M, R_M). We now require that the smoothed expanded shapes shall satisfy the orthonormality condition with respect to the updated analytical mass matrix in the form:

$$\Phi^T M \Phi = I \ . \tag{27}$$

In order to carry out the product in equation (27) the measured modes are expanded with respect to the unmeasured DOFs. The expansion is done by the so-called modal co-ordinate method [45] is done by expressing the expanded measured modal matrix Y_M by the analytical modal matrix $Y_A = [y_{o1} \cdots y_{oR}]_A$ according to

$$Y_M = Y_A \, C \tag{28}$$

After partitioning the modal matrices with respect to the measured DOFs (index c) and the unmeasured DOFs (index u) equation (28) yields

$$\begin{bmatrix} Y_{Mc} \\ Y_{Mu} \end{bmatrix} = \begin{bmatrix} Y_{Ac} \\ Y_{Au} \end{bmatrix} C \quad \text{with } C \text{ of order } (R_A, R_M) \quad . \tag{29}$$

The first row of equation (29) allows to solve for the unknown transformation matrix by minimising the norm $\left\| Y_{Mc} - Y_{Ac} \, C \right\|$ which results in

$$C = Y_{Ac}^{+} \, Y_{Mc} \tag{30}$$

("$^{+}$" indicates the pseudoinverse of the analytical (n_M, R_A) modal sub-matrix Y_{Ac}, n_M = no. of measured DOFs, R_A = no. of analytical modes). Using equation (30) the full measured modal matrix is expressed by equation (28). After introducing (26) and (28) and using the orthonormality condition, $Y_A^T M Y_A = I$, eq.(27) yields:

$$Q^T C^T C \, Q = I \tag{31}$$

Providing that the transformation matrix is symmetric $Q^T = Q$ then:

$$Q = (C^T C)^{-1/2} \tag{32}$$

Equation (33) is identical with the optimal orthogonalisation introduced by Baruch [25] which also includes the method of Targoff [24]. The smoothed shapes are finally obtained from eqs.(26) and (32) by

$$\Phi = \begin{bmatrix} \Phi_c \\ \Phi_u \end{bmatrix} = \begin{bmatrix} \mathbf{Y}_{Ac} \\ \mathbf{Y}_{Au} \end{bmatrix} \mathbf{C}(\mathbf{C}^T \mathbf{C})^{-1/2} \qquad (33)$$

This method was used in [25] and [26] to orthogonalise the mode shapes, and to update all terms of the stiffness matrix, in a global manner. We use this technique here after a given number of iteration steps, which allows to reduce the influence of the measurement errors and to build a test data set that is more consistent with the analytical model. It should be noted that the presented technique combines both, the mode shape smoothing and the expansion on the basis of the orthonormality condition. Numerically it does not require the use of the full mass matrix. The original measured mode shapes in the residual in eq.(24b) can now be replaced by the smoothes shapes taken at the measured DOF's:

$$\mathbf{Y}_{Mc} = [\mathbf{y}_{o,1} \cdots \mathbf{y}_{o,R_M}]_M \quad => \quad \Phi_c = [\varphi_1 \cdots \varphi_{R_M}]_c \ .$$

The **input error** is given by substituting the measured frequency response into the equation of motion. Since the number of measured DOF's is generally much smaller than the number of analytical DOF's it is necessary to expand the measured vector to full model size or to condense the model order down to the number of measured DOF's.

$$\varepsilon_F = \mathbf{f}_M - \hat{\mathbf{K}}_c(j\omega_M, p)\mathbf{y}_M \qquad (34a)$$

where
$$\hat{\mathbf{K}}_c(j\omega_M) = \hat{\mathbf{K}}_{ca}(j\omega_M) \ + \sum_s p_s \mathbf{S}_s \qquad (34b)$$
  := updated daynamic stiffness matrix dynamically condensed to
   $n_M < n$ measured DOF's,

$$\hat{\mathbf{K}}_{ca}(j\omega_M) = -\omega_M^2 \mathbf{M}_{ca} + \mathbf{K}_{ca} + j\omega_M \mathbf{D}_{ca} \qquad (34c)$$
  := condensed dynamic stiffness matrix at linearisation point a,

$\mathbf{S}_s = \mathbf{K}_i, \mathbf{M}_j$ or $\mathbf{D}_k$:= correction submatrices,

$\mathbf{y}_M(j\omega_M)$:= complex frequency response vector measured at n_M measurement
    DOF's

$\mathbf{f}_M$:= measured harmonic exciter forces. If $f_{Mi} = 1$ at exciter DOF i , $\mathbf{y}_M$ represents the i-th column of the frequency response matrix. With eqs.(34b,c) introduced the force residual at the measured DOF's can be expressed by

$$\varepsilon_F = \mathbf{r}_F - \mathbf{G}_F \Delta \mathbf{p} \qquad (35a)$$

where

$$\mathbf{r}_F = \mathbf{f}_M - \hat{\mathbf{K}}_{ca}(j\omega_M, p_a)\mathbf{y}_M := \text{residual vector at point a} \tag{35b}$$

$$\mathbf{G}_F = \left[\; ... - \omega_M^2 \mathbf{M}_{cj}\mathbf{y}_M ....\mathbf{K}_{ci}\mathbf{y}_M ...j\omega_M \mathbf{D}_{ck}\mathbf{y}_M ...\right]_a := \text{gradient matrix} \tag{35c}$$

Two condensation methods have most extensively been investigated in the past and are presented here. One is the dynamic condensation which includes the static condensation (also called the Guyan reduction) as a special case, the other is the modal expansion technique already presented in eqs.(29) and (30). Other techniques have been investigated like those described in [40].

The dynamically condensed matrices (index c) in the above equations are obtained by first introducing the parameterisations of eqs. (21) into the equation of motion and by partitioning the matrices with respect to the measured (index M) and unmeasured DOF's (index U):

$$\left(\begin{bmatrix} \hat{\mathbf{K}}_{MM} & \hat{\mathbf{K}}_{MU} \\ \hat{\mathbf{K}}_{UM} & \hat{\mathbf{K}}_{UU} \end{bmatrix}_A + \sum_s p_s \begin{bmatrix} \hat{\mathbf{K}}_{MM} & \hat{\mathbf{K}}_{MU} \\ \hat{\mathbf{K}}_{UM} & \hat{\mathbf{K}}_{UU} \end{bmatrix}_s \right) \cdot \begin{bmatrix} \mathbf{y}_M \\ \mathbf{y}_U \end{bmatrix} = \begin{bmatrix} \mathbf{f}_M \\ \mathbf{f}_U = 0 \end{bmatrix} \tag{36}$$

where the vector $\mathbf{y}_U$ contains the <u>unknown</u> response vector components at the unmeasured DOF's. The second row of this equation can be used to express the unmeasured vector $\mathbf{y}_U$ as a function of the correction parameters and the measured vector $\mathbf{y}_M$:

$$\mathbf{y}_U = \mathbf{T}(j\omega_M, p)\mathbf{y}_M = -\hat{\mathbf{K}}_{UU}^{-1}\hat{\mathbf{K}}_{UM}\mathbf{y}_M \tag{37}$$

The transformation matrix $\mathbf{T}$ is an function of the excitation frequency ω_M . For $\omega_M = 0$ this matrix represents the static condensation matrix (also called Guyan condensation).

The expanded full vector is given by

$$\mathbf{y} = \begin{bmatrix} \mathbf{y}_M \\ \mathbf{y}_U \end{bmatrix} = \begin{bmatrix} \mathbf{I} \\ \mathbf{T}(j\omega_M, p) \end{bmatrix} \mathbf{y}_M = \bar{\mathbf{T}}(j\omega_M, p)\mathbf{y}_M \tag{38}$$

where the (n, n_M) transformation matrix $\bar{\mathbf{T}}$ can be used to condense the system and the correction matrices according to

$$\mathbf{S}_c = \bar{\mathbf{T}}^T (\mathbf{S}_A + \sum_s p_s \mathbf{S}_s)\bar{\mathbf{T}} = \mathbf{S}_{Ac} + \sum_s p_s \mathbf{S}_{cs} \tag{39}$$

$\mathbf{S}_c = \mathbf{K}_c, \mathbf{M}_c$ or $\mathbf{D}_c :=$ condensed system matrices

Note: in the general damped case $\tilde{\mathbf{T}}$ is complex and $\tilde{\mathbf{T}}^T$ denotes complex-conjugate transpose.

This type of model reduction has several drawbacks. Looking at large order model typical for industrial applications the number of measured DOF's is much smaller than the model order: $n_M \ll n$. Since the transformation matrix depends on the current erroneous parameter estimate it becomes apparent that the sensitivity matrix in eq.(35c) will be polluted by a systematic error which can not be compensated by the LS method. In addition the numerical effort to decompose the $\hat{\mathbf{K}}_{UU}$ matrix in eq.(32) makes its application prohibitive for updating of large order models.

The second technique of order reduction also called the modal reduction technique has already been presented in eqs.(30) and (31). When applied to the frequency response vectors the transformation between the response at the unmeasured and the measured DOF's is given by:

$$y_U = \mathbf{T} y_M = \mathbf{Y}_{Au} \mathbf{Y}_{Ac}^{+} \, y_M \tag{40}$$

It should be noted that using this transformation to condense the system matrices according to eq.(40) the eigendata of the original and the condensed system are the same.
Another disadvantage results from the fact that the gradient matrix is polluted not only by the systematic condensation errors but also by measurement errors (y_M is used in $\mathbf{G}_F$, eq.(35c)) which both result in biased parameter estimates.

The **pseudo response** error residual is obtained from transforming the input force error to an output error by multiplying the force residual of eq.(34a) with the FRF matrix of initial model (index A). This is equivalent with exciting the initial model (assumed to exhibit modal damping for simplicity) with the arbitrary frequency ω_F :

$$\varepsilon_{PR} = \hat{\mathbf{K}}_{Ac}^{-1}(\omega_F)\varepsilon_F \tag{41}$$

where

$$\hat{\mathbf{K}}_{Ac}^{-1} = (-\omega_F^2 \mathbf{M}_{Ac} + j\omega_F \mathbf{D}_{Ac} + \mathbf{K}_{Ac})^{-1} = \sum_i y_{oAi} y_{oAi}^T /(\omega_{Ai}^2 - \omega_F^2 + j2\omega_{Ai}\omega_F\xi_F)$$

$$\tag{42}$$

:= condensed FRF matrix expressed by modal data of initial model (index A)
ω_F := filter frequency

$\hat{\mathbf{K}}_{Ac}^{-1}$ can also be interpreted as a dynamic filter. Applying the force residual vector on the initial system vibrating with the arbitrary filter frequency ω_F controls the magnification of the pseudo response error ε_{PR}. The idea behind this filtering is to reduce the bias of the estimate.

A **special case** is obtained by applying the residual force vector on the analytical system linearised at point 'a' using the measured excitation frequencies:

With $\hat{\mathbf{K}}_{Ac}^{-1} = \hat{\mathbf{K}}_{ca}^{-1}$ and $\omega_F = \omega_M$ introduced into eqs.(41) and (35a-c) the pseudo response residuals are calculated by

$$\varepsilon_{PR} = \hat{\mathbf{K}}_{ca}^{-1}(\omega_F)\varepsilon_F = \mathbf{r}_{PR} - \mathbf{G}_{PR}\Delta\mathbf{p} \tag{43a}$$

where

$$\mathbf{r}_{PR} = \mathbf{y}_{ca}(j\omega_M) - \mathbf{y}_M(j\omega_M) \tag{43b}$$
$$:= \text{residual response at point 'a'}$$

$$\mathbf{G}_{PR} = \hat{\mathbf{K}}_{ca}^{-1}[\ ...-\omega_M^2\mathbf{M}_{cj}\mathbf{y}_M....\mathbf{K}_{ci}\mathbf{y}_M...j\omega_M\mathbf{D}_{ck}\mathbf{y}_M...]_a \tag{43c}$$
$$:= \text{gradient matrix at measured DOF's .}$$

$$\hat{\mathbf{K}}_{ca}^{-1} = \mathbf{H}_{ca} = (-\omega_A^2\mathbf{M} + j\omega_A\mathbf{D} + \mathbf{K})^{-1}|_{p=p_a}$$
$$= \sum_i^R \mathbf{y}_{oci}\mathbf{y}_{oci}^T/(\omega_{oi}^2 - \omega_A^2 + j2\omega_{oi}\omega_A\xi_i)|_{p=p_a} \tag{43d}$$
$$:= \text{FRF matrix of analytical model expressed by } R <= N \text{ analytical modal}$$

quantities at linearisation point a (proportional modal damping values ξ_i assumed for simplicity) calculated only at the measured DOF's.

The force residual and the pseudo response technique have been investigated by several authors with respect to the bias and ill- conditioning problems mentioned and with respect to the optimal choice of the excitation frequencies above, e.g. in refs. [27]- [34] and [48].

Another special case investigated in [38] is obtained when the input force error is transformed to a static admissible displacement by replacing the dynamic stiffness matrix in eq.(41) by the static stiffness matrix $\hat{\mathbf{K}}_{Ac}^{-1} => \mathbf{K}_{ca}^{-1}$ and $\omega_F => 0$. The dynamic condensation matrix $\mathbf{T}$ of eq.(37) thus reduces to the static (Guyan) transformation matrix $\mathbf{T} = -\mathbf{K}_{UU}^{-1}\mathbf{K}_{UM}$.

The above **input error** and **pseudo response** equations also hold when measured modal data shall be used instead of the measured frequency response data. In this case the excitation force $\mathbf{f}_M$ is set to zero, the excitation frequency (Fourier variable) $-j\omega_M$ is replaced by the complex eigenvalue λ_{Mi} (Laplace variable) and the frequency response vector $\mathbf{y}_{Mi}$ is replaced by the complex mode vector. Undamped real modes can also be used when the damping matrices in the above equations are set to zero, the excitation frequency ω_M is replaced by the real eigenfrequency ω_{0i} and when the frequency response vector $\mathbf{y}_M$ is replaced by the measured real mode vector $\mathbf{y}_{0Mi}$.

When in the latter case the input error residual (34a) is multiplied by $\mathbf{y}_{0Mj}{}^T$ we arrive at the **orthogonality residual** $\mathbf{y}_{0Mj}^T \varepsilon_F = \mathbf{y}_{0Mj}^T \hat{\mathbf{K}}_c \mathbf{y}_{0Mi}$ allowing to enforce the orthogonality of the experimental modal matrix with the updated condensed mass and stiffness matrix.

The linearised **frequency response (FR) residuals** are obtained from the differences of the measured and the analytical FR at the reduced set of $n_M < n$ measured DOF's denoted by the index c. The sensitivity matrix of the **FR** residual is derived from differentiating the equation of motion (20a) with respect to the parameters.

$$\varepsilon_R = \mathbf{y}_M - \mathbf{y}_c = \mathbf{r}_{Ra} - \mathbf{G}_R \Delta \mathbf{p} \tag{44a}$$

where
$$\mathbf{r}_{Ra} = \mathbf{y}_M(j\omega_M) - \mathbf{y}_{ca}(j\omega_A) \tag{44b}$$

    := residual vector with test/ analysis differences of frequency response at
        linearisation point a and at excitation frequency $\omega_A = \omega_M$,

$\mathbf{y}_{ca}(j\omega_A) :=$ analytical complex frequency response vector at excitation
        frequency $\omega_A = \omega_M$,

$\mathbf{y}_M(j\omega_M) =$ frequency response vector measured at the same DOF's
        at excitation frequency ω_M,

$$\mathbf{G}_R = [...\partial \mathbf{y}_c / \partial p_s ...] = -\hat{\mathbf{K}}_{ca}^{-1} \left[... \frac{\partial \hat{\mathbf{K}}}{\partial p_s} \mathbf{y} \ ... \right]|_{p=p_a} \tag{44c}$$

$$= -\hat{\mathbf{K}}_{ca}^{-1} \left[... - \omega_A^2 \mathbf{M}_j \mathbf{y} \ .... \mathbf{K}_i \mathbf{y} \ ... j\omega_A \mathbf{D}_k \mathbf{y} \ ... \right]|_{p=p_a}$$

    := sensitivity matrix at measured DOF's at linearisation point a,

$\hat{\mathbf{K}}_{ca}^{-1} = \mathbf{H}_{ca} :=$ FRF matrix of analytical model according to eq.(43d).

The special case of using measured **static displacements** for parameter updating is also included in this formulation. In the static case the excitation frequencies ω_A

210

and ω_M are equal to zero, the residual vector of eq.(44b) includes the difference of the measured and analytical static displacements and the sensitivity matrix of eq.(44c) is reduced to $\mathbf{G}_R = -\mathbf{K}_{ca}^{-1}[\ ....\mathbf{K}_i\mathbf{y}\ ...]|_{p=p_a}$ where $\mathbf{K}_{ca}^{-1}$ and $\mathbf{y}$ denote the inverse static stiffness matrix and the analytical displacement vector at the linearisation point $p=p_a$.

Like the sensitivity matrix for the modal sensitivities the FR sensitivity matrix (44c) is not directly corrupted by measurement errors allowing (approximately) unbiased parameter estimates in contrast to the force and pseudo response residuals in eqs.(35c) and (43c) . In addition there is no need to expand the test vectors to the unmeasured DOF's. However, there is a crucial drawback in this formulation which has prevented its application to other than academic cases. Comparing measured and analytical FR functions like those of fig.7 it may be noticed that due to the shifts of the resonance peaks caused by the mismatch of the eigenfrequencies the test/analysis differences at a given excitation frequency become extremely large in particular in such cases where the ordering of the test and analysis eigenfrequencies is not the same (also called mode crossing) . The consequence is that the first order theory either fails to predict the updated model response or the rate of convergence is very low unless the test/analysis deviations are very small. Attempts have been made by several authors to overcome the problem, e.g. by using second order derivatives , artificial damping or eliminating or reducing the weight [32] of the data in the range of the resonances. Other approaches deliberately make use of the shift of the resonance peaks e.g. references [33], [34] and [39] and .

It is advocated to restrict updating to the resonance peaks taken at the analytical and experimental eigenfrequencies which are not identical due to mass and stiffness errors. The residuals therefore are the same as in eq.(44b) except that the analytical excitation frequencies are replaced by the analytical eigenfrequencies and the experimental excitation frequencies by the experimental eigenfrequencies, $\omega_A => \omega_o$ and $\omega_M => \omega_{Mo}$, which leads to the residual

$$\mathbf{r}_{Ra} = \mathbf{y}_M(j\omega_{Mo}) - \mathbf{y}_{ca}(j\omega_o) \tag{45}$$

Of course, the latter assumption means to extract the experimental eigenfrequencies prior to FRF updating which is in conflict with the motivation behind using FRF's for model updating instead of modal data aimed to avoid experimental modal analysis errors. However, it is believed that in practise when FRF measurements have been taken an experimental modal analysis will be performed anyway. Of course, the experimental modal data would also be useful to validate the FRF updating results. Also is must be expected that in cases when the accuracy of the test data and/or the elastodynamic mode assumptions will not allow a successful modal extraction then FRF updating will even be less successful due to the fact that the physical model includes more error sources than the curve fitting approach used in experimental modal analysis.

One advantage of FRF updating remains: the possibility to identify damping parameters. If it is assumed that the stiffness and mass parameters have been updated before by using the undamped modal residuals of eqs.(23) and (24) (Note: undamped experimental modal parameters can either be measured directly by phase resonance testing or by phase separation techniques like ISSPA [35] or by techniques transforming complex modes to real modes [36]) then the FRF residuals of eqs.(44) under consideration of eq.(45) could be applied directly with the damping parameters $\Delta p := \Delta \gamma$ being the only unkowns in this case. However, this possibility is very restricted because the local physical damping parameterisation of eq.(21c) is not available for most structural applications. This is the reason why the global modal damping approach is widely used in practise and why it is often retained in FRF based parameter updating.

The approach described in the following is directed to **update** initial estimates of **modal damping parameters**. Using the mass/stiffness updated models to calculate the frequency response one would of course use at first the damping values resulting from experimental modal analysis. If the fit of the eigenfrequencies and modes of the updated model to their experimental counterparts was perfect then the FRF's calculated with the updated model would be the same as the experimental FRF's used to extract the modal data, i.e. any updating of the modal damping parameters would not be necessary. Since in practical applications such a perfect fit cannot be obtained it seems desirable to have a procedure allowing to fit an initial modal damping estimate to the experimental FRF's.

Such an approach is described in the following. The residual used is the same as in eqs.(44a) and (45), i.e. the excitation frequencies are taken at the eigenfrequencies different in test and analysis:

$$\varepsilon_\xi = \mathbf{y}_M(\omega_{Mo}) - \mathbf{y}_c(\omega_o) = \mathbf{r}_{\xi a} - \mathbf{G}_\xi \Delta \xi \tag{46}$$

where

$$\mathbf{r}_{\xi a} = \mathbf{y}_M(j\omega_{Mo}) - \dot{\mathbf{y}}_{ca}(j\omega_o)$$

    := residual vector with test/ analysis differences of frequency response at linearisation point,

$\mathbf{y}_{ca}(j\omega_o)$:= complex frequency response vectors at excitation frequency equal to analytical eigenfrequency ω_o,

$\mathbf{y}_M(j\omega_{Mo})$:= frequency response vectors measured at the same DOF's at experimental excitation frequency equal to experimental eigenfrequency ω_{Mo}.

To calculate the sensitivity matrix of the modal damping the equation of motion

212

(20a) is first transformed to modal co-ordinates by the superposition of the undamped modes,

$$y_c = \sum_r^R y_{ocr} q_r \ , \tag{47a}$$

resulting in r = 1,...R decoupled equations if proportional damping is assumed:

$$(\omega_{or}^2 - \omega_{os}^2 + j\, 2\omega_{os}\omega_{or}\xi_r)\, q_r = \hat{K}_{g,rs}\, q_r = y_{or}^T f \ , \tag{47b}$$

ω_{os} := excitation frequencies taken at the analytical eigenfrequencies
s = 1,2...R <=n.

In case of non- proportional damping the modal equations are coupled by the non-zero off-diagonal modal damping terms $\mathbf{D}_{g,ij}^T = \mathbf{y}_{oi}^T \mathbf{D}\, \mathbf{y}_{oj} \neq 0$ for $i \neq j$. The proportional damping assumption is not necessary and was made here to simplify the derivation. Eqs.(47) derived with respect to the viscous modal damping parameter $\xi_r = D_{g,rr}/(2\omega_{or})$ yields

$$\mathbf{G}_\xi = \left[...\partial y_c(\omega_{os})/\partial \xi_r = y_{ocr}\frac{\partial q_r(\omega_{os})}{\partial \xi_r}... \right]_{\xi=\xi_a} \tag{48a}$$

where $\quad \dfrac{\partial q_r(\omega_{os})}{\partial \xi_r} = \dfrac{-j\, 2\, \omega_{os}\omega_{or}q_r(\omega_{os})}{\hat{K}_{g,rs}}. \tag{48b}$

(r ,s = 1,...R <= N number of active modes)

5.   UPDATING THE MODEL STRUCTURE

Model structure errors like those resulting from erroneous geometrical assumptions and from FE mesh discretisation errors can no longer be parameterised by substructure matrices according to eqs.(21) since they affect the mathematical structure of the model via topology and order. Even then the sensitivity approach as outlined in chapter 2 can be retained. Instead of calculating the eigenvalue and mode shape sensitivities in eqs.(23)-(25) via the parameterisation (21) they can approximately be calculated by finite differences in each iteration step by

$$\partial\lambda/\partial p_i = (\ \lambda(1+\varepsilon_{pi}) - \lambda\ )/\Delta p_i\ , \tag{49a}$$

$$\partial\varphi/p_i = (\ \varphi(1+\varepsilon_{pi}) - \varphi\ )/\Delta p_i\ . \tag{49b}$$

where ε_{pi} represents a small number, for example, $\varepsilon_{pi} = 0.1\%$. An example of updating erroneous mesh co-ordinates was presented in ref.[50].

The finite difference approach would also allow to update *mesh discretisation* parameters. The field of automatic mesh refinement which currently is based on minimising analytical energy norms of locally modified meshes could thus be based on experimental data.

6.   APPLICATION ASPECTS

In the following the different steps of updating industrial type complex structures are discussed when the model order can be as large as 100000 DOF's. Considering the non- unique correction parameter sets caused by the unavoidable idealisation and discretisation errors (not to forget the influence of random and systematic measurement errors which are discussed elsewhere, e.g. [1-2]) the problem of finding criteria to select the best initial model and the best parameter estimate must be solved in conjunction with an the assessment of the final model quality discussed in more detail in ref. [49b]. In the following this procedure is illustrated by an industrial application example. The aim was to validate the models of two aeroengine components (HPT, high pressure turbine casing and RBSS, rear bearing support structure, highlighted in Fig.1) as parts of the carcass of an aeroengine. Due to the complexity of such structures non-consistent model structures and parameterisations as well as experimental errors and incomplete test data can generally not be avoided.

Main goal of this industrial project was to improve the confidence in the predictions for an extremely complex model of the whole engine. The validation concept was based on updating the FE- models of engine components using experimental modal data of the components which allows to restrict the number of uncertain updating parameters of the whole engine model. The aim of model improvement for each individual component was to verify that the updated single models can improve the eigendynamics and the dynamic response predictions of the assembly. The first question to be answered was of how many modes each component must accurately predict in order to predict the assembly dynamics in a given frequency range. If the initial model idealisation and parameterisation was consistent one can expect that updating the parameters using experimental modal data of a limited frequency range would not only improve the analytical modal data in that range but would also improve the out-of-range data. Unfortunately this expectation can only seldom be fulfilled for industrial type structures. From an analytical component mode synthesis we concluded that about 20 component modes were necessary to predict the assembly dynamics in the desired frequency range, a number which governed the design of the experiments and also the design of the initial FE model.

214

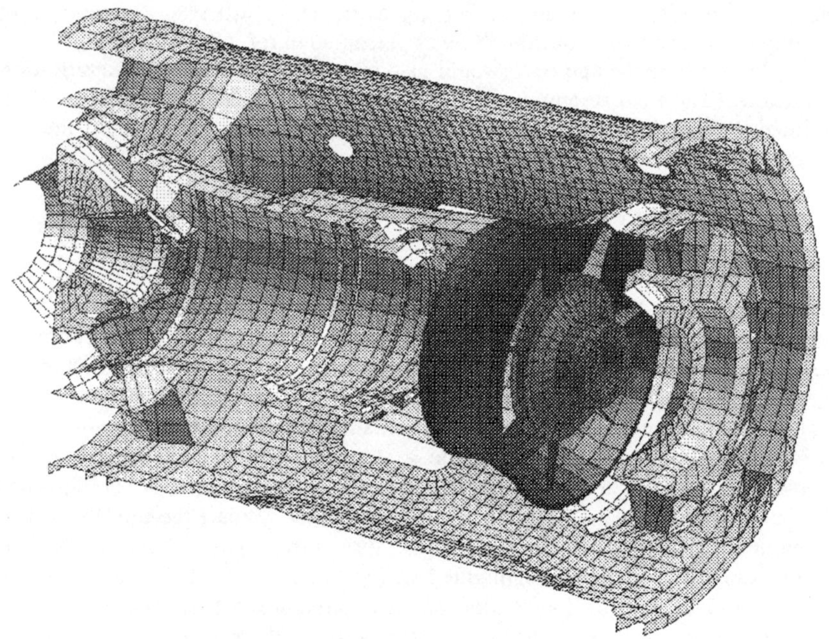

Fig. 1 Aeroengine carcass model with high pressure turbine casing (HPT) and rear bearing support structure (RBSS) highlighted

6.1   Update of the component models

Fig.2 shows the design and fig.3 the initial FE model of the HPT structure. 20 mode shapes could be correlated with their experimental counterparts with an average MAC - value of 91.1%. The average eigenfrequency error was -4%, the maximum error -6.6% and the error of the fundamental frequency was -2.5%. Usually this degree of accuracy can be considered to be sufficient for practical applications. Since due to the variety of parameters further improvement of the results by manual adjustment is no more feasible the code UPDATE_N developed at the University of Kassel for updating NASTRAN FE models using eigenfrequency and mode shape residuals was applied. Attempts to use a coarser mesh like that shown in fig.1 for the components failed because the test/analysis distance was too large.   The selection of the correction parameters was based on the sensitivity of the mode shapes and the eigenfrequencies with respect to small parameter variations. All parameters suspect of being uncertain had to be considered in the sensitivity analysis.

The evolution of the updating process is shown in Fig.4. Finally 21 analytical mode shapes correlated with the experimental shapes with an average MAC value of 92% and an average frequency error of -0.2%. The convergence of the frequency deviations, the parameter changes, the MAC values and the objective function expressed by the mean frequency error and the mean MAC error during the iterations is presented in Fig.4a-d. It should be noted that in contrast to the objective function some parameters in Fig.4b have not yet converged. This means that the diverging parameters may be changed without changing the modal behaviour in the frequency range since a convergence of the objective function can be stated from Fig.4d. The physical significance of such type of parameters must be questioned. In the present case these parameters represent shell thicknesses. Changing a thickness means to increase the membrane stiffness and the mass distribution simultaneously which tends to cancel the effect of the thickness change on the eigenfrequencies.

It should also be noted that even when the parameters have not converged each parameter set at a given iteration step represents an input data set which yields a valid FE solution. Of course, the solution is not unique in such cases. Other criteria can help to find the best solution. In the present case that solution was selected where the overall model mass and the measured mass was nearly equal.

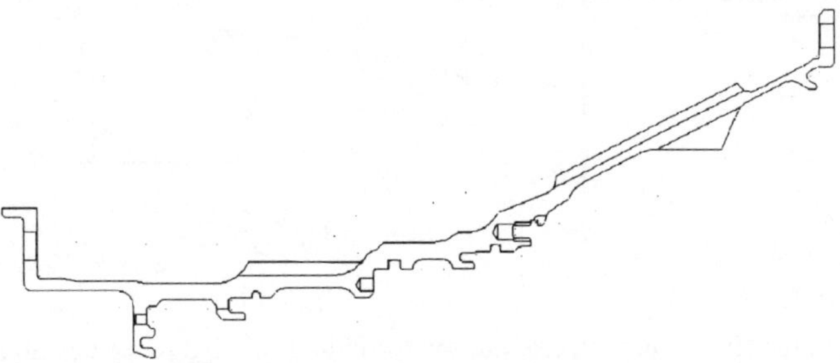

Fig. 2 HPT cross section as designed

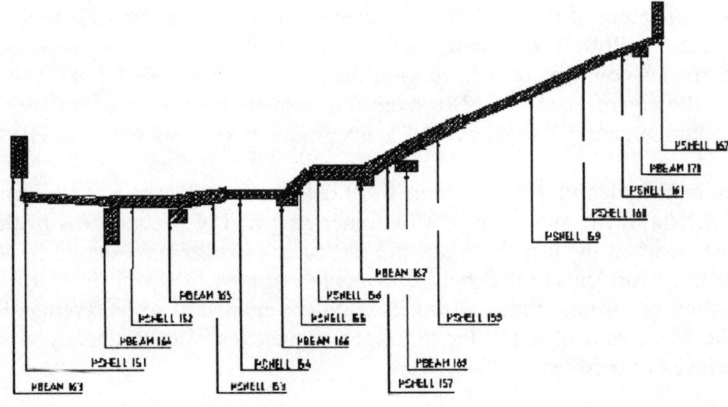

Fig. 3 FE model (NASTRAN)

216

Figs. 4a,d show the significant reduction of the 21 eigenfrequency errors before and after updating with the mean frequency error centred about zero. The quality of the final correlation results should be considered as above what can be expected from large scale industrial applications. In the present case this quality can be attributed to the quality of the initial modelling and also to the accuracy of the experimental modal data obtained under scientific laboratory conditions.

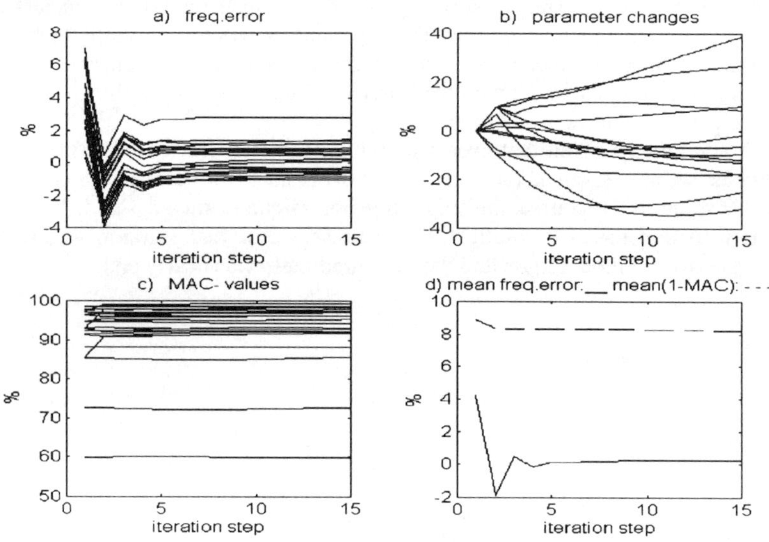

Fig. 4 Evolution of eigenfrequency errors, parameter changes, MAC - values and objective function during iteration of HPT update

Computational updating of the second component model, the rear bearing support structure (RBSS) followed the same principles. Initially 14 analytical and experimental modes could be correlated with an average of 81.8% for the MAC and -2.8% for the eigenfrequencies. However, the frequency error range between - 20% for the fundamental frequency and a maximum 6.3% was considered too large.

For computational updating 9 parameters were defined by the area and torsional moments of inertia of the spokes and of the outer ring and of the stiffness of the transition area between the circular plate and the outer cylinder represented by an equivalent Young's modulus of the shell elements in that area.

The correlation of 14 experimental modes could be increased to an average of 88.9% for the MAC and of 0.3% for the eigenfrequencies. The frequency error range was reduced to between -5.9% to 4.9%.

6.2 Assembly of HPT & RBSS components

The main goal of the present application was to improve the prediction capability of the assembly model by using *updated component* models. The test/analysis correlation of the assembly model composed of the not updated initial component models was poor. Under the assumption that the updating parameters were consistent with the source and the location of the real modelling uncertainties and under the assumption that the mathematical model structure was appropriate one has to expect that the assembly of the two updated components yields an improved model and should improve its prediction capability without further updating of the component parameters except for the uncertain parameters introduced at the bolted connection areas.

The results of using the *updated component* models confirmed that the prediction accuracy has been improved significantly compared with that of using the not updated initial component models. This gave an indication that the initial model structure and also the selected correction parameters fulfilled the consistency requirement. 21 modes were correlated with an average of 85.8% for the MAC and 5.5 % for the eigenfrequencies. However, since the frequency error range between 0.44% for the fundamental frequency and a maximum of 10.15 % still showed a positive bias and also because 3 experimental modes (nos. 4,7 and 16) could not be correlated at all computational updating was applied.

The correction parameters for updating the assembly model shown in fig.5 could be restricted to 7 parameters located in the bolted joints areas and represented by translational and rotational spring stiffnesses and the Young modulus of elasticity for the shell elements in the HPT/ spoke connection areas.

Typical results are shown in fig.6. 24 modes , that means 3 more modes compared to the previous case were correlated with an MAC average of 82.9% and an average frequency error of 0.9% in a band between −1.5 % and 4.6 %. The average MAC was a little bit lower than in the previous case which however can be tolerated in view of the fact that 3 more modes could be correlated.

A final model quality check was done by comparing the measured FRF's and those calculated with the updated assembly model together with the experimental damping values. The correlation of the envelope over all 82 measured DOF's in fig.7 and the correlation of the response plots for three selected DOF's in fig.8 looks quite satisfactory and meets the industrial requirement to ensure the response prediction up to 1000 Hz for the assembly.

Summary

In this article a summary of current procedures for model updating is presented. At first the mathematical background of the parameter estimation is given with respect to the numerical estimation techniques for solving the updating equations based on the iterative extended least squares technique. The model parameterisations defining the type and the location of the erroneous parameters are then used to construct the residuals formed by the test/analysis differences to be minimised. The residuals presented are formed by force and pseudo response equation errors, by eigenfrequency, antiresonance and mode shape errors and by frequency response errors. The procedures have been derived to handle incomplete test vectors, where the number of measured degrees of freedom (DOF) is much less than the DOF no. of the computational model. Some recent extensions of the estimation techniques with respect to handle systematic mode shape errors and to updating of modal damping values have also been included. Finally an application of updating the models of an aeroengine structure was used to demonstrate the most important updating steps.

Acknowledgement

The application in chapter 6 was supported by BMW Rolls-Royce, Germany.

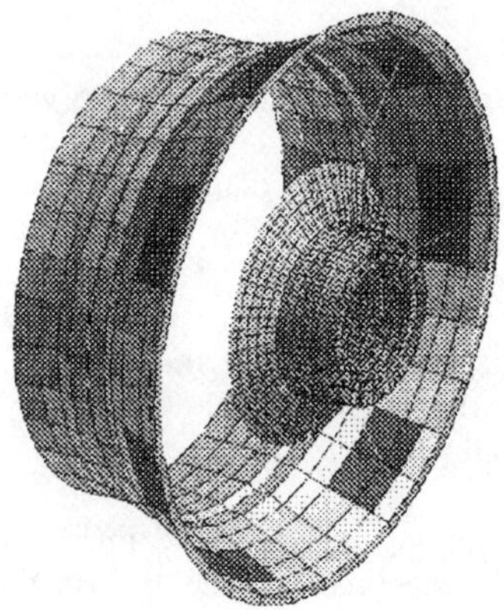

Fig. 5 HPT & RBSS Assembly Model

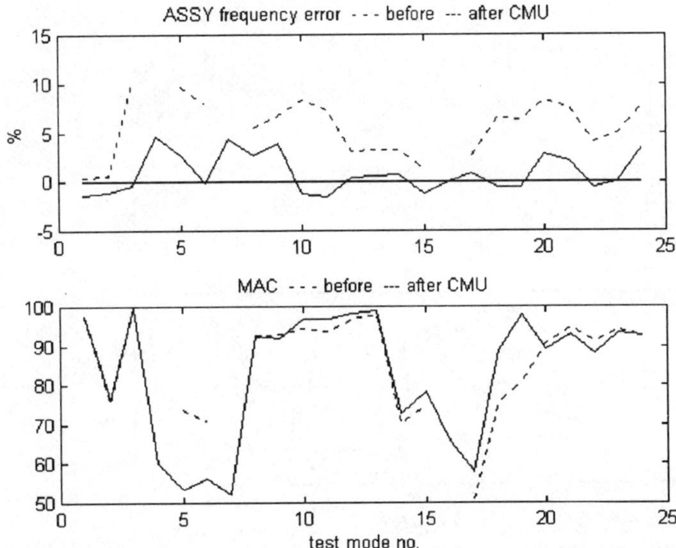

Fig. 6 Results before and after computational model updating of HPT& RBSS assembly

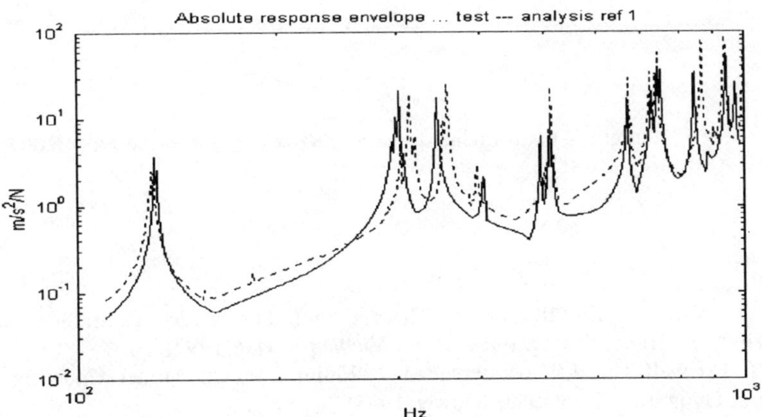

Fig. 7 Absolute response envelope over 82 measured DOF's

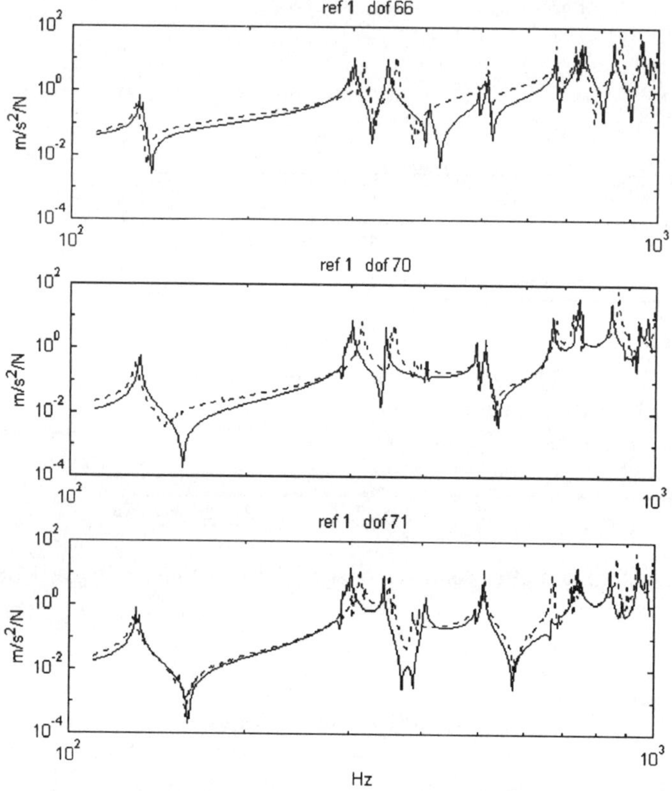

Fig. 8 Experimental (---) and updated model FRF's (___) at three representative DOF's

References

[1] H.G. Natke : Einführung in Theorie und Praxis der Zeitreihen und Modalanalyse. Braunschweig/Wiesbaden : Vieweg Verlag(1992)
[2] M.J. Friswell and J.E. Mottershead : Finite Element Model Updating in Structural Dynamics. Dordrecht: Kluwer(1995)
[3] G. Lallement: Localisation Techniques, Proc. of Workshop ' Structural Safety Evaluation Based on System Identification Approaches', Braunschweig/Wiesbaden , Vieweg Verlag(1988)
[4] Flores O. , Link M.: Localization Techniques for Parametric Updating of Dynamic Mathematical Models. Proc. of Int. Forum On Aeroelasticity and

Structural Dynamics 1993, Strassbourg, France Association Aeronautic et Astronautic de France (Hrsg), 75782 Paris (1993)

[5] H. Ahmadian, G.M.L. Gladwell and F. Ismail: Parameter Selection Strategies in Finite Element Model Updating, J.Vibration and Acoustics 119 (1994)

[6] J.W. Demmel: Applied Numerical Linear Algebra, SIAM, Philadelphia, PA,(1997)

[7] T. Soederstroem and P. Stoica: System Identification, Prentice Hall Int. UK (1989)

[8] M. Link : Updating of Analytical Models- Procedures and Experience. Proc. of Conf. on Modern Practice in Stress and Vibration Analysis, J.L. Wearing ed., Sheffield Academic Press, 35-52 (1993)

[9] A.N. Tikhonov and V.Y. Arsenin: Solutions of Ill-posed Problems, J. Wiley, New York (1977)

[10] H.G. Natke: On Regularization Methods Applied to the Error Localization of Mathematical Models. Proc. Int. Modal Analysis Conf. IMAC IX, Florence, Union College, Schenectady(1991)

[11] J.E. Mottershead and C.D. Foster: On the Treatment of Ill-Conditioning in Spatial Parameter Estimation from measured Vibration Data. Mechanical Systems and Signal Processing,Vol. 5,No.2,139-154 (1991)

[12] U. Prells : Eine Regularisierungsmethode für die lineare Fehlerlokalisierung von Modellen elastomechanischer Systeme , Dissertation, Univ. Hannover (1995)

[13] P. Ch. Hansen : Analysis of Discrete Ill-Posed Problems by Means of the L-Curve. Siam Review, Vol.34,No.4, pp. 561- 580,(1992)

[14] H. Ahmadian, J.E. Mottershead and M.I. Friswell: Regularisation Methods for Finite Element Model Updating, Mechanical Systems and Signal Processing, Vol.12, No.1 (1998)

[15] P. Ch. Hansen : Regularisation Tools. A MATLAB Package for Analysis and Solution of Discrete Ill- Posed Problems, Technical University of Denmark, Lyngby, Danmark

[16] H.G. Natke, D. Collmann and H. Zimmermann: Beitrag zur Korrektur des Rechenmodells eines elastomechanischen Systems anhand von Versuchsergebnissen. VDI- Bereichte 221, 23-32 (1974)

[17] H.G. Natke, G. Lallement and N. Cottin : Properties of Various Residuals within Updating of Mathematical Models.Inverse Problems in Engng., Vol.1, 329-348(1995)

[18] R. Fox and M. Kapoor: Rate of change of eigenvalues and eigenvectors, AIAA J., Vol.6, pp.2426- 2429 (1968)

[19] Th. R. Sutter, Ch. J. Camarda, J.L. Walsh and H. M. Adelman: Comparison of Several Methods for Calculating Vibration Mode Shape Derivatives. AIAA J., Vol.26, No.12(1988)

[20] G. Lallement and Q. Zhang: Selective Structural Modifications, Applications to the Problems of Eigensolution Sensitivity and Model Adjustment. Mech. Systems and Signal Processing, Vol.3, No.1 (1989)

[21] Balmes,E.: Efficient Sensitivity Analysis Based on Finite Element Model Reduction. Proc. of 16th Int. Modal Analysis Conf., IMAC XVI, Santa Barbara, USA, (1998)

[22] M. Link and J. Mardorf : The Role of Finite Element Idealisation and Test Data Errors in Model Updating. Proc. 2nd. Int. Conf. Structural Dynamics Modelling, NAFEMS, Glasgow, 493- 504 (1996)

[23] Link M.: Updating Analytical Models by Using Local and Global Parameters and Relaxed Optimisation Requirements, Mechanical Systems and Signal Processing , Vol.12, No.1(1998),

[24] W.P. Targoff: Orthogonality Check and Correction of Measured Modes. AIAA Journal, Vol. 14, No. 2 ,164 - 167(1976)

[25] M. Baruch and I.Y. Bar-Itzhack: Optimal Weighted Orthogonalisation of Measured Modes. AIAA Journal, Vol. 16, No.4, 346- 351 (1978)

[26]   M. Baruch : Optimal Correction of Mass and Stiffness Matrices using Measured Modes. AIAA Journal Vol. 20, No. 11 ,1623- 1626 (1982)

[27] M. Link: Localisation of Errors in Computational Models Using Dynamic Test Data. Proc. of the European Conf. on Struct. Dynamics, EURODYN '90, Structural Dynamics, W.B. Kraetzig et al. eds., A.A. Balkema (1991)

[28] Link M.:   Experiences with Different Procedures for Updating Structural Parameters of Analytical Models Using Test Data. Proc. Int. Modal Analysis Conf., IMAC X, San Diego, Union College, Schenectady, NY, USA (1992)

[29] C.-P. Fritzen and Th. Kiefer: Localisation and Correction of of Errors in Finite Element Models Based on Experimental Data. Proc. Int. Modal Analysis Conf., IMAC X, San Diego, Union College, Schenectady, NY, USA (1992)

[30] P.O. Larsson and P. Sas: Model Updating Based on Forced Vibration Testing Using Numerically Stable Formulation. Proc. Int. Modal Analysis Conf., IMAC X, San Diego, Union College, Schenectady, NY, USA (1992)

[31] W. D'Ambrogio and A. Fregolent: On the Use of Consistent and Significant Information to Reduce Ill- Conditioning in Dynamic Model Updating. Mechanical Systems and Signal Processing , Vol.12, No.1(1998)

[32] S.R. Ibrahim, W. Teichert and O. Brunner: Frequency Response Function FE Model Updating Using Multi Perturbed Analytical Models and Information Density Matrix.Proc. of 16th Int. Modal Analysis Conf., IMAC 15, Santa Barbara, USA, (1998)

[33] R. Pascual, J.C. Golinval and M. Razeto: A Frequency Domain Correlation Technique for Model Correlation and Updating. Proc. of 15th Int. Modal Analysis Conf., IMAC 15, Santa Orlando, USA (1997)

[34] S. Cogan, D. Lenoir and G. Lallement: An Improved Frequency Response Residual for Model Correction. Proc. of 14th Int. Modal Analysis Conf., IMAC 14, Dearborn, USA (1996)

[35] M. Link,   M. Weiland and J. Moreno Barragan: Direct Physical Matrix Identification as Compared to Phase Resonance Testing. Proc. of 5th Int. Modal Analysis Conf., IMAC 5, London, UK(1987)

[36] N. Niedbal: Analytical Determination of Real Normal Modes from Measured Complex Modes. AIAA Paper 84-0995 (1984)

[37] G.M.L. Gladwell and H. Ahmadian: Generic Element Matrices Suitable for Finite Element Model Updating. Mechanical Systems and Signal Processing , Vol.9 (1995)

[38] P. Ladeveze and M. Reynier: A localisation Method of Stiffness Errors for Adjustment of Finite Element Models, ASME, Vibration Analysis Techniques and Applications, Vol. 18, No.4 (1989)

[39]   W. Heylen and S. Lammens: FRAC, Aconsistent Way of Comparing Frequency Response Functions. Proc. of the international Conf. on Identification in Engineering Systems, Univ. of Wales, Swansea (1996)

[40] HP. Gysin: Expansion, the Achilles Heel of FE Model Updating . Proc. of ISMA 15, International Seminar on Modal Analysis, KU Leuven, Belgium (1990)

[41]   G. Lallement and S. Cogan: Reconciliation Between Measured and Calculated Dynamic Behaviours: Enlargement of the Knowledge Space. Proc. of the 10th Int. Modal Analysis Conf., IMAC 10, San Diego, USA (1992)

[42] D.A. Rade and G. Lallement: A Strategy for the Enrichment of Experimental Data as Applied to an Inverse Eigensensitivity- Based Finite Element Model Updating Method. Mech. Systems and Signal Processing, Vol.12, pp.293-307 (1998).

[43] J.E. Mottershead: On the Zeros of Structural Frequency Response Functions and their Application to Model Assessment and Updating. Proc. of 16th Int. Modal Analysis Conf., IMAC 16, Santa Barbara, USA (1998)

[44] W. D'Ambrogio and A. Fregolent: New Figures of Merit for Non- Modal Test- Analysis Correlation. Proc. of ISMA 23, International Conf. on Noise and Vibration Engineering, KU Leuven, Belgium (1998)

[45]   J. Lipkens and U. Vandeurzen: The Use of Smoothing Techniques for Structural Modification Applications. Proc. of ISMA 12, International Seminar on Modal Analysis, KU Leuven, Belgium (1987)

[46] S.W. Smith and Ch. A. Beattie: Multiple- Constraint Matrix Updates for Structural Identification. Proc. of International Conf. on Structural Dynamic Modelling, Test, Analysis and Correlation , DTA/ NAFEMS(Publ.), Milton Keynes, UK (1993)

[47]   B. Caesar: Update and Identification of Dynamic Mathematical Models. Proc. of 4th Int. Modal Analysis Conf., IMAC 4, Los Angeles, USA (1986)

[48]   R.M. Lin and D. Ewins: Model Updating Using FRF Data, Proc. of ISMA 15, International Conf. on Noise and Vibration Engineering, KU Leuven, Belgium (1990)

[49](a) M. Link: Updating of Analytical Models- Basic Procedures and Extensions.
(b) M. Link and G. Hanke: Model Quality Assessment and Model Updating.
In: Modal Analysis and Testing, Proceedings of NATO Advanced Study Institute (N.M.M. Maia and J.M. Montalvao e Silva Eds.), Kluwer Acad. Publ., Dordrecht, The Netherlands (to be published 1999)

[50] M. Link and Th. Graetsch: Assessment of Model Updating Results in the Presence of Model Structure and Parameterisation Errors. Proc. Int. Conference on Identification in Engineering Systems, Univ. of Wales, Swansea(1999)

AN OVERVIEW OF NONLINEAR SYSTEM DYNAMICS

Ali H. Nayfeh
University Distinguished Professor
Haider N. Arafat
Postdoctoral Research Associate

Department of Engineering Science and Mechanics, MC 0219, Virginia Polytechnic Institute and State University, Blacksburg, VA 24061, U.S.A.

Abstract. An overview of nonlinear dynamical systems and the influence of nonlinearities on their responses is presented. The types of solutions nonlinear systems may possess, the bifurcations these solutions may undergo, and the consequences of these bifurcations on the responses of such systems are explained by way of several examples. Furthermore, the types of resonances that may occur in nonlinear systems and the importance of modal interactions in nonlinearly coupled multi-degree-of-freedom systems are discussed.
Keywords: Nonlinear systems, resonances, modal interactions, bifurcations.

1. INTRODUCTION

The dynamic behavior of most systems is inherently nonlinear, and hence one needs a nonlinear theory to accurately model their responses. Only when the motion is very small that linear theory may be accurate for modeling. Continuous systems, such as structures and fluids, have infinite degrees of freedom, and therefore they are modeled by partial-differential equations. Beam, plate, and shell theories are all models of flexible structures. The Navier-Stokes equations, the Euler equations, and Laplace's equation are all models of fluids. Discrete systems, on the other hand, have finite degrees of freedom, and hence they are modeled by ordinary-differential equations of the form $\dot{x} = F(x,t)$. Many systems are hybrid, consisting of both continuous and discrete parts. Moreover, if the evolution of a system occurs at discrete points in time, then the system is modeled by a set of algebraic equations (or a map) of the form $x_{k+1} = F(x_k)$ instead of differential equations. The influence of nonlinearity on the behavior of a system is common in many engineering and science fields. This includes mechanical, electrical, thermal, nuclear, chemical, and biological systems. However, we concentrate here on the importance of nonlinearities in structural systems.

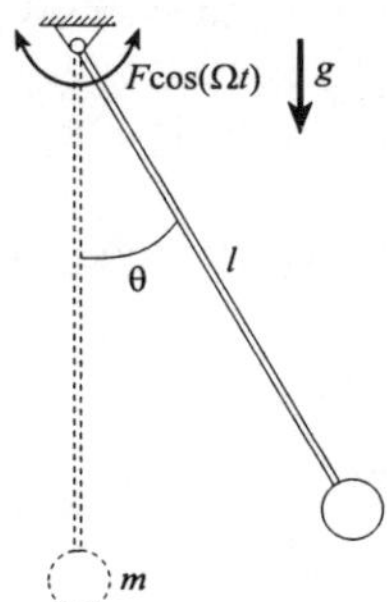

Fig. 1 An externally excited simple pendulum.

The sources of nonlinearities include the geometry, inertia, material, boundary conditions, and damping. Geometric nonlinearities result when the system undergoes large displacements and/or deflections. The dynamics of a harmonically excited simple pendulum of length l and mass m, as shown in Figure 1, are governed by

$$ml^2\ddot{\theta} + mgl\sin\theta + f(\theta,\dot{\theta}) = F\cos\Omega t \tag{1}$$

where the overdot denotes differentiation with respect to time t, $\theta(t)$ is the angular displacement, and $f(\theta,\dot{\theta})$ is a damping term, which may be nonlinear. Equation (1) is transcendental and hence nonlinear. The term $mgl\sin\theta$ is a geometric nonlinear term that models the large angular motions of the pendulum.

The equation of motion governing large bending deflections $v(s,t)$ of an inextensional beam (e.g., cantilever) according to the Euler-Bernoulli beam theory is given by

$$\rho A\ddot{v} + c\dot{v} + EIv^{iv} = -\frac{1}{2}\rho A\left\{v'\int_L^s\left[\frac{\partial^2}{\partial t^2}\int_0^s v'^2 ds\right]ds\right\}'$$
$$- EI[v'(v'v'')']' + F\cos\Omega t \tag{2}$$

where the prime denotes differentiation with respect to the arclength s, ρA is the mass per unit length, c is the coefficient of viscous damping, L is the length of the beam, and EI is the bending stiffness. In Eq. (2), the first two terms on the right-hand side are nonlinear, which are cubic in $v(s,t)$: the first is due to *nonlinear inertia* and the second is due to *nonlinear curvature*. If one considers, instead, the response of an extensional beam (e.g., hinged-clamped), then the governing equation of motion is

$$\rho A\ddot{v} + c\dot{v} + EIv^{iv} = EA\left[P(t) + \frac{1}{2L}\int_0^L v'^2 ds\right]v'' + F\cos\Omega t \tag{3}$$

where $P(t)$ is an axial force. In this case, the geometric nonlinearity, given by the second term between brackets, is due to *stretching* of the mid-surface.

The beams modeled by Eqs. (2) and (3) are assumed to be made of a material that obeys Hooke's stress-strain relations (i.e., elastic material), which are linear. However, for some materials and/or under certain forcing levels, the behavior exhibited may be plastic and/or viscoplastic. In these cases, nonlinear stress-strain relations may be necessary to accurately model the response.

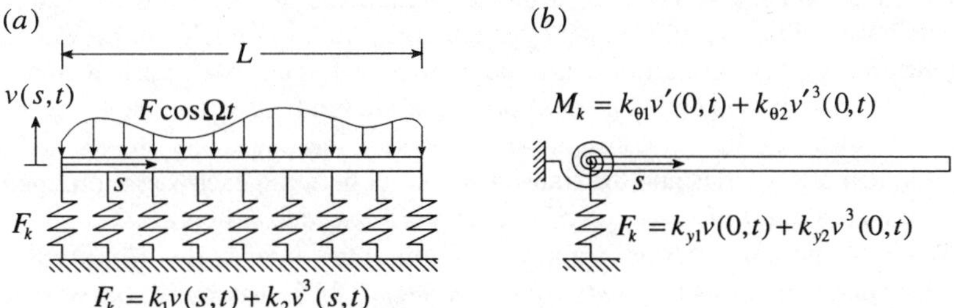

Fig. 2 (a) Free-free beam resting on a nonlinear Winkler-type foundation and (b) free-free beam with nonlinear boundary conditions.

Nonlinearities also result when elements attached to the boundary of a structure are nonlinear. For example, if we consider a linear beam resting on a nonlinear Winkler-type foundation, as shown in Figure 2a, then the equation of motion and boundary conditions are

$$\rho A\ddot{v} + EIv^{iv} + c\dot{v} + k_1 v + k_2 v^3 = F\cos\Omega t \tag{4}$$

$$v = 0 \text{ and } v' = 0 \text{ at } s = 0 \quad \text{and} \quad v'' = 0 \text{ and } v''' = 0 \text{ at } s = L \qquad (5)$$

where the k_i are constant coefficients. Nonlinearities can also appear in the boundary conditions. For example, the boundary conditions at the end $s = 0$ for the beam shown in Figure 2b are given by $EIv'' = k_{\theta 1}v' + k_{\theta 2}v'^3$ and $EIv''' = -k_{y1}v - k_{y2}v^3$.

Lastly, damping can play a major role in the response of a system. In addition to the linear viscous damping that is present in Eqs. (2)-(4), many systems are influenced by one or more of the following nonlinear dampings: Coulomb (or dry) friction where $f(v, \dot{v}) \propto \text{sgn}(\dot{v})$; aerodynamic damping, where for high Reynolds numbers, $f(v, \dot{v}) \propto \dot{v}|\dot{v}|$ and, for moderate Reynolds numbers, $f(v, \dot{v}) \propto \dot{v}|\dot{v}|^\alpha$, $0 < \alpha < 1$; hysteretic damping; and negative damping where $f(v, \dot{v}) = -c\dot{v} + h(v, \dot{v})\dot{v}$ and $h(v, \dot{v})$ is positive definite.

There are several ways one can classify the excitations acting on a system. Going by the way they appear in the equations of motion, we have *direct* (or additive) excitations and *parametric* (or multiplicative) excitations. Unlike direct excitations, the coefficients of parametric excitations are functions of the states of the system. For example, in Eq. (3), the term $F\cos\Omega t$ is a direct excitation, while the term $P(t)$ is a parametric excitation because it multiplies the term EAv''. A system that is directly excited must have nontrivial solutions whereas a parametrically excited system can exhibit a trivial response. Excitations are also classified as *ideal* and *nonideal*. *Ideal* excitations are produced by an infinite source of energy relative to the energy of the system. *Nonideal* excitations, on the other hand, have a limited source of energy that affects and is affected by the system's response. Ideal excitations may be deterministic or random in nature. Examples of deterministic excitations include harmonic and multifrequency excitations, both of which can be *stationary* or *nonstationary*. An excitation whose amplitude or frequency varies at a small, but finite, rate with time is called *nonstationary*. If we hold either the amplitude or the frequency of an excitation constant while we vary the other very slowly and wait until the response settles and the transients die out between each variation, then the excitation is called *stationary* and this process is referred to as being *quasistationary*.

When the nonlinear terms in a model are the same order as the linear terms, we describe such a model as being *strongly nonlinear*. This is usually typical of systems that exhibit large finite motions like dynamic rigid-body systems. If the nonlinear terms in a model are relatively small compared to the linear terms, then we describe such a model as being *weakly nonlinear*. This is typical of systems that exhibit small finite-amplitude motions like many elastic structures. When the motion of a system is infinitesimally small, it is common to neglect the nonlinearities altogether and describe its dynamics by a *linear* model. However, caution must be taken when doing so because, in real life, there are many factors that can easily and unexpectedly cause a linearly behaving system to exhibit nonlinear behavior. In such a situation, a linear model of the system would be inadequate and misleading.

As a measure of the strength of the nonlinear terms, we introduce a small dimensionless parameter $\epsilon \ll 1$ in the model. As an example, we consider

$$\ddot{u} + \omega^2 u + \epsilon\mu\dot{u} + \epsilon\alpha\dot{u}^3 = \epsilon F \cos\Omega t \tag{6}$$

where $u(t)$ is a generalized coordinate of the system and ω is its associated natural frequency; Ω is the excitation frequency; and α, μ, and F are constant coefficients of the cubic nonlinear damping term, linear viscous damping term, and excitation term, respectively. Assuming that $\alpha = O(1)$ (i.e., α is of order 1), Eq. (6) is classified as being *weakly nonlinear*. If $\alpha = O(\epsilon^{-1})$, the nonlinear terms are the same order as the linear terms and Eq. (6) is now classified as being *strongly nonlinear*. Moreover, if $\alpha = 0$, Eq. (6) is linear. In the same manner, we describe the influence of the damping and excitation. For example, if $\mu = O(1)$ and $F = O(1)$, we say that Eq. (6) is weakly damped and weakly excited, and if $\mu = O(\epsilon^{-1})$ and $F = O(\epsilon^{-1})$, we say that Eq. (6) is strongly damped and strongly excited.

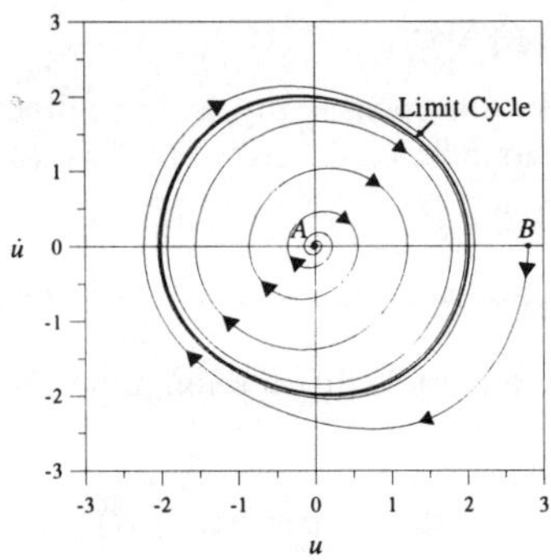

Fig. 3 A phase portrait demonstrating the behavior of a self-excited system modeled by the Rayleigh equation.

When $\mu < 0$ and $\alpha > 0$, Eq. (6) is referred to as Rayleigh's equation; it is an example of a *self-excited* system. That is, even if the system is not externally excited (i.e., $F = 0$), its response can still be nontrivial. For small values of $u(t)$, the negative linear damping dominates the term $f(u, \dot{u}) = \mu\dot{u} + \alpha\dot{u}^3$, making the system negatively damped and causing the solution to grow. For large values of $u(t)$, the positive nonlinear damping dominates $f(u, \dot{u})$, making the system positively damped and causing the solution to decay. Both solutions are attracted to a limit cycle, as shown in Figure 3.

By in large, closed-form analytic solutions to nonlinear systems do not exist. Therefore, numerical methods, such as the finite element, finite difference, and boundary element methods, may be needed to obtain a solution, especially for strongly nonlinear systems. For weakly nonlinear systems, one can use *perturbation methods* to obtain asymptotically uniform approximations of the solutions. The advantage of having an analytical solution, even if it is just an approximation, over a numerical solution is that the former provides the flexibility to investigate the response under many conditions and for a wide range of system parameters. Analytical solutions also can give insight as to the reasons a system behaves in a certain manner under certain conditions. Perturbation methods include the method of averaging, the Lindstedt-Poincaré technique, the method of multiple scales, the method of normal forms, the method of renormalization, the method of harmonic balance, the method of matched asymptotic expansions, and averaging by using Lie series and transforms. These methods may be applied directly to a set of partial-differential equations, when considering a continuous system, or to a set of ordinary-differential equations, when considering a discrete system or a discretized model of a continuous system. One can also apply perturbation methods to the Lagrangian and virtual work associated with a system.

2. NONLINEAR SYSTEM RESPONSES AND BIFURCATIONS

There are many practical systems possessing cubic nonlinearities, including the beam resting on a Winkler-type foundation described by Eqs. (4) and (5), that can be modeled by the Duffing equation

$$\ddot{u} + \omega^2 u + 2\epsilon\mu\dot{u} + \epsilon\alpha u^3 = \epsilon F \cos\Omega t \tag{7}$$

Setting $\alpha = 0$ in Eq. (7), one finds that the longtime forced response of the linear system is given by

$$u(t) = a\cos(\omega t + \beta) \tag{8}$$

The full solution of Eq. (7) includes the free response. However, because the system is viscously damped, the free response dies out, leaving the particular solution given by Eq. (8). The amplitude a and phase β of the response are constants in time and are given by

$$a = \frac{\frac{\epsilon F}{\omega^2}}{\sqrt{\left[1 - \left(\frac{\Omega}{\omega}\right)^2\right]^2 + \left[2\zeta\left(\frac{\Omega}{\omega}\right)\right]^2}} \quad \text{and} \quad \beta = -\tan^{-1}\left[\frac{2\zeta\left(\frac{\Omega}{\omega}\right)}{1 - \left(\frac{\Omega}{\omega}\right)^2}\right] \tag{9}$$

where $\zeta = \frac{\epsilon\mu}{\omega}$ is the damping ratio.

To describe the nearness of the forcing frequency Ω to the system's linear natural frequency ω, we introduce the parameter σ such that

$$\Omega = \omega(1 + \epsilon\sigma) \quad \text{or} \quad \frac{\Omega}{\omega} = 1 + \epsilon\sigma \tag{10}$$

When σ deviates slightly from zero, we refer to it as a *frequency detuning parameter*. Taking $\frac{\epsilon F}{\omega^2} = 1$, we show in Figure 4a the amplitude a as a function of $\epsilon\sigma$ for different

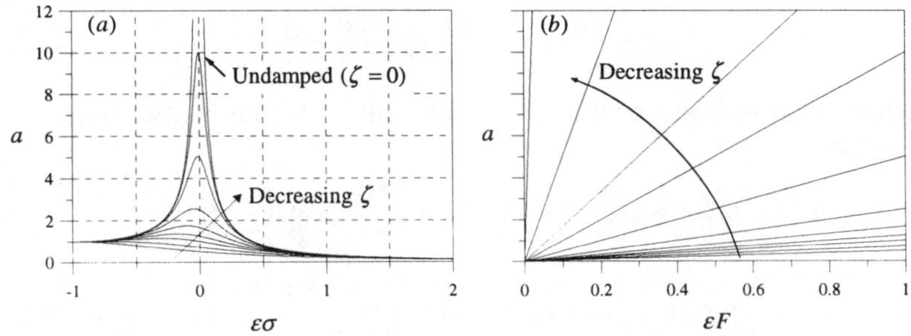

Fig. 4 (a) Frequency-response and (b) force-response curves for the linearized system for different values of viscous damping.

damping ratios ζ. When $\epsilon\sigma$ is close to zero, and hence $\Omega \approx \omega$, the amplitude a increases (and approaches ∞ for the undamped system). This behavior is called *resonance*. When $\sigma = 0$, $\Omega = \omega$ and the system is said to being excited at *primary resonance*. The influence of increasing the amount of viscous damping is to reduce the response amplitude. As σ becomes of order $O(\epsilon^{-1})$ and varies away from zero, the amplitude a monotonically decreases. In Figure 4b, we show the amplitude a as a function of the forcing amplitude ϵF for different damping ratios when $\sigma = 0$. As the excitation amplitude F increases, the response amplitude a monotonically increases in a linear fashion.

Next, we investigate the solution of Eq. (7) when $\alpha \neq 0$. Using a perturbation technique, such as the method of multiple scales or averaging, one obtains a first-order approximation of the solution of Eq. (7) when $\Omega \approx \omega$ (i.e., primary resonance) in the form

$$u(t) = a(t)\cos[\omega t + \beta(t)] + O(\epsilon) \tag{11}$$

where the a and β are now time-varying functions governed by

$$\dot{a} = -\epsilon\mu a + \frac{\epsilon F}{2\omega}\sin(\epsilon\sigma\omega t - \beta) \tag{12}$$

$$a\dot{\beta} = \frac{3\epsilon\alpha}{8\omega}a^3 - \frac{\epsilon F}{2\omega}\cos(\epsilon\sigma\omega t - \beta) \tag{13}$$

Equations (12) and (13) are nonlinear ordinary-differential equations; they are called *modulation* or *evolution* equations because they govern the slow dynamics of the system. Because Eqs. (12) and (13) are explicit functions of time t, they are called a *nonautonomous* set of modulation equations. One can introduce the transformation $\gamma(t) = \epsilon\sigma\omega t - \beta(t)$ into Eqs. (12) and (13) and obtain

$$\dot{a} = -\epsilon\mu a + \frac{\epsilon F}{2\omega}\sin\gamma \tag{14}$$

$$a\dot{\gamma} = \epsilon\sigma\omega a - \frac{3\epsilon\alpha}{8\omega}a^3 + \frac{\epsilon F}{2\omega}\cos\gamma \tag{15}$$

which is an *autonomous* set of modulation equations. Moreover, the solution of the system becomes

$$u(t) = a(t)\cos[\omega t + \beta(t)] + O(\epsilon) = a(t)\cos[\Omega t - \gamma(t)] + O(\epsilon) \tag{16}$$

Equations (14) and (15) are an example of a two-dimensional *dynamical system*. For the undamped and unforced system, we set $\mu = 0$ and $F = 0$ in Eqs. (14) and (15) and obtain

$$\dot{a} = 0 \ \longrightarrow \ a = a_0 \qquad \text{and} \qquad \dot{\gamma} = \epsilon\sigma\omega - \frac{3\epsilon\alpha}{8\omega}a_0^2 \tag{17}$$

where a_0 is constant. Then, the nonlinear natural frequency of the system, defined by $\omega_{nl} \equiv \omega + \dot{\beta}$, is equal to

$$\omega_{nl} = \omega\left(1 + \frac{3\epsilon\alpha}{8\omega^2}a_0^2\right) + \cdots \tag{18}$$

From Eq. (18), we note that the nonlinear natural frequency ω_{nl} is a function of the square of the free-response amplitude a_0 of the system. Hence, the nonlinearity introduces a *shift* in the natural frequency of the system. When $\alpha > 0$, the nonlinear frequency increases with vibration amplitude, and one speaks of a *hardening-type* nonlinearity. On the other hand, when $\alpha < 0$, the nonlinear frequency decreases with vibration amplitude, and one speaks of a *softening-type* nonlinearity.

For the damped and forced system, it is of interest to determine the solutions of the modulation equations because they characterize the behavior of the response. *Constant* solutions of a general dynamical system are called *equilibrium solutions* or *fixed points*. An equilibrium solution corresponds to a *periodic* or *limit-cycle* response of the system (i.e., $u(t)$ is periodic). Equilibrium solutions are determined by setting $\dot{a} = 0$ and $\dot{\gamma} = 0$ in the modulation equations. The stability of an equilibrium solution is determined by calculating the eigenvalues of the Jacobian matrix corresponding to Eqs. (14) and (15) evaluated at the equilibrium point. When all of the eigenvalues have negative real parts, the equilibrium solution is *asymptotically stable* and is either a *stable focus* or a *node*. On the other hand, when one or more of the eigenvalues have positive real parts, the equilibrium solution is *unstable*. If some of the eigenvalues are real and positive while the rest are negative, then the unstable equilibrium solution is called a *saddle*. If some of the eigenvalues are complex conjugates with positive real parts while the rest are negative, then the unstable fixed point is called an *unstable focus*. This does not happen for this system.

A graph whereby a state of the equilibrium solution, such as a, is plotted as a function of a certain (bifurcation) parameter is called a *bifurcation diagram*. An example is shown in Figure 5. When the bifurcation parameter used is the excitation

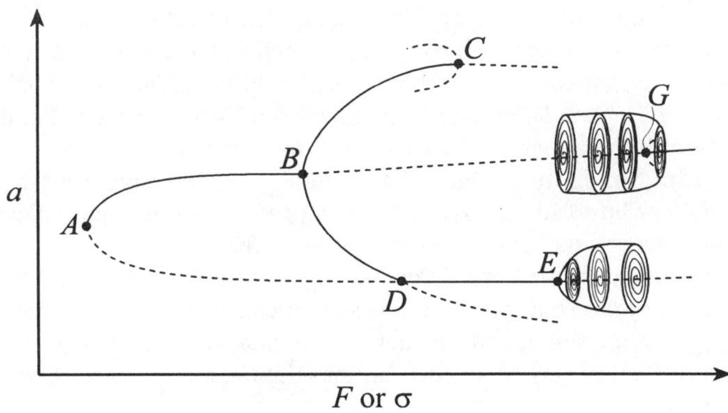

Fig. 5 Sketch of a bifurcation diagram showing some of the possible static and dynamic bifurcations. Point A is a saddle-node bifurcation, points B and C are supercritical and subcritical pitchfork bifurcations, respectively, point D is a transcritical bifurcation, and points E and G are supercritical and subcritical Hopf bifurcations. Solid lines (—) denote stable equilibrium solutions and dashed lines (– – –) denote unstable equilibrium solutions.

amplitude F, the graph is referred to as a *force-response curve*; and when the bifurcation parameter used is a measure of the excitation frequency Ω (e.g., $\sigma \equiv \frac{\Omega}{\omega} - 1$), then the graph is referred to as a *frequency-response curve*. For an n-dimensional system, as we slowly vary a parameter, such as F or σ, a stable equilibrium solution loses stability if either a real eigenvalue or a pair of complex-conjugate eigenvalues crosses from the left-half to the right-half of the complex plane. At the crossing point, we say that the system undergoes a *bifurcation* because a qualitative change in the character of the system dynamics occurs. If the loss of stability results from a real eigenvalue crossing from the left- to the right-half plane, the bifurcation is called *static*. Static bifurcations are either *saddle-node*, *pitchfork*, or *transcritical* bifurcations. When two branches of equilibrium solutions, one stable and the other unstable, meet each other at a point of vertical tangency, such as point A in Figure 5, we call point A a *saddle-node* bifurcation. If a single stable branch of equilibrium solutions loses stability at a point and splits into three branches (two of which are stable), such as point B, we call point B a *supercritical pitchfork* bifurcation. If a single unstable branch of equilibrium solutions regains stability at a point and splits into three branches (two of which are unstable), such as point C, then we call that point a *subcritical pitchfork* bifurcation. Lastly, if two branches intersect each other and exchange their stability, such as point D, we call point D a *transcritical* bifurcation. Saddle-node and pitchfork bifurcations are common in systems having cubic and/or quadratic nonlinearities and transcritical bifurcations occur mostly in systems having quadratic nonlinearities. These bifurcations occur in two- and higher-dimensional dynamical systems.

234

If a pair of complex-conjugate eigenvalues transversely crosses the imaginary axis from the left-half to the right-half of the complex plane, as a parameter is varied, we say that the system lost stability through a *Hopf* bifurcation. A Hopf bifurcation is referred to as a *dynamic bifurcation* because it results in a limit cycle. For dissipative systems, Hopf bifurcations do not occur in two-dimensional systems, such as Eqs. (14) and (15). Similar to pitchfork bifurcations, there are *supercritical Hopf* and *subcritical Hopf* bifurcations. When the Hopf bifurcation is supercritical, such as point E in Figure 5, the transition from an equilibrium solution to a limit cycle is gradual, resulting in a small-amplitude limit cycle near the Hopf bifurcation. On the other hand, when the Hopf bifurcation is subcritical, such as point G, the transition from the equilibrium solution to the limit cycle is sudden and results in a jump to a far away solution, which may be an equilibrium solution, a limit cycle, a quasiperiodic solution, or chaos.

In Figure 5, the dashed lines denote unstable equilibrium solutions and the solid lines denote stable equilibrium solutions. As we sweep past points A, C, and G, a jump to either another branch of equilibrium solutions or to a branch of dynamic solutions occurs. Furthermore, regions where multiple stable equilibrium solutions coexist are possible, as in the region between points B and D. In these regions, the initial conditions play an important role in the response of the system. Moreover, it is possible for the presence of regions where no stable equilibrium solutions exist, as in the region left of point A and the region between points E and G. Usually, the response of the system in such regions exhibits very rich and complex dynamics.

To determine the equilibrium solutions of Eqs. (14) and (15), we set $\dot{a} = 0$ and $\dot{\gamma} = 0$ and obtain

$$\frac{F}{2\omega}\sin\gamma = \mu a \tag{19}$$

$$\frac{F}{2\omega}\cos\gamma = -\left(\sigma\omega - \frac{3\alpha}{8\omega}a^2\right)a \tag{20}$$

Squaring Eqs. (19) and (20), adding the results, and solving for either σ or F, we obtain

$$\sigma = \frac{3\alpha}{8\omega^2}a^2 \pm \frac{1}{\omega}\sqrt{\left(\frac{F}{2\omega a}\right)^2 - \mu^2} \quad \text{or} \quad F = 2\omega a\sqrt{\mu^2 + \left(\sigma\omega - \frac{3\alpha}{8\omega}a^2\right)^2} \tag{21}$$

In Figure 6, we present typical frequency-response curves when $\alpha < 0$ and $\alpha > 0$. Unlike for the linear system (i.e., $\alpha = 0$) that is shown in Figure 4, where a increases monotonically as $\sigma \to 0$, the curves in Figure 6 are bent to the left when $\alpha < 0$ and to the right when $\alpha > 0$. The bending of the curves to the left indicates that the nonlinearity is of the *softening-spring* type, whereas the bending of the curves to the right indicates that the nonlinearity is of the *hardening-spring* type. Taking the case $\alpha < 0$, starting at a point on the lower branch at about $\sigma = -1$, and slowly increasing σ, we find that the solution loses stability through the saddle-node bifurcation SN_1.

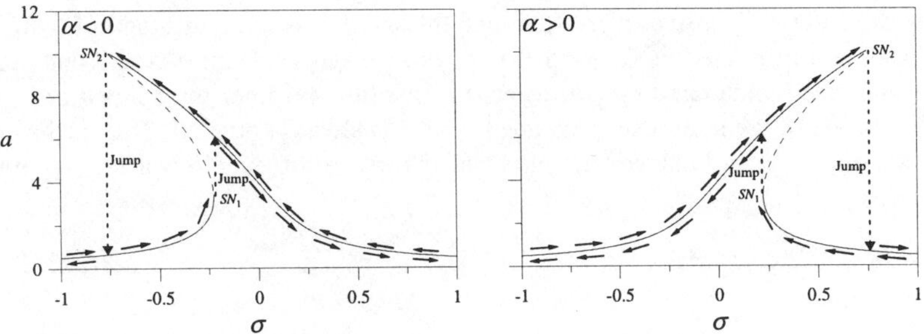

Fig. 6 Typical frequency-response curves for the Duffing equation near primary resonance when $\alpha < 0$ and $\alpha > 0$.

The result is a jump to the upper branch. If we continue to sweep to the right, the amplitude of the response monotonically decreases and becomes small as σ exceeds 1. On the other hand, if we slowly decrease σ, we find that the response amplitude increases monotonically until it encounters the saddle-node bifurcation SN_2, resulting in a jump to the lower branch and a sudden change in the response amplitude. Qualitatively, a similar behavior occurs when $\alpha > 0$. Therefore, unlike the linearized system, the response of the nonlinear system can lose stability, exhibit jumps, coexist with other stable solutions, and be sensitive to initial conditions.

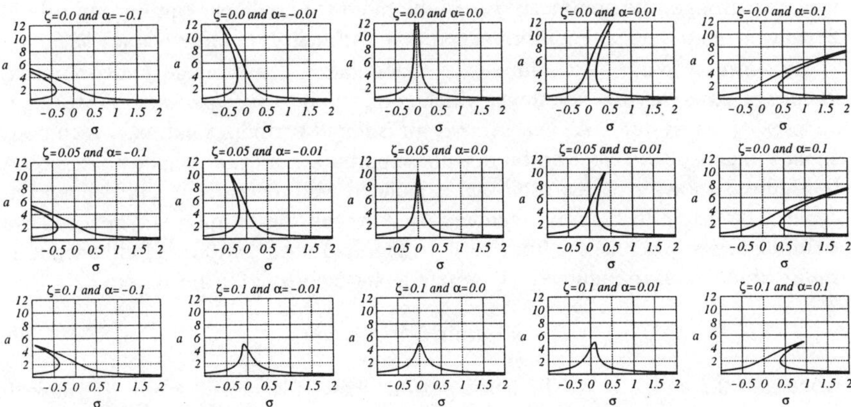

Fig. 7 Frequency-response curves for the Duffing equation for different values of α and ζ when $\frac{\epsilon F}{\omega^2} = 1$.

The magnitude of bending of the response curves depends on the strength of the nonlinearity α, the value of the forcing amplitude F, and the degree of damping μ. In Figure 7, we show the influence of the nonlinearity α and damping ratio $\zeta = \frac{\epsilon \mu}{\omega}$ on

236

the frequency-response curves. Although the stability is not indicated in Figure 7, it follows that presented in Figure 6. Moreover, we can see from Figure 7 that, unlike the linearized undamped system where the solution becomes unbounded as $\sigma \to 0$, the solution of the nonlinear undamped system is always bounded. That is, the effect of the nonlinearity balances the effect of the excitation even when no damping is present.

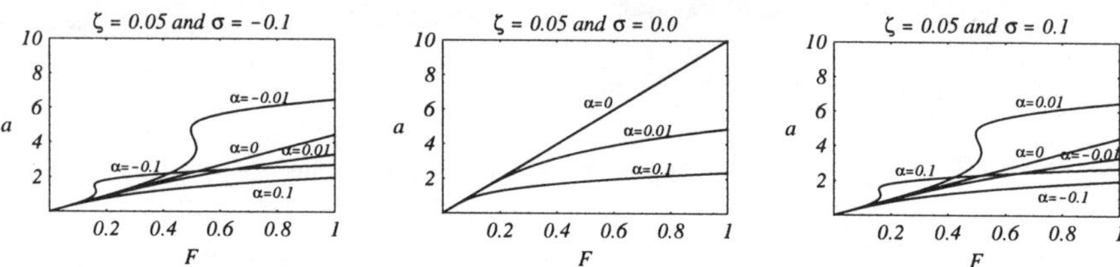

Fig. 8 Force-response curves for the Duffing equation for different values of α and σ when $\zeta = 0.05$.

In Figure 8, we present typical force-response curves for the Duffing equation when $\zeta = 0.05$ for different values of α and σ. We note that, depending on the values of α and σ, the equilibrium solution can either be always stable and monotonically increasing or it can lose stability through a saddle-node bifurcation, resulting in a jump up or a jump down to another branch of stable equilibrium solutions. Either way, the stronger the nonlinearity α is, the more the nonlinear equilibrium solution deviates from the linear equilibrium solution, especially for larger values of F.

In addition to constant solutions, the modulation equations generally admit *dynamic* (or time-varying) solutions, which correspond to *quasiperiodic* or *chaotic* responses of the system. A *limit cycle* is an *isolated* periodic solution, which corresponds to an isolated closed orbit in a phase portrait. For example, the phase portrait in Figure 3 shows a closed orbit that is isolated, and therefore it represents a limit cycle. The origin in Figure 3 (denoted by A) repels the solution and hence it is an *unstable focus*, whereas the limit cycle in Figure 3 is an *attractor*, and hence it is a *stable* limit cycle. In Figure 9, we present a phase portrait for the system

$$\ddot{u} - \mu \sin \dot{u} + \omega^2 u = 0 \tag{22}$$

when $\mu = 0.5$ and $\omega = 1$. The phase portrait shows five closed orbits. Because the orbits are isolated, they are limit cycles; three of them are stable and the other two lying in between are unstable. Limit cycles can result from an equilibrium solution undergoing a Hopf bifurcation or they can be found isolated.

The stability of limit cycles is determined by using *Floquet theory*. Similar to equilibrium solutions where the eigenvalues of the Jacobian matrix determine their stability, in applying Floquet theory, one calculates the eigenvalues of a characteristic matrix called the *monodromy matrix*. The eigenvalues ρ_i of the monodromy matrix are referred to as *Floquet* or *characteristic multipliers*. If all of the Floquet multi-

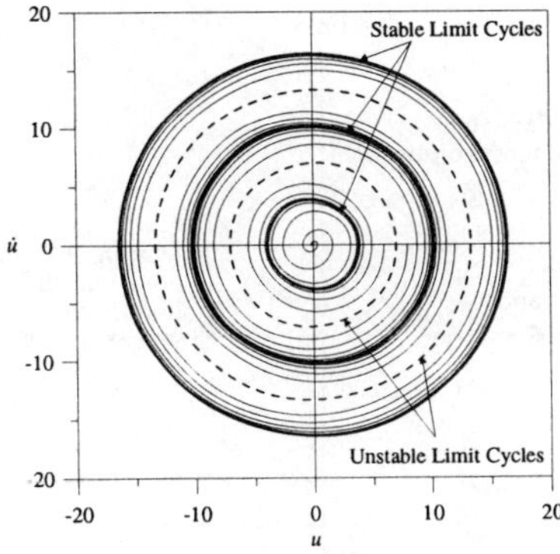

Fig. 9 Phase portrait for the system $\ddot{u} - \mu\sin\dot{u} + \omega^2 u = 0$, $\mu = 0.5$ and $\omega = 1$, demonstrating stable and unstable limit cycles.

pliers lie within the unit circle, then the limit cycle is *stable*. On the other hand, if one or more of the Floquet multipliers lie outside the unit circle, then the limit cycle is *unstable*. We note that when dealing with an autonomous system, there is always one Floquet multiplier that equals unity, but it has no influence on the stability of the limit cycle. Once a stable limit cycle is determined, one can trace its behavior as a bifurcation parameter is slowly varied through the use of continuation schemes, such as the arclength and pseudo-arclength algorithms.

Much like equilibrium solutions, limit cycles also can undergo bifurcations. If one of the Floquet multipliers leaves the unit circle through $+1$, then the limit cycle loses stability through one of the following: a *cyclic-fold bifurcation*, a *symmetry-breaking bifurcation*, or a *transcritical bifurcation*. Similar to the saddle-node bifurcation for equilibrium solutions, the *cyclic-fold bifurcation* occurs when two branches of limit cycles, one stable and the other unstable, meet and destroy each other; the result is a jump to a far away attractor. A *symmetry-breaking bifurcation* causes a symmetric limit cycle to lose stability, resulting in a stable asymmetric limit cycle. The *transcritical bifurcation* for limit cycles is similar to that for equilibrium solutions in that an exchange of stability between the two limit cycles occurs. If one of the Floquet multipliers leaves the unit circle through -1, then the limit cycle loses stability through a *period-doubling* (or *flip*) *bifurcation*. Hence, if a stable limit cycle of period T loses stability through a *period-doubling bifurcation*, then the resulting stable limit cycle has a period of $2T$. If the new limit cycle undergoes another period-doubling bifurcation, then the resulting stable limit cycle has a period of $4T$. Continuous repetition of the bifurcation results in stable limit cycles with periods of $8T$, $16T$, $32T$, and so on. We note that, for a limit cycle to undergo a period-doubling bifurcation, it is necessary for it to be asymmetric. Lastly, if a pair of complex Floquet multipliers leaves the unit circle, the limit cycle loses stability through a *secondary Hopf* (or *Neimark*) *bifurcation*. In this case, a stable *two-period*

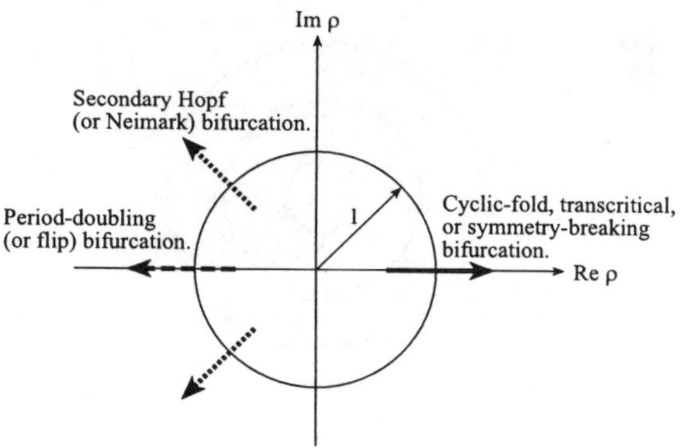

Fig. 10 A sketch showing the possible bifurcations of a limit cycle as a Floquet multiplier leaves the unit circle.

quasiperiodic solution (or a *torus*) results. In Figure 10, we show a sketch of these scenarios.

A *quasiperiodic* solution is characterized by two or more incommensurate frequencies. Two frequencies ω_1 and ω_2 are *incommensurate* if the ratio ω_1/ω_2 is an irrational number. Quasiperiodic motions can result when a limit cycle undergoes a secondary Hopf (or Neimark) bifurcation, and unlike limit cycles, which form a closed orbit in a phase portrait, quasiperiodic solutions do not form a closed orbit. Limit-cycle solutions of the modulations equations result, in general, in a quasiperiodic response of the corresponding system. For example, the solutions $u_1(t) = \cos 2t \cos 5t$ and $u_2(t) = \cos 2.1t \cos 5.2t$ are periodic solutions, whereas the solution $u_3(t) = \cos \sqrt{5}t \cos \sqrt{31}t$ is quasiperiodic. This is because for u_1 and u_2, respectively, $\omega_1/\omega_2 = 2/5$ and $\omega_1/\omega_2 = 2.1/5.2 = 21/52$ are rational, whereas for u_3, $\omega_1/\omega_2 = \sqrt{5/31}$ is an irrational number.

In addition to time histories and phase portraits, two tools that are useful in analyzing data are *Poincaré sections* and *Fourier spectra*. *Poincaré* sections are phase portraits that are constructed by stroboscopely sampling a set of data at times $t = m\Delta t$, $m = 0, 1, 2, \cdots$. A period-k periodic motion corresponds to k points in the Poincaré section; a quasiperiodic motion corresponds to a closed curve in the Poincaré section; and a chaotic attractor corresponds to a large cluster of points in the Poincaré section. The *Fourier spectra* convert a set of data to a collection of amplitude peaks at each frequency. The Fourier spectrum of a periodic motion of period T consists of a single basic frequency $f = \frac{1}{T}$. For a *symmetric* periodic solution, the Fourier spectrum contains only *odd* harmonics of f, and for an *asymmetric* periodic solution, the Fourier spectrum contains both *odd* and *even* harmonics of f. For a periodic solution of period nT, $n = 2, 3, 4, 5, \cdots$, in addition to the odd and even harmonics of f, the Fourier spectrum should contain $\frac{1}{n}$-order harmonics of f. When the spectrum has k basic (i.e., k incommensurate) frequencies, the corresponding motion is k-period quasiperiodic. Lastly, when the spectrum shows a broadband character, the corresponding motion is chaotic. Usually, it is necessary to use two or more tools to adequately determine the character of a motion.

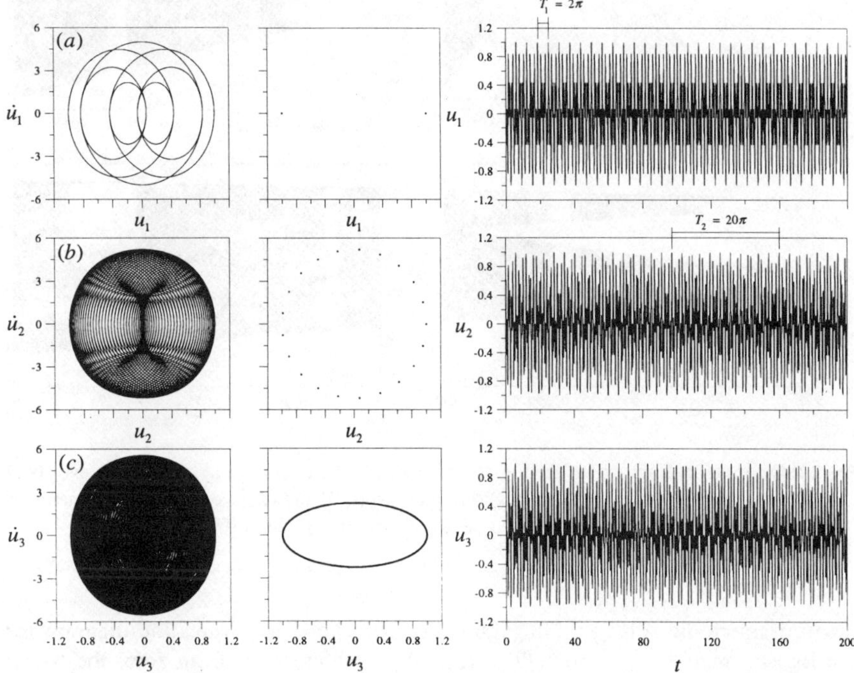

Fig. 11 Phase portraits, Poincaré sections, and time histories for (a) the periodic motion $u_1 = \cos 2t \cos 5t$, (b) the long-period periodic (or phase-locked) motion $u_2(t) = \cos 2.1t \cos 5.2t$, and (c) the quasiperiodic motion $u_3(t) = \cos\sqrt{5}t \cos\sqrt{31}t$.

In Figure 11, we present phase portraits, Poincaré sections, and time histories for u_1 (part a), u_2 (part b), and u_3 (part c). Although it is clear from the phase portrait and time history in part (a) that u_1 is periodic, the characters of the solutions u_2 and u_3 are hard to determine from such tools alone. Therefore, more information, by way of the Poincaré section, is needed. The Poincaré sections were constructed by sampling points at periods of $2\pi/\omega_1$, where $\omega_1 = 2, 2.1$, and $\sqrt{5}$. The Poincaré sections in parts (a) and (b) show discrete number of points, indicating that the solutions u_1 and u_2 are periodic. Moreover, because of the large number of points in part (b), the solution u_2 is usually referred to as a *phase-locked* periodic motion; it has a large period. As for u_3, the Poincaré section in part (c) shows a closed curve, indicating that u_3 is a quasiperiodic motion, as predicted earlier.

A *chaotic* solution is a bounded solution that is not a fixed point, or a limit cycle,

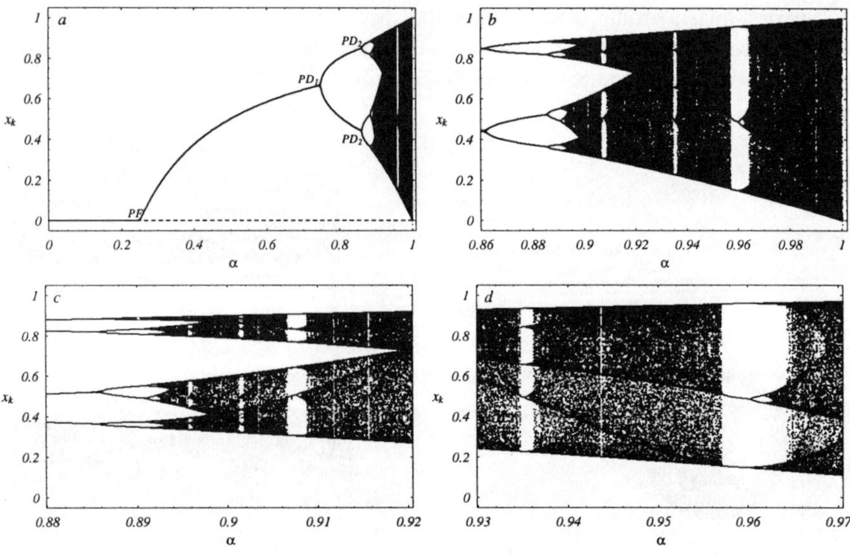

Fig. 12 Bifurcation diagram for the logistic map $x_{k+1} = 4\alpha x_k(1 - x_k)$. In part (a), we show the entire bifurcation diagram and in parts (b), (c), and (d), we show enlargements of the regions $\alpha \in [0.86, 1]$, $\alpha \in [0.88, 0.92]$, and $\alpha \in [0.93, 0.97]$, respectively.

or a quasiperiodic solution. In Figure 12a, we show the bifurcation diagram for the logistic map $x_{k+1} = 4\alpha x_k(1 - x_k)$. As we increase α from zero, the trivial solution loses stability through the pitchfork bifurcation PF at $\alpha = 0.25$, resulting in a stable nontrivial solution. As α exceeds 0.75, the nontrivial solution undergoes a period-doubling bifurcation PD_1, resulting in two stable solutions. Continuing to increase α, the solutions undergo a sequence of period-doubling bifurcations, which culminate in chaos. This behavior is called a *period-doubling route to chaos*.

Zooming in on different regions of the bifurcation diagram, we show in Figures 12b-d clearer views of these sequences of period-doubling bifurcations. We note from Figures 12b-d that there are windows within the chaotic region where periodic solutions exist. For example, windows of period-three, period-five, and period-six solutions exist at $\alpha \approx 0.958$, $\alpha \approx 0.935$, and $\alpha \approx 0.907$, respectively. These windows usually result when the solution becomes *intermittent* and/or undergoes an *interior crisis*. A solution that is *intermittent* exhibits long spans of periodic behavior interrupted by chaotic bursts. An intermittency that is associated with a cyclic-fold bifurcation is called *type I intermittency*. *Type II* and *type III intermittencies* are associated with subcritical Hopf and subcritical period-doubling bifurcations, respectively. In Figures 12c and 12d, we note that as α approaches the critical values α_i at which the windows of periodic solutions start, there appears to be areas in the chaotic attractor that are more densely populated. These areas seem to be located near the branches of the impending periodic solutions and are a result of intermittent behavior of the attractor. Intermittency is a possible *route to chaos*. An *interior crisis* occurs when the chaotic attractor collides with an unstable equilibrium or periodic solution, resulting in a larger chaotic attractor. As we continue to increase α, we find that the attractor continues to exist until α approaches 1. At $\alpha = 1$, the chaotic attractor collides with the unstable trivial solution, resulting in its destruction; this is an example of a *boundary* or an *exterior crisis*. Exterior crises usually result in the destruction of a chaotic attractor and the solution tends to either a bounded attractor (i.e., an equilibrium solution or a dynamic solution) or to an unbounded solution. Increasing α past 1, we find that the solution of the logistic map becomes unbounded.

In addition to interior and exterior crises, an *attractor-merging crisis* may occur in a system with symmetries if, as a bifurcation parameter is varied, two or more chaotic attractors merge together to form a single larger chaotic attractor. Interior and attractor-merging crises are referred to as *explosive bifurcations* because they result in larger attractors. However, the reverse is also true, in which case the attractor may shrink or split. Boundary crises are referred to as *dangerous* or *bluesky catastrophe bifurcations* because of the sudden disappearance of the attractor.

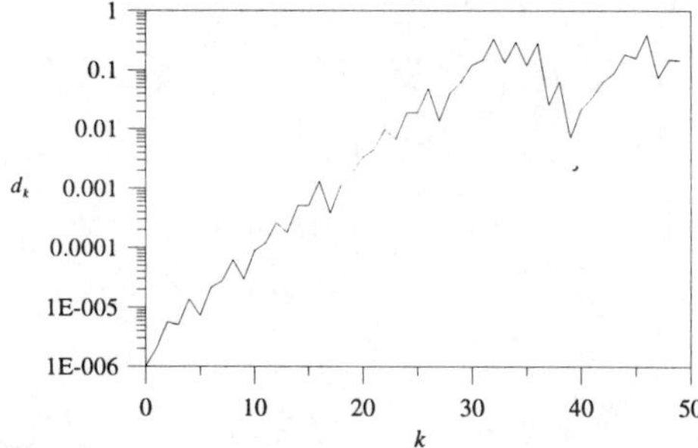

Fig. 13 A plot illustrating the sensitivity of the chaotic behavior of the logistic map to initial conditions when $\alpha = 0.92$.

We note the structure of the solutions shown in Figure 12 as we zoom in closer and closer on a region remains the same. Attractors that exhibit such *fractal* behavior are called *strange attractors*. A characteristics of chaotic attractors is their sensitivity to initial conditions. At $\alpha = 0.92$, where x_k is chaotic, we take two initial conditions, 0.875661 and 0.875660, which are separated by an amount $d_0 = 10^{-6}$. In Figure 13, we show the separation distance d_k between the two solutions as we iterate on the map fifty times. When $k = 50$, $d_k \approx 0.36 \gg d_0$, which illustrates the sensitivity of this chaotic solution to initial conditions. Another characteristic of chaotic attractors is the presence of a broadband signal in the Fourier spectrum. We note that quasiperiodic and chaotic solutions are grouped under the terms *almost-periodic* and *aperiodic* solutions, respectively.

In Figure 14, we show a sketch of a bifurcation diagram of a limit cycle that oc-

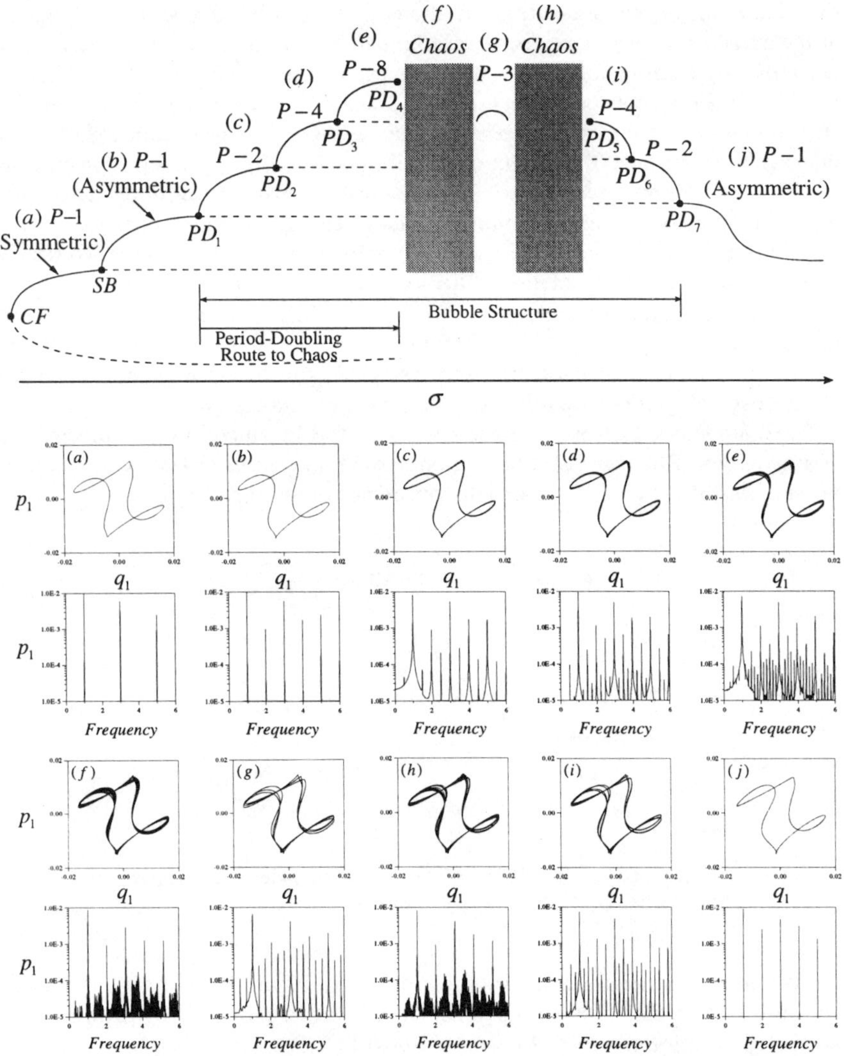

Fig. 14 A Schematic bifurcation diagram of a limit cycle and the corresponding two-dimensional projections of the phase portraits onto the p_1q_1-plane and FFTs. P-i denotes a period-i limit cycle, CF denotes a cyclic-fold bifurcation, SB denotes a symmetry-breaking bifurcation, and PD_i denotes the ith period-doubling bifurcation.

curs in a two-degree-of-freedom system (i.e., four-dimensional dynamical system) in terms of the parameter σ. Because a limit cycle consists of an infinite number of points that form a closed orbit, we usually show the behavior of a single point on the limit cycle as we slowly vary the bifurcation parameter σ. Therefore, Figure 14 is not the actual bifurcation diagram, which may be three- or higher-dimensional depending on the degree of freedom of the system, but rather a qualitatively representative schematic of it. In addition, we present in Figure 14 two-dimensional projections of the phase portraits onto the $p_1 q_1$-plane and the corresponding FFTs (fast Fourier transforms). The p_i and q_i are Cartesian coordinates that are related to the amplitudes a_i by $a_i = \sqrt{p_i^2 + q_i^2}$. We start at part (a) with a period-one limit cycle. The corresponding Fourier spectrum shows only odd harmonics, indicating that this limit cycle is symmetric. As we decrease σ, the limit cycle loses stability through the cyclic-fold bifurcation CF. The response may jump to a far away attractor, which may be an equilibrium solution, another limit cycle, or chaos, or it may be intermittent of type I.

As we increase σ, the symmetric period-one limit cycle loses stability through the symmetry-breaking bifurcation SB; the result is an asymmetric period-one limit cycle, as evidenced by the presence of both odd and even harmonic in the Fourier spectrum in part (b). Increasing σ further, we find that the asymmetric period-one limit cycle undergoes a sequence of period-doubling bifurcations, resulting in period-two, period-four, and period-eight limit cycles, as shown in parts (c)-(e). In addition to odd and even harmonics, the corresponding Fourier spectra in parts (c)-(e) show $\frac{1}{2}$, $\frac{1}{4}$, and $\frac{1}{8}$ harmonics, respectively. As we increase σ further, the sequence of period-doubling bifurcations quickly leads to the chaotic attractor shown in parts (f) and (h). The corresponding Fourier spectra are broadband, which is a characteristic of chaos. The behavior exhibited in parts (b)-(f) is another example of a period-doubling route to chaos. For the system described in Figure 14, we find that inside the chaotic region (i.e., parts (f) and (h)), there exists a small window of period-three limit cycles, as shown in part (g). The corresponding Fourier spectrum shows harmonics of order $\frac{1}{3}$.

If we increase σ further, the chaotic attractor undergoes a sequence of *reverse* period-doubling bifurcations, resulting once again in an asymmetric period-one limit cycle, as shown in parts (i)-(j). The sequence of period-doubling and reverse period-doubling bifurcations shown in parts (b)-(j) is referred to as a *bubble structure*. We note that period-doubling sequences do not always culminate in chaos and it is possible for a system to exhibit a bubble structure free of chaos.

Other ways a system can become chaotic include the following: quasiperiodic routes, such as the Ruelle-Takens scenario, doubling of a torus, or breakdown of a torus; Melnikov theory, such as homoclinic and hetroclinic tangles; and bifurcations of homoclinic orbits. However, the explanations of such routes to chaos are beyond the scope of this paper.

3. SINGLE-DEGREE-OF-FREEDOM SYSTEMS

As previously discussed, the excitation acting on a system can be modeled as being either direct (or additive) or parametric (or multiplicative). Next, we discuss the influence of these excitations on the response of single-degree-of-freedom systems, such as the Duffing equation described by Eq. (7).

3.1. Direct Excitations

Let us consider the following directly excited nonlinear single-degree-of-freedom system:

$$\ddot{u} + \omega^2 u + \epsilon f(u, \dot{u}) + \epsilon N(u, \dot{u}) = F \cos \Omega t \tag{23}$$

We seek a *straightforward expansion* of the solution of Eq. (23) in the form

$$u(t) = u_0(t) + \epsilon u_1(t) + \epsilon^2 u_2(t) + \cdots \tag{24}$$

Substituting Eq. (24) into Eq. (23) and equating coefficients of like powers of ϵ, we obtain a set of second-order nonhomogeneous ordinary-differential equations for the u_i in the form

$$\ddot{u}_0 + \omega^2 u_0 = F \cos \Omega t \tag{25}$$

$$\ddot{u}_1 + \omega^2 u_1 = -f_0(u_0, \dot{u}_0) - N_0(u_0, \dot{u}_0) \tag{26}$$

$$\ddot{u}_2 + \omega^2 u_2 = -f_1(u_0, \dot{u}_0, u_1, \dot{u}_1) - N_1(u_0, \dot{u}_0, u_1, \dot{u}_1) \tag{27}$$

$$\vdots$$

Depending on the nonlinearity $N(u, \dot{u})$ and the order to which the expansion is carried out, one finds that the solutions for the u_i may contain coefficients that can be troublesome. For example, terms like

$$\frac{1}{\Omega^2 - \omega^2}, \quad \frac{1}{\Omega^2 - n^2 \omega^2}, \quad \text{and} \quad \frac{1}{n^2 \Omega^2 - \omega^2}, \quad n = 2, 3, \cdots \tag{28}$$

are very large when $\Omega \approx \omega$, $\Omega \approx n\omega$, and $n\Omega \approx \omega$, respectively, causing the straightforward expansion of the solution to breakdown and lose uniformity. The denominators of such coefficients are called *small-divisor terms* and relationships like $\Omega \approx \omega$, $\Omega \approx n\omega$, and $n\Omega \approx \omega$ are called *resonances*.

The resonance $\Omega \approx \omega$ is referred to as *primary resonance* because it appears in the first term of the expansion (i.e., u_0), and therefore a system directly excited near primary resonance can exhibit large-amplitude responses even when the excitation amplitude is very small. The resonances $\Omega \approx n\omega$ and $n\Omega \approx \omega$ are referred to as *secondary resonances* because they appear in the higher-order terms of the expansion (i.e., $u_1, u_2, \cdots$). For secondary resonances, a relatively larger excitation amplitude may be necessary in order to produce large-amplitude responses. Secondary resonances of the form $\Omega \approx n\omega$ are called *subharmonic resonances of order $\frac{1}{n}$* and those of the form $n\Omega \approx \omega$ are called *superharmonic resonances of order n*, where $n = 2, 3, \cdots$. For example, if $N(u, \dot{u})$ is a cubic nonlinearity, then the resonances $\Omega \approx 3\omega$ and $3\Omega \approx \omega$ may occur. Moreover, depending on the order of the expansion, resonances such as $\Omega \approx \frac{2}{3}\omega$ and $\frac{2}{3}\Omega \approx \omega$ are also possible. Similarly, for systems with quadratic nonlinearities, the resonances $\Omega \approx 2\omega$, $2\Omega \approx \omega$, $\Omega \approx 4\omega$, and $4\Omega \approx \omega$ may occur.

3.2. Parametric Excitations

In addition to direct (or additive) excitations, there are systems that are parametrically excited. In such cases, the excitation appears as a time-dependent coefficient (or parameter) in the equation of motion. Unlike directly excited systems, where an excitation with $\Omega \approx \omega$ (primary resonance) produces the most significant resonance, in parametrically excited systems, an excitation with $\Omega \approx 2\omega$ produces the most significant resonance. That is, a small-amplitude parametric forcing can produce a very large-amplitude response when $\Omega \approx 2\omega$. Therefore, the resonance $\Omega \approx 2\omega$ is called *principal parametric resonance*. In addition, $\Omega \approx \omega$ is called *fundamental parametric resonance*. There are many systems that are influenced by parametric excitations. For example, the bending response of a hinged-clamped beam excited along its axis, as described by Eq. 3 when $F = 0$. Another example is the motion of a simple pendulum attached to a vibrating base; it is described by

$$ml^2\ddot{\theta} + ml\,[g - \ddot{y}(t)]\sin\theta - ml\ddot{x}(t)\cos\theta = 0 \tag{29}$$

where $x(t)$ and $y(t)$ are prescribed horizontal and vertical displacements of the base.

3.3. Multifrequency Excitations

Many systems are influenced by several excitation sources. In such cases, it is necessary to include all of the sources in the model. For example, the system

$$\ddot{u} + \omega^2 u + \epsilon f(u, \dot{u}) + \epsilon N(u, \dot{u}) = \sum_{k=1}^{K} F_k \cos\Omega_k t \tag{30}$$

where $\Omega_k > \Omega_{k-1}$ is said to be directly forced by a *multifrequency excitation*. In such a case, in addition to subharmonic and superharmonic resonances, one needs to account for combination resonances. For example, for a two-frequency (Ω_1 and Ω_2) excitation and if $N(u, \dot{u})$ is a cubic nonlinearity, combination resonances of the form $\omega \approx |2\Omega_2 \pm \Omega_1|$, $\omega \approx |\Omega_2 \pm 2\Omega_1|$, and $\omega \approx \frac{1}{2}(\Omega_2 \pm \Omega_1)$ may occur. For a three-frequency (Ω_1, Ω_2, and Ω_3) excitation, combination resonances of the form $\omega \approx |\pm\Omega_1 \pm \Omega_2 \pm \Omega_3|$ may occur. Moreover, it is possible for two or more resonances to occur simultaneously.

4. MULTI-DEGREE-OF-FREEDOM SYSTEMS: MODAL INTERACTIONS

Many systems have more than one degree of freedom, and therefore their responses are modeled by sets of nonlinear equations, which may be algebraic equations, integral equations, ordinary-differential equations, partial-differential equations, or combinations of all four types. These equations of motion are usually coupled, either linearly or nonlinearly, and therefore the behavior of one state of the system may influence the behavior of other states. Here, we concern ourselves with systems that are nonlinearly coupled but linearly uncoupled.

4.1. Internal (Autoparametric) Resonances

Internal resonances may occur when two or more natural frequencies of a system are commensurate or nearly commensurate. Two frequencies ω_1 and ω_2 are commensurate if the ratio ω_1/ω_2 is rational. For example, consider the following model of a two-degree-of-freedom system:

$$\ddot{u}_1 + \omega_1^2 u_1 + \epsilon c_1 \dot{u}_1 + \epsilon N_1(u_1, u_2) = 0 \tag{31}$$

$$\ddot{u}_2 + \omega_2^2 u_2 + \epsilon c_2 \dot{u}_2 + \epsilon N_2(u_1, u_2) = \epsilon f \cos \Omega t \tag{32}$$

where the $u_i(t)$ are generalized coordinates of the system and the ω_i are their associated natural frequencies; the c_i are damping coefficients; and the N_i are nonlinear terms that couple the two modes. Some of the systems that may be modeled by Eqs. (31) and (32) include the motion of an elastic pendulum, the motion of a double pendulum, the motion of a ship undergoing pitch and roll, and the nonplanar motion of a two-mode discretized model of a cantilever beam.

In Eqs. (31) and (32), the coordinate u_2 is directly excited, whereas the coordinate u_1 is not; depending on Ω, u_2 may be excited at primary or secondary resonance. Therefore, because the system is damped, one may expect that, after some time, u_1 dies out and the response of the system is described only by u_2. However, this is not always true. Depending on the nonlinearities and system parameters, it is possible for the response to contain components from both u_1 and u_2. This is because, in addition to external resonances, many systems are *internally resonant*. For the system described by Eqs. (31) and (32), if $\omega_2 \approx \omega_1$, we say that the system may possess a *one-to-one* internal resonance. Spherical pendulums, strings, and near-square or near-circular cross-section beams are systems that inherently possess one-to-one internal resonances, which may result in nonplanar motions. If $\omega_2 \approx 2\omega_1$ or $\omega_2 \approx 3\omega_1$, then we say that the system possesses *two-to-one* and *three-to-one* internal resonances, respectively. Ships undergoing pitch and roll motions have been observed to exhibit two-to-one internal resonances. Hinged-clamped beams, as described by Eq. (3), and buckled beams are some of the systems that possess three-to-one internal resonances.

The response of a cantilever beam undergoing bending in the two orthogonal directions y and z, as shown in Figure 15, is governed by the following equations of motion:

$$\rho A \ddot{v} + c_v \dot{v} + D_\zeta v^{iv} = (D_\eta - D_\zeta) \left[w'' \int_L^s v'' w'' ds - w''' \int_0^s v'' w' ds \right]'$$

$$- \frac{(D_\eta - D_\zeta)^2}{D_\xi} \left[w'' \int_0^s \int_L^s v'' w'' ds ds \right]'' - D_\zeta \left[v' (v'v'' + w'w'')' \right]'$$

$$- \frac{1}{2} \rho A \left\{ v' \int_L^s \left[\frac{\partial^2}{\partial t^2} \int_0^s (v'^2 + w'^2) ds \right] ds \right\}' \tag{33}$$

$$\rho A \ddot{w} + c_w \dot{w} + D_\eta w^{iv} = -(D_\eta - D_\zeta) \left[v'' \int_L^s v'' w'' ds - v''' \int_0^s v' w'' ds \right]'$$

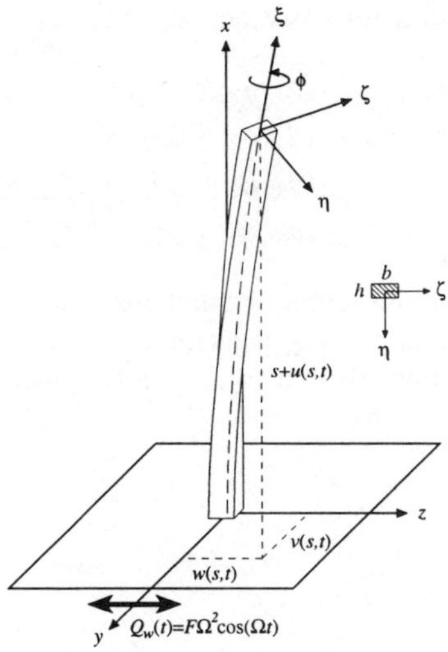

Fig. 15 A schematic of an externally excited cantilever beam undergoing nonplanar motion.

$$-\frac{(D_\eta - D_\zeta)^2}{D_\xi}\left[v''\int_0^s\int_L^s v''w''\,ds\,ds\right]'' - D_\eta\left[w'\left(v'v'' + w'w''\right)'\right]'$$

$$-\frac{1}{2}\rho A\left\{w'\int_L^s\left[\frac{\partial^2}{\partial t^2}\int_0^s\left(v'^2 + w'^2\right)ds\right]ds\right\}' + F\Omega^2\cos\Omega t \qquad (34)$$

where $v(s,t)$ and $w(s,t)$ are the bending deflections in the y- and z-directions; D_η, D_ζ, and D_ξ are the bending and torsional rigidities; and c_v and c_w are the coefficients of viscous damping. Equations (33) and (34), which are nonlinearly coupled, are an extension of Eq. (2) which models the planar motion of a cantilever beam. The nonlinearities arise from the geometry and the inertia.

Let ω_{vm} and ω_{wn}, $m, n = 1, 2, 3, \cdots$, be the natural frequencies of the mth mode $\Phi_{vm}(s)$ in the y-direction and the nth mode $\Phi_{wn}(s)$ in the z-direction and consider the case where the nth mode in the z-direction is directly excited near primary resonance (i.e., $\Omega \approx \omega_{wn}$). Using a perturbation method, such as averaging or multiple scales, one obtains a first-order expansion of the solution of Eqs. (33) and (34) when $\omega_{vm} \approx \omega_{wn}$ (i.e., one-to-one internal resonance) in the form

$$v(s,t) \approx \epsilon\Phi_{vm}(s)a_1(t)\cos[\omega_{vm}t + \beta_1(t)] \qquad (35)$$

$$w(s,t) \approx \epsilon\Phi_{wn}(s)a_2(t)\cos[\omega_{wn}t + \beta_2(t)] \qquad (36)$$

where we assumed that, because of the presence of damping in the system, all modes that are not activated, either directly by the excitation or indirectly by the internal resonance, die out with time. The amplitudes $a_i(t)$ and phases $\beta_i(t)$ are governed by

248

an autonomous set of modulation equations of the form

$$\dot{a}_1 = \epsilon g_{11}(a_1, a_2, \gamma_1, \gamma_2) \tag{37}$$

$$a_1\dot{\gamma}_1 = \epsilon g_{12}(a_1, a_2, \gamma_1, \gamma_2) \tag{38}$$

$$\dot{a}_2 = \epsilon g_{21}(a_1, a_2, \gamma_1, \gamma_2) + \epsilon f_{21}(\gamma_2) \tag{39}$$

$$a_2\dot{\gamma}_2 = \epsilon g_{22}(a_1, a_2, \gamma_1, \gamma_2) + \epsilon f_{22}(\gamma_2) \tag{40}$$

where the $\gamma_i \propto \beta_i$, the g_{ij} are nonlinear functions in the a_i and γ_i, and the f_{ij} are functions associated with the forcing. Equations (37)-(40) are nonlinearly coupled ordinary-differential equations that govern the slow dynamics of the two interacting modes.

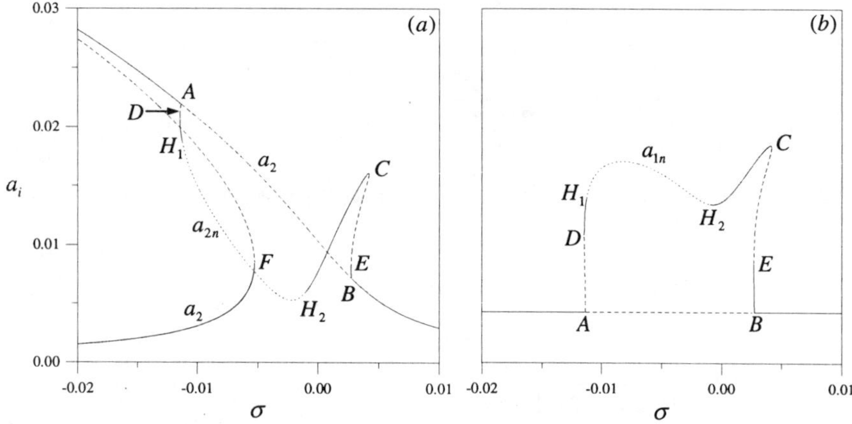

Fig. 16   Frequency-response curves for a near-square cross-section cantilever beam when the second bending mode in the z-direction is excited near primary resonance and in the presence of a one-to-one internal resonance. a_2 ($a_1 = 0$) denotes the planar (or single-mode) response amplitude; a_{1n}, a_{2n} denote the nonplanar (or two-mode) response amplitudes; solid lines (—) denote stable solutions; dashed lines (– – –) denote saddles; and dotted lines ($\cdots$) denote unstable foci.

If we assume that the beam has a near-square cross-section, then the one-to-one internal resonance $\omega_{vm} \approx \omega_{wn}$ occurs when $m = n = 1, 2, 3, \cdots$. In Figure 16, we show the frequency-response curves for a near-square cross-section cantilever beam when the second bending mode in the z-direction (i.e., $n = m = 2$) is excited near primary resonance; that is $\Omega = \omega_{w2} + \epsilon\sigma$ and $\omega_{w2} \approx \omega_{v2}$. In Figure 16, a_1 and a_{1n} denote the amplitudes of the bending deflection $v(s,t)$ and a_2 and a_{2n} denote the amplitudes of the bending deflection $w(s,t)$. The a_i are the amplitudes of the single-mode responses of the beam; that is, when $a_2 \neq 0$, $a_1 = 0$ and the oscillations are planar. The a_{in} are the amplitudes of the two-mode responses of the beam; that is, $a_{1n} \neq 0$, $a_{2n} \neq 0$, and the oscillations are nonplanar. The solution $a_1 \neq$

0 and $a_2 = 0$ does not occur in this system because the $w(s,t)$-motion is directly excited, and hence a_2 must be nontrivial. It does, however, occur when the beam is parametrically excited along its elastic axis. In this case, Eqs. (37) and (38) would have the additional terms $\epsilon a_1 f_{11}(\gamma_1)$ and $\epsilon a_1 f_{12}(\gamma_1)$ and the terms $\epsilon f_{21}(\gamma_2)$ and $\epsilon f_{22}(\gamma_2)$ in Eqs. (39) and (40) would be replaced by $\epsilon a_2 f_{21}(\gamma_2)$ and $\epsilon a_2 f_{22}(\gamma_2)$.

Similar to the Duffing equation when $\alpha < 0$ (see Figure 6), the planar-response curves in Figure 16a are bent to the left, indicating that the effective nonlinearity for the second bending mode is softening. This is because, for the second and higher modes, the influence of the inertia nonlinearity, which is softening, dominates the influence of the geometric nonlinearity, which is hardening. We note that, for the first bending mode, the effective nonlinearity is hardening because the influence of the geometric nonlinearity dominates the influence of the inertia nonlinearity, and hence the planar-response curves would be bent to the right.

However, unlike Figure 6 where the upper branch of equilibrium solutions is always stable, the upper branch of the single-mode equilibrium solution a_2 in Figure 16a loses stability through two pitchfork bifurcations at points A and B, resulting in the nonplanar equilibrium solutions a_{2n} and a_{1n} (shown in Figure 16b). These latter equilibrium solutions correspond to the beam exhibiting simple whirling motions. The pitchfork bifurcation at point A is subcritical and a jump is associated with it as σ is slowly increased, whereas the pitchfork bifurcation at point B is supercritical and the transition from planar to nonplanar responses is gradual as σ is slowly decreased. The points C, D, E, and F are saddle-node bifurcations, which result in the solution jumping to a far away attractor which may be an equilibrium solution, a limit cycle, or an aperiodic solution. Furthermore, the nonplanar equilibrium solutions in Figure 16 undergo two Hopf bifurcations H_1 and H_2 as σ is slowly varied. These Hopf bifurcations result in limit cycles, which correspond to the beam exhibiting unsteady (or beating-type) whirling motions.

From Figure 16, we note that there are regions where multiple solutions (equilibrium and dynamic, planar and nonplanar) coexist, and hence, for a given set of parameters and depending on the initial conditions, the oscillatory response of the beam may be planar or it may be nonplanar, exhibiting whirling, unsteady whirling, or chaos. Moreover, we note that, in Figure 16, equilibrium solutions do not exist in the region between the saddle-node bifurcation F and the Hopf bifurcation H_2, and hence one expects the response of the beam to be nonplanar, exhibiting unsteady whirling or chaos. In such regions, one expects the system to contain very rich and complex dynamics. In fact, the bifurcation diagram presented in Figure 14 is one part of several dynamical solutions present in the region between points F and H_2.

An interesting phenomenon that can appear in forced systems having quadratic nonlinearities and possessing a two-to-one internal resonance is *saturation*. For example, consider the system

$$\ddot{u}_1 + \omega_1^2 u_1 + 2\epsilon\mu_1\dot{u}_1 + \epsilon\alpha_1 u_1 u_2 = \epsilon F_1 \cos\Omega_1 t \tag{41}$$

$$\ddot{u}_2 + \omega_2^2 u_2 + 2\epsilon\mu_2\dot{u}_2 + \epsilon\alpha_2 u_1^2 = \epsilon F_2 \cos\Omega_2 t \tag{42}$$

250

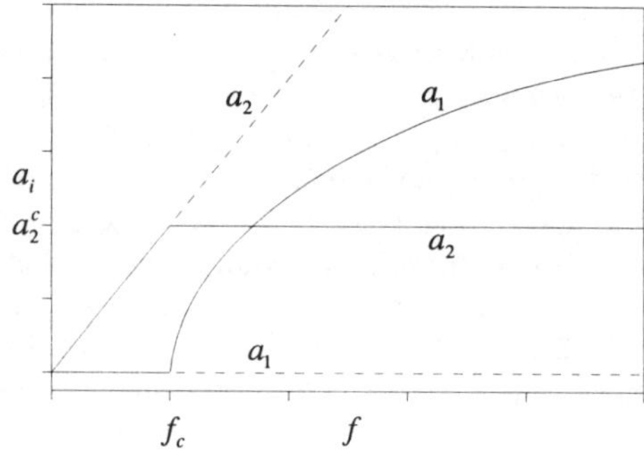

Fig. 17 Force-response curves of a system exhibiting the saturation phenomenon.

where the α_i are constants and have the same sign. When $F_1 = 0$, $\Omega_2 \approx \omega_2$, and $\omega_2 \approx 2\omega_1$ (i.e., two-to-one internal resonance), we show in Figure 17 typical force-response curves. In Figure 17, a_1 is the amplitude of the indirectly excited mode u_1 and a_2 is the amplitude of the directly excited mode u_2. From Figure 17, we note that as the forcing amplitude f ($\propto F_2$) is increased from zero, the response consists of only the u_2 mode whose amplitude increases linearly with f. However, when f reaches the critical value f_c, the amplitude a_2 reaches an upper bound a_2^c and spills the extra energy it receives to the u_1 mode; that is, the u_2 mode becomes *saturated*. As f is increased further, the amplitude a_1 continues to increase monotonically while the amplitude a_2 remains constant. Eventually, the amplitude a_1 of the indirectly excited mode, as a result of the internal resonance, becomes larger than the amplitude a_2 of the directly excited mode.

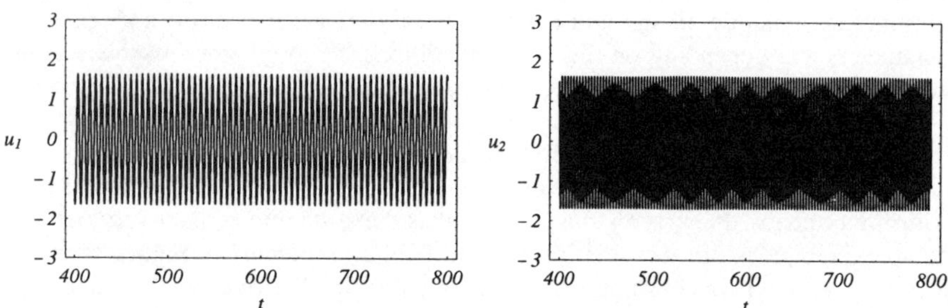

Fig. 18 Time histories for the system governed by Eqs. (41) and (42) when $F_1 = 0.2$, $F_2 = 0$, $\Omega_1 = \omega_1 = 1$, and $\omega_2 = 2\omega_1 = 2$, $\mu_1 = \mu_2 = 0.02$, and $\alpha_1 = \alpha_2 = 0.1$; it shows that the modes u_1 and u_2 are periodic.

If one takes $F_2 = 0$, $\Omega_1 \approx \omega_1$, and $\omega_2 \approx 2\omega_1$, another phenomenon that a system governed by Eqs. (41) and (42) may exhibit is a *continuous exchange of energy* between the interacting modes. Depending on the system parameters, periodic solutions of Eqs. (41) and (42) may not exist, and instead solutions where energy is continually channeled between u_1 and u_2 may occur. For example, taking $F_2 = 0$, $\Omega_1 = \omega_1 = 1$, $\omega_2 = 2\omega_1 = 2$, $\mu_1 = \mu_2 = 0.02$, and $\alpha_1 = \alpha_2 = 0.1$, Eqs. (41) and (42) were integrated in time for two values of F_1. When $F_1 = 0.2$, the time histories in Figure 18 show that the system response is periodic. However, when $F_1 = 0.3$, the time histories in Figure 19 show that the system response is aperiodic.

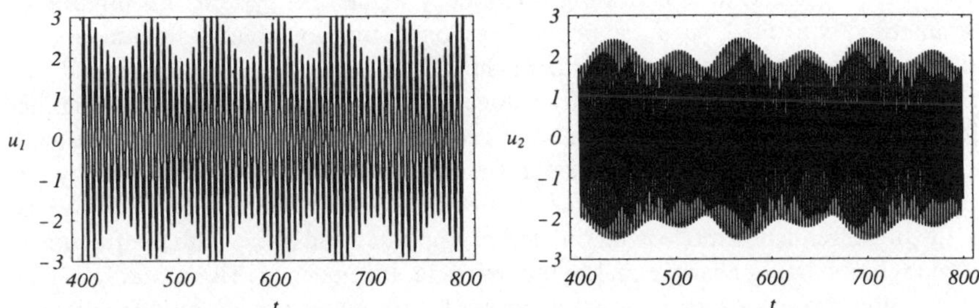

Fig. 19 Time histories for the system governed by Eqs. (41) and (42) when $F_1 = 0.3$, $F_2 = 0$, $\Omega_1 = \omega_1 = 1$, and $\omega_2 = 2\omega_1 = 2$, $\mu_1 = \mu_2 = 0.02$, and $\alpha_1 = \alpha_2 = 0.1$; it shows that the modes u_1 and u_2 continuously exchange energy.

4.2. Combination Internal Resonances

In systems that have three or more degrees of freedom, *combination internal resonances* may occur. For example, the three-degree-of-freedom system

$$\ddot{u}_1 + \omega_1^2 u_1 + \epsilon c_1 \dot{u}_1 + \epsilon N_1(u_1, u_2, u_3) = 0 \tag{43}$$

$$\ddot{u}_2 + \omega_2^2 u_2 + \epsilon c_2 \dot{u}_2 + \epsilon N_2(u_1, u_2, u_3) = 0 \tag{44}$$

$$\ddot{u}_3 + \omega_3^2 u_3 + \epsilon c_3 \dot{u}_3 + \epsilon N_3(u_1, u_2, u_3) = \epsilon f \cos \Omega t \tag{45}$$

may possess combination internal resonances of the form $\omega_3 \approx \omega_1 \pm \omega_2$ if the nonlinearities are quadratic, and of the forms $\omega_3 \approx |2\omega_1 \pm \omega_2|$ and $\omega_3 \approx \frac{1}{2}(\omega_1 \pm \omega_2)$ if the nonlinearities are cubic.

4.3. Combination External and Combination Parametric Resonances

Systems that are excited by combination resonances may exhibit modal interactions, even when the natural frequencies are incommensurate (i.e., the system is not autoparametric). Referring back to the system governed by Eqs. (31) and (32) where the forcing is direct, when the nonlinearities are quadratic, resonances of the form $\Omega \approx \omega_1 \pm \omega_2$ may occur, and when the nonlinearities are cubic, resonances of the form $\Omega \approx |2\omega_1 \pm \omega_2|$ and $\Omega \approx \frac{1}{2}(\omega_1 \pm \omega_2)$ may occur. Such resonances are called

252

combination external resonances. On the other hand, if the forcing in Eq. (32) were parametric (e.g., $u_2 f \cos \Omega t$), then *combination parametric resonances* of the form $\Omega \approx \omega_1 \pm \omega_2$ may occur. We note that if $\omega_2 \gg \omega_1$, then a system excited by a combination resonance, such as $\Omega \approx \omega_1 + \omega_2$, may be an example of a high-frequency excitation resulting in a large-amplitude low-frequency response.

4.4. Energy Transfer between Widely-Spaced Modes

Recent experiments, in which a high-frequency mode of a system was directly or parametrically excited, have documented responses that contained a significant contribution from a low-frequency mode of the system, even though the high and low natural frequencies are not internally resonant and the excitation is not a combination resonance. The spectra for these experiments contain peaks at both the high and low natural frequencies. Moreover, the peak at the high natural frequency has sidebands that are separated by an amount equal to the low natural frequency. That is, the high-frequency mode is modulated in amplitude and phase with the frequency of modulation being nearly equal to the low natural frequency. Therefore, in such a case, a high-frequency excitation activates a low-frequency mode by a mechanism that is neither an autoparametric resonance nor a combination-type resonance. This mechanism is sometimes dubbed *zero-to-one resonance.*

5. SUMMARY

We presented a brief overview of nonlinear systems, the possible sources of the nonlinearities, and their importance on the responses. We showed that nonlinear systems can exhibit many behaviors that can not be explained by linear theory. Some of these behaviors include multiplicity of responses, jumps, saturation, modal interactions, and almost-periodic (quasiperiodic) and aperiodic (chaotic) system responses. Hence, when modeling a dynamical system, the influence of nonlinearities must be taken into consideration.

Often, one assumes that because a system is excited at a high frequency, its response will be small and possibly negligible; hence, modeling the system using linear theory is adequate. Several experimental and theoretical results clearly demonstrate that this is not always true. Rather, the energy of an excited high-frequency low-amplitude mode may be transferred to a low-frequency high-amplitude mode of the system. The transfer mechanism may be either resonant, such as a three-to-one internal resonance or a combination-type resonance, or nonresonant. Then, because the system's overall response contains significant contributions from both modes, it may have a large amplitude. In such a case, one cannot ignore the influence of the nonlinearities and model the system linearly.

Although, most of the examples and results we presented correspond to mechanical and structural systems, we note that nonlinearities influence many other systems in the engineering and science fields. Therefore, solutions of these systems may also exhibit some of the nonlinear behaviors discussed.

ACKNOWLEDGEMENT

This work was supported by the Air Force Office of Scientific Research (AFSOR) under Grant No. F49620-98-1-0393, the Army Research Office (ARO) under Grant No. DAAG55-98-1-0210, and the Office of Naval Research (ONR) under the MURI Grant No. N00014-96-1-1123.

REFERENCES

1. Arafat, H. N., 1999, *Nonlinear Response of Cantilever Beams*, Dissertation, Virginia Polytechnic Institute and State University, Blacksburg, VA.

2. Arafat, H. N., and Nayfeh, A. H., 1998, "Investigation of Subcombination Internal Resonances in Cantilever Beams," *Shock and Vibration*, Vol. 5, pp. 289-296.

3. Arafat, H. N., and Nayfeh, A. H., 1999, "Nonlinear Bending-Torsion Oscillations of Cantilever Beams to Combination Parametric Excitations," in *Proceedings of the 17th International Modal Analysis Conference, IMAC XVII*, Orlando, Florida, February 8-11, pp. 1203-1209.

4. Arafat, H. N., and Nayfeh, A. H., 2000, "Nonlinear Response of Symmetrically Laminated Composite Cantilever Beams," in *Proceedings of the 41st AIAA/ASME/ASCE/ AHS/ASC Structures, Structural Dynamics, and Materials Conference and Exhibit*, Atlanta, Georgia, April 3-6, AIAA Paper # 2000-1471.

5. Arafat, H. N., and Nayfeh, A. H., 2000, "Non-Linear Non-Planar Dynamics of Directly Excited Cantilever Metallic Beams," *Journal of Sound and Vibration*, submitted for publication.

6. Arafat, H. N., Nayfeh, A. H., and Chin, C., 1998, "Nonlinear Nonplanar Dynamics of Parametrically Excited Cantilever Beams," *Nonlinear Dynamics*, Vol. 15, pp. 31-61.

7. Ariaratnam, S. T., 1986, "Parametric Resonance," in *Proceedings of the Tenth U.S. National Congress of Applied Mechanics*, Austin, Texas, June 16-20, pp. 181-185.

8. Asmis, K. G., and Tso, W. K., 1972, "Combination and Internal Resonance in a Non-linear Two-Degrees-of-Freedom System," *Journal of Applied Mechanics*, Vol. 39, pp. 832-834.

9. Billah, K. Y., and Scanlan, R. H., 1991, "Resonance, Tacoma Narrows Bridge Failure, and Undergraduate Physics Textbooks," *American Journal of Applied Physics*, Vol. 59, pp. 118-124.

10. Bolotin, V. V., 1964, *The Dynamic Stability of Elastic Systems*, Holden-Day, San Francisco, California.

11. Cartmell, M. P., 1990, *Introduction to Linear, Parametric and Nonlinear Vibrations*, Chapman and Hall, London.

12. Cartmell, M. P., and Roberts, J. W., 1987, "Simultaneous Combination Resonances in a Parametrically Excited Cantilever Beam," *Strain*, Vol. 23, pp. 117-126.

13. Chin, C., and Nayfeh, A. H., 1996, "Bifurcations and Chaos in Externally Excited Circular Cylindrical Shells," *Journal of Applied Mechanics*, Vol. 63, pp. 565-574.

14. Chin, C., Nayfeh, A. H., and Mook, D. T., 1995, "The Response of a Nonlinear System With a Nonsemisimple One-to-One Resonance to a Combination Parametric Resonance," *International Journal of Bifurcation and Chaos*, Vol. 5, pp. 971-982.

15. Crespo da Silva, M. R. M., and Glynn, C. C., 1978, "Nonlinear Flexural-Flexural-Torsional Dynamics of Inextensional Beams. I. Equations of Motion," *Journal of Structural Mechanics*, Vol. 6, pp. 437-448.

16. Crespo da Silva, M. R. M., and Glynn, C. C., 1978, "Nonlinear Flexural-Flexural-

Torsional Dynamics of Inextensional Beams. II. Forced Motions," *Journal of Structural Mechanics*, Vol. 6, pp. 449-461.

17. Dugundji, J., and Mukhopadhyay, V., 1973, "Lateral Bending-Torsion Vibrations of a Thin Beam Under Parametric Excitation," *Journal of Applied Mechanics*, Vol. 40, pp. 693-698.

18. Evan-Iwanowski, R. M., 1965, "On the Parametric Response of Structures," *Applied Mechanics Reviews*, Vol. 18, pp. 699-702.

19. Feng, Z. C., and Leal, L. G., 1994, "Symmetries of the Amplitude Equations of an Inextensional Beam With Internal Resonance," *Journal of Applied Mechanics*, Vol. 62, pp. 235-238.

20. Glendinning, P., 1994, *Stability, Instability and Chaos: an Introduction to the Theory of Nonlinear Differential Equations*, Cambridge University Press, New York.

21. Hayashi, C., 1985, *Nonlinear Oscillations in Physical Systems*, Princeton University Press, Princeton, New Jersey.

22. Meirovitch, L., 1997, *Principles and Techniques of Vibrations*, Prentice Hall, Upper Saddle River, New Jersey.

23. Mettler, E., 1967, "Stability and Vibration Problems of Mechanical Systems Under Harmonic Excitation," in *Dynamic Stability of Structures*, Herrmann, G., ed., Pergamon Press, New York, pp. 169-183.

24. Mook, D. T., HaQuang, N., and Plaut, R. H., 1986, "The Influence of an Internal Resonance on Non-Linear Structural Vibrations under Combination Resonance Conditions," *Journal of Sound and Vibration*, Vol. 104, pp. 229-241.

25. Moon, F. C., 1992, *Chaotic and Fractal Dynamics*, Wiley-Interscience, New York.

26. Nayfeh, A. H., 1973, *Perturbation Methods*, Wiley-Interscience, New York.

27. Nayfeh, A. H., 1981, *Introduction to Perturbation Methods*, Wiley-Interscience, New York.

28. Nayfeh, A. H., 1983, "Response of Two-Degree-of-Freedom Systems to Multifrequency Parametric Excitations," *Journal of Sound and Vibration*, Vol. 88, pp. 1-10.

29. Nayfeh, A. H., 1983, "Parametrically Excited Multidegree-of-Freedom Systems with Repeated Frequencies," *Journal of Sound and Vibration*, Vol. 88, pp. 145-150.

30. Nayfeh, A. H., 1983, "Combination Resonances in the Non-Linear Response of Bowed Structures to a Harmonic Excitation," *Journal of Sound and Vibration*, Vol. 90, pp. 457-470.

31. Nayfeh, A. H., 1985, "The Response of Non-Linear Single-Degree-of-Freedom Systems to Multifrequency Excitations," *Journal of Sound and Vibration*, Vol. 102, pp. 403-414.

32. Nayfeh, A. H., 1997, "On the Discretization of Weekly Nonlinear Spatially Continuous Systems," in *Differential Equations and Chaos Lectures on Selected Topics*, Ibragimov, N. H., Mahomed, F. M., Mason, D. P., and Sherwell, D., eds., pp 3-39.

33. Nayfeh, A. H., 2000, *Nonlinear Interactions*, Wiley-Interscience, New York.

34. Nayfeh, A. H., and Arafat, H. N., 1998, "Nonlinear Response of Cantilever Beams to Combination and Subcombination Resonances," *Shock and Vibration*, Vol. 5, pp. 277-288.

35. Nayfeh, A. H., and Balachandran, B., 1989, "Modal Interactions in Dynamical and Structural Systems," *Applied Mechanics Reviews*, Vol. 42, pp. S175-S201.

36. Nayfeh, A. H., and Balachandran, B., 1995, *Applied Nonlinear Dynamics*, Wiley-Interscience, New York.

37. Nayfeh, A. H., and Chin, C-M., 1995, "Nonlinear Interactions in a Parametrically Ex-

cited System with Widely Spaced Frequencies," *Nonlinear Dynamics*, Vol. 7, pp. 195-216.

38. Nayfeh, A. H., and Jebril, A. E. S., 1987, "The Response of Two-Degree-of-Freedom Systems with Quadratic and Cubic Non-Linearities to Multifrequency Parametric Excitations," *Journal of Sound and Vibration*, Vol. 115, pp. 83-101.

39. Nayfeh, A. H., and Mook, D. T., 1978, "A Saturation Phenomenon in the Forced Response of Systems With Quadratic Nonlinearities," in *Proceedings of the VIIIth International Conference on Nonlinear Oscillations*, Prague, Czech Republic.

40. Nayfeh, A. H., and Mook, D. T., 1979, *Nonlinear Oscillations*, Wiley-Interscience, New York.

41. Nayfeh, A. H., and Mook, D. T., 1995, "Energy Transfer from High-Frequency to Low-Frequency Modes in Structures," in *ASME Special Combined Issue of the Journal of Mechanical Design and Journal of Vibration and Acoustics, 50th Anniversary of the Design Engineering Division*, Vol. 117, pp. 186-195.

42. Nayfeh, A. H., Mook, D. T., and Nayfeh, J. F., 1987, "Some Aspects of Modal Interactions in the Response of Beams," in *Proceedings of the AIAA 28th Structures, Structural Dynamics and Materials Conference*, Monterey, California, April 6-8.

43. Nayfeh, A. H., and Pai, P. F., 1989, "Non-Linear Non-Planar Parametric Responses of an Inextensional Beam," *International Journal of Non-Linear Mechanics*, Vol. 24, pp. 139-158.

44. Nayfeh, A. H., and Zavodney, L. D., 1986, "The Response of Two-Degree-of-Freedom Systems with Quadratic Non-Linearities to a Combination Parametric Resonance," *Journal of Sound and Vibration*, Vol. 107, pp. 329-250.

45. Nayfeh, S. A., and Nayfeh, A. H., 1992, "Energy Transfer from High- to Low-Frequency Modes in Flexible Structures," *Nonlinear Vibrations*, ASME DE-Vol. 50/AMD-Vol. 144, pp. 89-97.

46. Nayfeh, S. A., and Nayfeh, A. H., 1993, "Nonlinear Interactions Between Two Widely Spaced Modes - External Excitation," *International Journal of Bifurcation and Chaos*, Vol. 3, pp. 417-427.

47. Nayfeh, S. A., and Nayfeh, A. H., 1994, "Energy Transfer from High- to Low-Frequency Modes in a Flexible Structure via Modulation," *Journal of Vibration and Acoustics*, Vol. 116, pp. 203-207.

48. Oueini, S. S., 1999, *Techniques for Controlling Structural Vibrations*, Dissertation, Virginia Polytechnic Institute and State University, Blacksburg, Virginia.

49. Pai, P. F., and Nayfeh, A. H., 1990, "Non-Linear Non-Planar Oscillations of a Cantilever Beam Under Lateral Base Excitations," *International Journal of Non-Linear Mechanics*, Vol. 25, pp. 455-474.

50. Popovic, P., Nayfeh, A. H., Oh, K., and Nayfeh, S. A., 1995, "An Experimental Investigation of Energy Transfer from a High-Frequency Mode to a Low Frequency Mode in a Flexible Structure," Journal of Vibration and Control, Vol. 1, pp. 115-128.

51. Rayleigh, J. W. S., 1945, *The Theory of Sound*, Vol. 1, Dover, New York.

52. Sathyamoorthy, M., 1982, "Nonlinear Analysis of Beams, Part I: A Survey of Recent Advances," *Shock and Vibration Digest*, Vol. 14, pp. 19-35.

53. Sathyamoorthy, M., 1982, "Nonlinear Analysis of Beams, Part II: Finite Element Methods," *Shock and Vibration Digest*, Vol. 14, pp. 7-18.

54. Sathyamoorthy, M., 1997, *Nonlinear Analysis of Structures*, CRC Press, Boca Raton, Florida.

55. Seydel, R., 1994, *Practical Bifurcation and Stability Analysis*, Springer-Verlag, New

256

York.

56. Shyu, I.-M. K., Mook, D. T., and Plaut, R. H., 1993, "Whirling of a Forced Cantilevered Beam with Static Deflection. III: Passage through Resonance," *Nonlinear Dynamics*, Vol. 4, pp. 461-481.

57. Srinivasan, P., 1995, *Nonlinear Mechanical Vibrations*, Wiley, New York.

58. Streit, D. A., Bajaj, A. K., and Krousgrill, C. M., 1988, "Combination Parametric Resonance Leading to Periodic and Chaotic Response in Two-Degree-of-Freedom Systems with Quadratic Nonlinearities," *Journal of Sound and Vibration*, Vol. 124, pp. 297-314.

59. Tabaddor, M., and Nayfeh, A. H., 1997, "An Experimental Investigation of Multimode Responses in a Cantilever Beam," *Journal of Vibration and Acoustics*, Vol. 119, pp. 532-538.

60. Tezak, E. G., Nayfeh, A. H., and Mook, D. T., 1982, "Parametrically Excited Non-Linear Multidegree-of Freedom Systems with Repeated Natural Frequencies," *Journal of Sound and Vibration*, Vol. 85, pp. 459-472.

61. Thompson, J. M. T., and Stewart, H. B., 1986, *Nonlinear Dynamics and Chaos*, Wiley-Interscience, Chichester, England.

62. Tso, W. K., and Asmis, K. G., 1974, "Multiple Parametric Resonance in a Non-Linear Two Degree of Freedom System," *International Journal of Non-Linear Mechanics*, Vol. 9, pp. 269-277.

63. Verhulst, F., 1989, *Nonlinear Differential Equations and Dynamical Systems*, Springer-Verlag, Berlin.

64. von Kármán, T., 1940, "The Engineer Grapples with Nonlinear Problems," *Bulletin of the American Mathematical Society*, Vol. 46, pp. 615-683.

65. Yamamoto, T., Yasuda, K., and Tei, N., 1981, "Summed and Differential Harmonic Oscillations in a Slender Beam," *Bulletin of the JSME*, Vol. 24, pp. 1214-1222.

66. Yamamoto, T., Yasuda, K., and Tei, N., 1982, "Super Summed and Differential Harmonic Oscillations in a Slender Beam," *Bulletin of the JSME*, Vol. 25, pp. 959-968.

67. Yasuda, K., and Hayashi, N., 1982, "Summed and Differential Oscillations in a Circular Plate," *Bulletin of the JSME*, Vol. 25, pp. 1582-1590.

68. Yasuda, K., Kato, M., and Masuda, H., 1993, "Single-Mode and Multimode Combination Tones in a Nonlinear Beam," *JSME International Journal, Series C*, Vol. 36, pp. 17-25.

69. Zaretzky, C. L., and Crespo da Silva, M. R. M., 1994, "Experimental Investigation of Non-Linear Modal Coupling in the Response of Cantilever Beams," *Journal of Sound and Vibration*, Vol. 174, pp. 145-167.

70. Zavodney, L. D., 1987, "Can the Modal Analyst Afford to Be Ignorant of Nonlinear Vibration Phenomena?," in *Proceedings of the Fifth International Modal Analysis Conference*, London, England, April 6-9, pp. 154-159.

Random Vibrations: Assessment of the State of the Art

Thomas L. Paez
Experimental Structural Dynamics Department
Sandia National Laboratories
Albuquerque, New Mexico

ABSTRACT
The study of random vibration is the phenomenon wherein random excitation applied to a mechanical system induces random response. We summarize the state of the art in random vibration analysis and testing, commenting on history, linear and nonlinear analysis, the analysis of large-scale systems, and probabilistic structural testing.

1. INTRODUCTION
Random vibrations deals with the probabilistic analysis of the response of structures with potentially random parameters and initial conditions and with random excitation and potentially random boundary conditions, and it deals with the simulation of random environments in the laboratory for the testing of real structures. Both random vibration analysis and testing are used for the design, optimization, and reliability assessment of structures and for other purposes.

The activities undertaken in analysis and testing in random vibrations can be succinctly described using a simple, representative dynamic equation of equilibrium. Consider the following (scalar or vector) representation.

$$\dot{x} = g(x, y, a) \qquad x(0) = x_0, t \geq 0 \tag{1}$$

The quantity x represents system response; y represents system excitation; a represents system parameters; the dot denotes differentiation with respect to time; and $g(.)$ is the deterministic functional form that relates the former quantities to the response derivative.

Traditional random vibrations specifies the form of $g(.)$, takes a as a scalar or vector of constants, specifies y as a random process, and analyzes the probabilistic character of the response random process x. Many aspects of this problem have been solved for the case where $g(.)$ is a linear function of the excitation, and the excitation is a stationary, Gaussian random process. When the excitation is nonstationary, the expression for the response can be written, but it cannot always be solved easily in closed form for the desired response characteristics. For the case where $g(.)$ is not a linear function of the excitation, many means have been developed for characterizing the response random process. None is completely general, and practically all involve some form of approximation.

Many large-scale structural analysis computer codes implement the Gaussian excitation/linear response solution mentioned above. However, not much else is typically analyzed in commercial finite element codes. One can always resort to a Monte Carlo approach for the analysis of practically any system including nonlinear systems and systems with random structural parameters, a. However, depending on the information desired and the complexity of the structure, the number of random quantities or the number of simulations may be significantly limited.

Two techniques for the analysis of large-scale structures with random parameters have been developed: reliability-based techniques and stochastic finite element techniques. The former takes the parameters, a, to be random variables with known joint probability distributions and uses this information to characterize the probabilistic response of structural systems. It provides solutions that are not traditional random vibration solutions. Stochastic finite element analysis is a relatively new technique. It permits system parameters, a, to be represented approximately as random fields and approximates the response, x, as a random field.

The laboratory testing of structural systems is the link between random vibration theory and the practical excitation of structures in the field. Stationary random vibration testing in the laboratory operates in two phases, with the assumption that the electromechanical testing system and the structure under test have parameters, a, that are either constant or slowly varying functions of time. In the first phase of operation, the random vibration test seeks a preliminary identification of the system parameters, a, in the linear framework. Having performed this identification, the second phase seeks a random input, y, to excite a response, x, with a preestablished spectral density. As the test progresses, the system parameters, a, are periodically updated, and the spectral density of the input, y, is modified. Though satisfactory algorithms for the control of single-shaker, stationary random vibration tests have been developed, more development is required, particularly in the control of nonstationary tests, the generation of non-Gaussian environments, the testing of nonlinear systems, and the operation of multi-shaker tests.

In this contribution we will briefly summarize the history of analytical technique development in random vibrations; then we will discuss the random vibrations of linear systems, the topic on which the great majority of research and development effort has historically been placed. Next we will discuss the random vibration of nonlinear systems. Following this the state of the art in the random vibration analysis of large-scale structures using numerical codes will be described. Finally, the state of the art in probabilistic experimental structural dynamics will be described. The contribution concludes with a summary.

2. HISTORY

It is difficult to identify precisely the paper or the event that marked the beginning of the field of random vibrations analysis. Lord Rayleigh wrote a paper in the late 1800s (see Rayleigh, 1880) considering a problem that is a very much idealized and specialized case of random structural response. Two later papers (see Rayleigh, 1919a, 1919b) were extensions of the earlier one and treated more practical random vibration subjects, but by that time several other papers that must be considered treatments of random vibrations had been published.

Around the turn of the century, Einstein (1905) constructed a framework for analyzing the Brownian movement, the random oscillation of particles suspended in a fluid medium and caused by the molecular motion postulated by the kinetic theory of matter. His framework is a special form of what was later to become known as the Fokker-Planck equation governing the probability density function (PDF) of particle motion and relating it to mechanical system parameters. Einstein augmented his initial study with more, related investigations (all reprinted in Einstein, 1956) considering, among other things, the problem of parameter identification.

According to Gnedenko (1997), Smoluchowski (1916) generalized Einstein's analysis and performed Brownian motion experiments to verify the predictions of the theory. Smoluchowski was first to write the Fokker-Planck equation for systems in which a displacement-related force restrains the mass. The single-degree-of-freedom system in which linear damping and stiffness are present was called the case of the "harmonically bound" particle. Planck (1927) and Fokker (1913) started with the discrete space/discrete time framework of the random walk and offered arguments regarding the relative and limiting values of various parameters in the model. They found that a limiting form of the random walk model is the partial differential equation developed by Einstein to characterize the Brownian motion.

In a paper written a few years later, Wiener (1923) overcame a problem with previous analyses - namely, that the solution of a differential equation excited by ideal white noise does not normally possess as many derivatives as appear in the equation of dynamic equilibrium. He did this by rewriting the governing equation in stochastic differential form. Doob (1942) later formalized this procedure.

An important paper written in 1930 by Wiener (1930) defined the concept of the spectral density function, and he attributed the fundamental idea to Schuster (1906). He showed that the Fourier transform of the correlation function exists when the underlying random process is stationary and called that quantity the spectral density. These, of course, are concepts that are essential to the theory of random vibration today.

Numerous other investigators extended the theme introduced by Einstein in the following decades. Uhlenbeck and Ornstein (1930) developed the probability distribution of white noise-excited response at small times. Wang and Uhlenbeck (1945) provided a general derivation of the Fokker-Planck equation from the Chapman, Kolmogorov, Smoluchowski equation.

Some papers by S. O. Rice (1944, 1945) appear to be the first to introduce ideas on how to develop probabilistic measures of some special and important characteristics of system response. He developed (1) equations governing the rate of zero crossings and crossings of a preestablished level by a random process; (2) the probability distribution of maxima of a Gaussian random process; (3) the probability distribution of the envelope of a stationary, narrowband Gaussian random process; (4) the probability distribution of the output of many (zero memory) nonlinear devices; and many other subjects.

3. ANALYSIS OF LINEAR SYSTEMS

Linear system analysis is the single area in random vibrations where a substantial number of problems have been solved and the solutions used in a wide variety of practical applications. In fact, outside of research and development projects where new techniques for solving random vibrations problems are being sought, perhaps 95 percent or more of current efforts in random vibration problem solution involve the analysis of linear systems excited by stationary inputs. The fundamental techniques for analyzing the response of linear systems with constant coefficients to stationary random excitation were given in Crandall (1958). His paper appears to be the first to express structural response moments in the time and frequency domains, in the form used widely today. He cites Laning and Battin (1956), a text on control systems, as the source of the fundamental statistical input/output relations that he presents. They, in turn, refer to an article by Phillips (1947), dealing with servomechanisms. He refers to the classical paper of Wiener (1930) as the source of the fundamental mathematics (autocorrelation/spectral density relation) in his paper, though it appears to be Phillips who derived the spectral input/output relation. Many authors rederived and generalized these results. Among them are Crandall (1963); Crandall and Mark (1963); Robson (1964); Lin (1967); Elishakoff (1983); Nigam (1983); Newland (1984); Bolotin (1984); Augusti, Baratta and Casciati (1984); Ibrahim (1985); Yang (1986); Schueller and Shinozuka (1987); Roberts and Spanos (1990); Soong and Grigoriu (1993); and Wirsching, Paez, and Ortiz (1995).

To summarize these results, consider a stable, linear system that is discretely modeled with n degrees-of-freedom. Let the system excitation be a stationary, Gaussian, vector random process denoted $\{Y(t), -\infty < t < \infty\}$. There are n elements in the vector random process, each a scalar random process that represents the excitation at a degree of freedom. Because the excitation is assumed stationary, it possesses a spectral density; denote this $S_{YY}(\omega), -\infty < \omega < \infty$. The ith row j[th] column element of this matrix is the cross-spectral density function between the excitation random processes at the ith and j[th] degrees of freedom. The matrix is Hermitian at every frequency, ω. The diagonal elements of the matrix are the autospectral density functions of the excitations. These are real, symmetric, non-negative functions of the frequency, ω.

Because the system is linear, it possesses a frequency response function. Let $H(\omega), -\infty < \omega < \infty$, denote the matrix frequency response function of the system under consideration; this matrix has dimension $n \times n$. The ith row j[th] column element of this matrix is the frequency response function of motion at the ith degree of freedom excited by input at the jth degree of freedom.

Let the response be the vector random process denoted $\{X(t), -\infty < t < \infty\}$, and let the response have the spectral density $S_{XX}(\omega), -\infty < \omega < \infty$. The vector random process contains n elements, and its spectral density has dimension $n \times n$. Then the spectral density of the response can be expressed:

$$S_{XX}(\omega) = H(\omega)S_{YY}(\omega)H'(\omega) \qquad -\infty < \omega < \infty, \tag{2}$$

where the prime denotes the operations of transposition and complex conjugation. The response vector random process is Gaussian distributed because the system was assumed linear, and a linear function of a Gaussian random process is also Gaussian. If the excitation has a nonzero mean, then the mean of the response can be easily established. The response mean and the second order statistics provided in Eq. (2) are sufficient to completely specify the response random process.

In the special case where the input excitation is a stationary random process but the initial conditions of the system are specified at a particular time, the system response is nonstationary from the time of initiation to the time when the system achieves the stationary state. The responses at all degrees of freedom approach the stationary state at an exponential rate that is a function of the system damping. The responses to initial and boundary conditions that are random processes can also be written directly.

The moments and probability distributions of many important measures of structural response were derived by Rice (1944, 1945). Among them are the average rate at which a random process crosses a barrier, the probability distribution of the maxima of a random process, and the probability distribution of the envelope of a narrowband random process. The first of these, in combination with a Poisson process assumption, is used to approximate the first passage probability distribution, even now. (See Coleman, 1959.)

When the system under consideration is modeled as continuous, we can write an integral expression equivalent to Eq. (2) for the response spectral density. However, it

262

is typically more difficult to use because it requires knowledge of the system's Green's function and yields information about the response spectral density only through integration of what may be complicated functions. The continuous approach is used infrequently in this age when spatially discrete finite element models are available and widely used.

All real random excitations are nonstationary in nature. The stationarity assumption is an idealization that is used frequently and sometimes quite accurately. When the duration of an input is short with respect to the fundamental period of the structure being excited, then it is appropriate to treat the input and the response it excites as nonstationary. It is relatively simple to write the expression for the random response of a linear system excited by a nonstationary excitation. It is

$$X(t) = \int_0^\infty h(t-\tau)Y(\tau)d\tau \qquad X(0) = \dot{X}(0) = 0, t \geq 0, \tag{3}$$

where the excitation and response random processes are assumed to exist in some well-defined sense, and the initial time and initial conditions have arbitrarily been assumed zero. The quantity $h(t)$ is the system impulse response function; $\{X(t), t \geq 0\}$ is the response random process; and $\{Y(t), t \geq 0\}$ is the excitation random process. It is usually a matter of interest to establish the mean and autocorrelation function of the response, and one would hope to be able to do this for realistic excitation random processes. Because of our preoccupation with white noise excitation in the stationary excitation case, we might hope to specify the input as a filtered and modulated white noise or as a sum of filtered and modulated, band-limited white noise components. For example, we might hope to define a scalar excitation random process as

$$Y(t) = m(t)L[w(t)] \qquad t \geq 0, \tag{4}$$

where $m(t)$ is an appropriately specified modulating function, $w(t)$ is an ideal or band-limited white noise, and $L[.]$ is a filter operating on the white noise. However, it turns out to be very difficult to evaluate the required integrals to make this approach work. We can evaluate the integrals numerically, of course, but then we lose the advantage of relating changes in the input parametric model to the effects they cause in the response statistics.

To overcome this problem, an approach has been developed over the past decade, or so, that simplifies (in some sense) the computation of nonstationary response moments. It has been recently summarized by Masri, Smyth, and Traina (1998). They first establish a representation for the excitation nonstationary random process using a truncated Karhunen-Loeve expansion (Karhunen, 1947, Loeve, 1948). This expresses the excitation random process, $\{Y(t), t \geq 0\}$, as a finite sum of terms involving the eigenvalues and eigenvectors of the excitation autocorrelation function and a sequence of uncorrelated random variables. Each eigenvector is then represented as a series in Chebychev polynomials. The resulting expression for the excitation random process is used in Eq. (3), and an expression for the response random process can be developed because of the simple form of the Chebychev polynomials. Moments of the response can be computed, and if the probability distribution of the underlying uncorrelated random variables (from the Karhunen-Loeve expansion) has been derived, then the probability distribution of the response and its measures can be established. However, when the probability distribution of the excitation is non-Gaussian, this may be a difficult problem.

An important problem related to both the stationary and nonstationary response of linear systems is the first passage/peak response problem. It is typically treated by modeling a system response (or envelope of the response) crossing process as a Poisson random process. The approach is accurate at high barrier levels but less so at low levels. More work is required in this area. Pandey and Ariaratnam (1996) is an example of a recent paper in this area.

An interesting method for bounding the peak response probability distribution (and, therefore, the first passage probability distribution) of a linear system was introduced by Koopmans, Qualls, and Yao (1973). It is based on an inequality described by Drenick (1970), which relates peak response to input energy and system impulse response function. They use this relation to derive an upper bound on the probability distribution of peak response of a linear system, and they specialize the analysis to the Gaussian excitation case. Their bound on the peak response probability distribution has the advantage that it does not rely upon any assumptions regarding the probability distribution of response random process peaks. A tighter bound on the response of a linear system was proposed by Shinozuka (1970) soon after the one described above. This bound can also be used to establish a bound on the peak response probability distribution (see Rojwithya, 1980).

The techniques described in the previous paragraph can be used in several other contexts. For example, they can be used to bound the peak response probability distribution of large-scale systems, they can be used in the experimental context, and they can be extended to the nonlinear case. Some of these will be discussed later.

The problem of non-Gaussian excitation and response is also important because many real excitations are non-Gaussian, and almost all real structural responses are nonlinear and non-Gaussian, to some extent. The problem of non-Gaussian excitations has not been treated to near the depth of the Gaussian problem. The problem of nonlinear response will be treated in the following section.

4. NONLINEAR SYSTEM ANALYSIS AND CHAOTIC VIBRATIONS

Practically all real structural systems display some degree of nonlinear behavior when subjected to realistic environments. When the types of nonlinearities that a system displays are continuous with input level or when the level of response is mostly below the threshold at which nonlinear behavior commences, then, under many circumstances, it may be reasonable to model the system as linear, ignoring the nonlinearity in the system response behavior. However, when it is necessary to understand detailed structural response behavior at full level or near a failure threshold, then a nonlinear analysis of the system must be performed. There are several approaches available for dealing with nonlinear systems. Some of them are summarized in the following.

The response quantities of interest in a random vibration analysis vary according to the application. When only quantities such as the root-mean-square (RMS) structural response levels are required from an analysis, then equivalent linearization approaches might be used to investigate structural behavior. The subject of equivalent linearization has been widely investigated and is discussed, for example, in Roberts and Spanos (1990) and Lin (1967). Another linearization technique that accounts for the potential for non-Gaussian response is given by Iyengar and Roy (1996). The fundamental objective of equivalent linearization techniques is to establish the "best" linear approximation to a nonlinear mathematical model of a system or to an actual physical nonlinear system.

When the critical characteristics of nonlinear structural response are not preserved in a linearized system, then a nonlinear model is required for random vibration analysis. The techniques available for constructing a framework, identifying a nonlinear model, and analyzing its behavior depend on whether data are available for use in the system identification.

When data are available for the identification of a nonlinear model, then random vibration analyses can be performed by, first, identifying the parameters of the nonlinear model, second, generating realizations of excitation from the desired random source, then third, using the model to compute the response to arbitrary random excitation. This is a Monte Carlo approach. Several models have been used in this framework. For example, Hunter (1997, 1998) and Hunter and Theiler (1992) have utilized a local linear modeling approach to predict the response of a system known only through measured realizations of its excitation and response. They assume that finite length excitation vectors and their associated response vectors, grouped by a measure of excitation level, can be used to establish a local linear, state space model of the system. When an excitation vector is near the group of vectors used to create the linear model, then the model can be used to predict response. Of course, the modeling technique needs to be efficient because it is essential to identify the system with minimal data. They have chosen to use the canonical variate analysis technique (Larimore, 1983) for state space analysis. An implicit assumption is that the system under consideration is time invariant. Further, it is required that the known inputs and responses occupy regions of the hyperspace to be visited by the future, predicted excitations and responses of the system.

A similar approach identifies the parameters of an artificial neural network (ANN) to model system behavior. The ANN can be trained directly using measured input and response data, or it can be trained using transformed data. For example, data obtained in the process of training a local linear model, as the one mentioned above, can be used to train an ANN. Urbina, Hunter, and Paez (1998) and Paez and Hunter (1997) are examples of this type of analysis. The reason for using an ANN when the model that produces the training data is available is that the ANN is typically much more efficient than the former model, sometimes up to a few orders of magnitude more efficient. Chance, Worden, and Tomlinson (1998) and Bailer-Jones, MacKay, and Withers (1998) describe other ANNs that are suitable for random vibration analysis.

A very general model for nonlinear structural system behavior is the Volterra series model (see Volterra, 1959, Marmarelis and Marmarelis, 1978, Schetzen, 1980, Roy and Spanos, 1989). It is a generalization of the convolution model for linear structural response, and it is particularly good at representing harmonic distortion phenomena. (Harmonic distortion phenomena are connected to nonlinear response behavior, and occur frequently in real structural systems under test.) The Volterra model has the form

$$x(t) = h_0 + \int_0^\infty h_1(t-\tau)y(\tau)d\tau$$

$$+ \int_0^\infty d\tau_1 \int_0^\infty d\tau_2\, h_2(t-\tau_1, t-\tau_2)y(\tau_1)y(\tau_2)$$

$$+ \int_0^\infty d\tau_1 \int_0^\infty d\tau_2 \int_0^\infty d\tau_3\, h_3(t-\tau_1, t-\tau_2, t-\tau_3)y(\tau_1)y(\tau_2)y(\tau_3)$$

$$+ \dots$$

$$t \geq 0, \qquad (5)$$

where $x(t)$ is the system response, $y(t)$ is the system input excitation, and the $h_0, h_1(t_1), h_2(t_1,t_2), h_3(t_1,t_2,t_3), \dots$, are the system kernels. The quantity h_0 is a constant; $h_1(t)$ is analogous to the linear system impulse response function; and the $h_i(t_1, \dots, t_i), i = 2, 3, \dots$, are the higher- order kernels that encapsulate the nonlinear system behavior. Volterra argued that the input/output relation of a nonlinear, time-invariant, finite memory, analytic system can be expressed as in Eq. (5).

Identification of the kernels can, in principle, be done in the time or frequency domain. Identification is discussed in Marmarelis and Marmarelis (1978) and Schetzen (1980). When the objective is to identify the truncated Volterra series for a system known, for example, by its differential equation, then one set of identification procedures might be followed. For example, realizations of the input and response might be generated from the differential equation, then used to identify the system kernels. When data from an experimental system are available to identify the system kernels, then system identification can be done, especially directly in the frequency domain. However, the higher-order kernels require a particularly large amount of data to identify (Hunter and Paez, 1987).

The Weiner series model for nonlinear structural system behavior is a generalization of the Volterra series to the case where the system inputs are wide-band, Gaussian random processes (see Wiener, 1942, 1958; Marmarelis and Marmarelis, 1978; Schetzen, 1980). Specifically, Wiener created a functional series model for nonlinear system response to random excitation, where each functional in the series is orthogonal to all lower-order functionals in the series. The functionals appear like the ones in Eq. (5), except that each term has added terms involving the same order kernel whose purpose is to enforce the orthogonality.

In the same way that a general, nonlinear model for system behavior can be specified, as above, a specific, parametric nonlinear model can also be specified. In fact, when some characteristic or characteristics of a system are known through observation of the physical system or through knowledge of the form of its differential equation, that information can often be used to specify a parametric form. The problem then becomes one of identifying the parameters of the model, and this is almost always easier than identifying system kernels of a Volterra model. For example, Zhu and Lei (1997) discuss the identification of nonlinear models. Ghanbari and Dunne (1998) consider a parametric model for damping.

A class of models that appears to be a special case of the Volterra series is that suggested by Bendat (1982, 1990, 1998) and Bendat and Piersol (1983). These comprise a set of robust and easy-to-use techniques, requiring response data. Bendat suggests that nonlinear structural models can be constructed by combining series of elements like those shown in Figure 1. In the diagram the first and final computational elements, denoted ZMNL, are zero memory, nonlinear operators. A ZMNL operator computes a function of its input at time t, to establish the output at time t. For example, some simple ZMNL operations on the signal $y = y(t)$ are:

$$g_1(y) = y, \quad g_2(y) = y^2, \quad g_3(y) = y|y|, \quad g_4(y) = y^3. \tag{6}$$

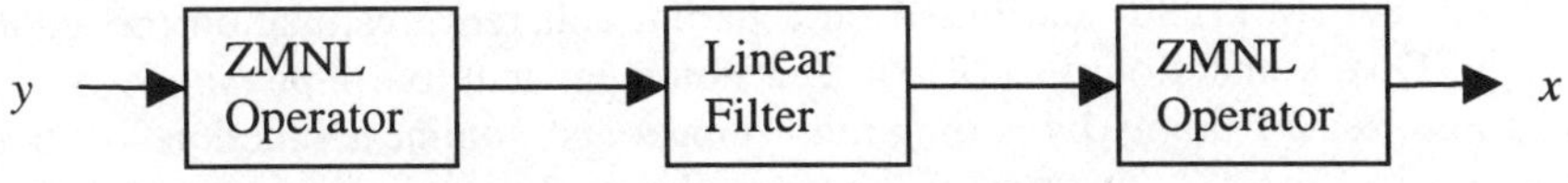

Figure 1. A series of operations on the signal $y = y(t)$, yielding $x = x(t)$.

The middle element in Figure 1 is linear filter operating on its input to yield a filtered output. For example, if we denote the Fourier transform of the filter input as $\eta(f), -\infty < f < \infty$, and the Fourier transform of the filter output as $\xi(f), -\infty < f < \infty$, then the filtering operation can be expressed as:

$$\xi(f) = H(f)\eta(f) \qquad -\infty < f < \infty, \tag{7}$$

where $H(f)$ is the filter transfer function. Bendat suggests that only one ZMNL operation normally be used in a series like the one shown in Figure 1 (though two such operations can be used).

Multiple series elements, as the ones shown in Figure 1, are used to model a structure input/output relation. For example, Figure 2 shows a three-element model that seeks to express the output as a function of linear, quadratic, and cubic terms.

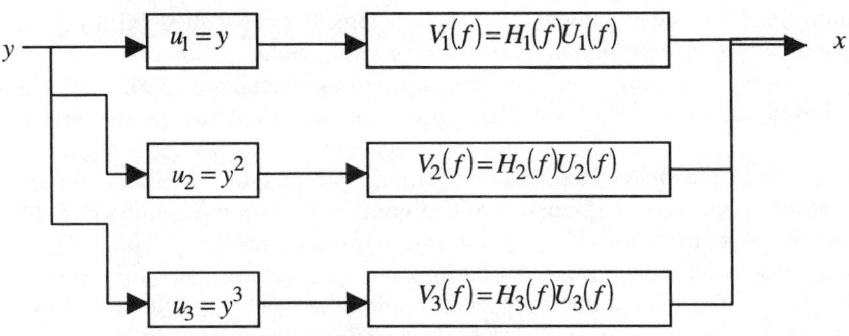

Figure 2. A nonlinear input/output model, expressing the response as a function of linear, quadratic, and cubic terms.

A model like the one shown in Figure 2 can be identified using a sequential application of the principles used in linear system identification.

The treatment developed above is simply a special use of the multiple input/single output system identification procedure first outlined in the mechanical system literature by Bendat and Piersol (1986) and based on a partial coherence estimation procedure developed by Dodds and Robson (1975). The nonlinear multiple input/single output problem can be treated simply by adding more inputs and nonlinear functions of them to the input set in Figure 2. The model shown in Figure 2 can also be augmented and modified via the inclusion of other nonlinear input forms.

In addition to models like the one shown in Figure 2, other types of models can be specified. For example, the ZMNL operation might follow the filtering operation. System identification follows the same general sequential procedure described above, in this case, except that the signal must be run from right to left through the ZMNL operators to infer the filter outputs for $H_2(f)$ and beyond. Richards and Singh (1998) and Zeldin and Spanos (1998a) consider the modeling of nonlinear systems with this type of model.

We showed in the previous section that there are techniques that can be used to bound the response of a linear system - the so-called least favorable response techniques - and, therefore, be used to find probability distributions that bound the peak response probability distribution. It so happens that Drenick (1977) published a paper showing that an analogous bound can be developed for nonlinear systems and that the bound on nonlinear system response is related to the behavior of a linearized version of the system. It appears that this result has never been used to establish a peak response probability distribution for nonlinear systems, though it could be.

A collection of methods known as reliability-based techniques have been used to solve problems not normally associated with random vibrations. However, recent investigations and the development of a finite element code called NESSUS have demonstrated that some problems connected with random vibrations, including nonlinear response problems, can be solved by these techniques. The subject of reliability-based techniques and the NESSUS code are discussed in the following section.

Researchers in the fields of random vibrations and chaotic vibrations appear, at least until recently, to have religiously avoided commenting on one anothers' fields in the frameworks of their own studies. This is an interesting omission in view of the fact that there appear to be many strong similarities between system behaviors in the two fields. Of course, the generation of pseudo-random numbers on digital computers during Monte Carlo analyses and random vibration tests uses carefully crafted procedures to generate numeric sequences that are simply chaotic realizations. Therefore, computed and experimental responses are, in these cases, simply functions of chaotic inputs. Still a few texts, including the ones by Ruelle (1991) and Schroeder (1991), have discussed these relations. The latter text discusses, among many other interesting subjects, the fractal character of Brownian motion. Among the many invariants used to characterize chaotic systems is the PDF. The PDF of a chaotic process can be estimated based on experimental data. Sums of chaotic processes obey the Central Limit Theorem. Further, many low-order chaotic processes transform into high-order chaotic, or random, processes with the variation of a fundamental parameter. These and many other related subjects are discussed by Eubank and Farmer (1990). Some other papers that discuss the relations between chaos and random vibrations are the ones by Lin and Yim (1996a, 1996b) and Feng and Pfeiffer (1998). Gregory and Paez (1990) and Paez and Gregory (1990) consider a chaotic system for its potential use in generating high frequency/high-level random environments.

Finally, we mention that a frequently used tool for the analysis of nonlinear random vibrations is the Fokker-Planck equation. The Fokker-Planck equation can be written for the transition probabilities of many structural systems. However, a closed form solution for system behavior, in the most general form, cannot always be practically obtained. Measures of system behavior, as the stationary state response probability distribution or moments of the response, can be obtained more easily. An early study in this area is that of Caughey (1963). Other examples of studies that involve the use of the Fokker-Planck equation are those by Parssinen (1998) and Jing and Young (1990).

5. ANALYSIS OF LARGE-SCALE SYSTEMS IN NUMERICAL ANALYSIS CODES/STOCHASTIC SYSTEM ANALYSIS

One of the areas of study in random vibrations that connects theoretical analysis techniques to the common application of random vibration in analysis, design, and diagnosis is the assessment of response character with computer codes meant to accommodate models of large systems. Many practical systems are analyzed using finite element codes today. Most commercial codes include the capability to perform the analysis symbolized in Eq. (2) for linear systems. Beyond this commercial finite element codes can be used to perform Monte Carlo analyses where the excitation, initial conditions, boundary conditions, and material properties are varied randomly, but such analyses must usually be performed simply by using the code as a deterministic function analyzer, and this may be difficult. Engelhardt (1999) shows how one code is being automated to permit Monte Carlo analysis. Other tools could be directly incorporated into existing finite element codes. For example, capability to analyze response of structures excited by nonstationary random excitations (Masri, Smyth, Traina, 1998). In fact, though, this would require very substantial augmentation of existing codes.

The material and/or geometric properties of many systems must be treated as random when they affect structural response at the same level as inputs, initial conditions, and boundary conditions. The analysis of such systems is more complex than the analysis of systems whose random response is caused only by random excitation. As noted above, deterministic finite element tools can be used to perform probabilistic analysis of systems with both random excitations and random structural properties using a Monte Carlo approach. However, as the random character of excitation and system properties becomes more complex, the number of Monte Carlo

270

simulations required for convergence of statistical measures of the response increases and may become prohibitive.

In view of this, there has been a continuing interest in the development of nonsampling techniques. An early approach to the analysis of complex structures was the perturbation method. The use of this method requires that coefficients in the system governing equations be separated into two parts: a mean, or deterministic part, and a random part. The random part of the governing equation is taken as a perturbation to the mean part; components of the response are matched to like-ordered components of the excitation, in the traditional way; and the equations of motion are solved. Some problems are that, first, only the first term in the perturbation can usually be maintained for reasons of computational complexity, thereby limiting potential accuracy of the series representation of the response, and second, convergence cannot typically be proven. Therefore, other techniques are required.

In the 1980s an entire class of techniques became available with the development of the reliability-based techniques, and these can be used for the analysis of large-scale probabilistic structural dynamics problems. Among these are first and second order reliability methods (FORM/SORM) (see Madsen, Krenk and Lind, 1986); fast probability integration (FPI) techniques (see Wu and Wirsching, 1987a); and the advanced mean value (AMV) technique (see Wu and Wirsching, 1987b).

The FORM is typical of other reliability-based techniques, and it seeks to approximately evaluate points on the cumulative probability distribution function (CDF) of a measure of the output of an arbitrary, deterministic function of a set of random variables. The function need not be explicitly defined; it may be a mathematical analysis computer code - e.g., a finite element code. The capability to approximately evaluate the CDF of a function of several random variables implies that we can characterize a measure of the response of a system excited by a dynamic excitation and, perhaps, with random system characteristics. The problem is posed in the following way. Let Y be an n-vector of random variables with known joint PDF, $f_Y(y)$; and let X be a scalar, deterministic function of Y:

$$X = g(Y) . \tag{8}$$

We seek the CDF of X, $F_X(x)$, at a set of points $x_j, j = 1, \ldots, N$. The CDF is defined

$$
\begin{aligned}
F_X(x) &= P(X \le x) \\
&= \int dy_1 \ldots \int_{g(y) \le x} dy_n \, f_Y(y) \cdot
\end{aligned}
\tag{9}
$$

The n-fold integral on the right of Eq. (9) is typically difficult to evaluate for arbitrary joint PDF, $f_Y(y)$. To simplify the approximation of the integral, we define a compound Rosenblatt transform (see Rosenblatt, 1952) from the domain of the PDF of the n-vector Y to the domain of the PDF of n uncorrelated, standard normal random variables denoted by the vector Z. We denote the transformation

$$Z = T(Y), \tag{10}$$

and note that it is based on the marginal CDF of Y_1, and the conditional CDFs of the Y_j given $Y_{j-1},\ldots,Y_1$, for $j = 2,\ldots,n$. The transformation is monotone increasing in each variable when the random variables in the vector Y are continuous valued; therefore, the transformation is invertible. Based on this, Eq. (9) can be rewritten

$$F_X(x) = \int dz_1 \ldots \int dz_n \, f_Z(z). \tag{11}$$
$$g\left(T^{-1}(z)\right) \leq x$$

The integrand is much simpler in this expression than Eq. (9). To permit the approximate evaluation of the integral, we find the location, z^*, on the constraint

$$g\left(T^{-1}(z)\right) = x, \tag{12}$$

where the norm of z, $\|z\|$, is minimum. This is the so-called design point, and we denote this distance to the origin as $\|z^*\| = \beta$. At this point we approximate the constraint, $g\left(T^{-1}(z)\right) \leq x$, with a version that is linearized in z, $\gamma_L(z) \leq x$; replace the integration limit in Eq. (11) with this approximate one; and evaluate the integral.

$$F_X(x) = \int dz_1 \ldots \int dz_n \, f_Z(z) = \Phi(\pm\beta), \tag{13}$$
$$\gamma_L(z) \leq x$$

where $\Phi(.)$ is the CDF of a standard normal random variable. The plus sign is taken on β when x is greater than the mean of $g(Y)$, and the minus sign is taken otherwise. The integral can be evaluated because it is the CDF of a weighted sum of uncorrelated, standard normal random variables; and any weighted sum of uncorrelated, standard normal random variables is simply a normal random variable. By repeating the procedure described in Eqs. (8) through (13) for $x = x_j, j = 1,...,n$, we approximately evaluate the CDF of the random variable, X, at these abscissa values.

Second order reliability methods work in precisely the same way as the FORM except that the linear approximation to the integration limit in Eq. (13) is replaced by a quadratic approximation. The advanced mean value technique works in the same way as the FORM except that we start the analysis with a linear approximation to $X = g(Y)$, and after completing the analysis make a correction that renders the result quite accurate. An iterative and more accurate, yet still very efficient, version of the AMV technique is available. (See Red-Horse and Paez, 1998.)

Normally we consider the solution of a random vibrations problem to be a probabilistic characterization of the temporal response. From this we normally derive the probabilistic character of critical response measures, like peak values, RMS values, first passage times, etc. The approaches described above yield results in a different form. Here we obtain directly the probabilistic character of response measures, and normally we cannot obtain the entire temporal response characterization. If the analyst requires the probabilistic characterization of the entire response, then one of the techniques would need to be adapted for that purpose. This appears not to have been done, though the Karhunen-Loeve approach for excitation modeling, specified above, connected with nonstationary response analysis, may be useful in this connection.

One example of a reliability-based finite element code is the computer code called NESSUS (Numerical Evaluation of Stochastic Structures Under Stress). It was developed at Southwest Research Institute (see NESSUS, 1996). It is based on the AMV technique and has been implemented in connection with multiple finite element codes including Sandia National Laboraories' PRONTO code (See Taylor and Flanagan, 1987). The code is implemented as a "wrap-around" in the sense that the finite element code is only called to perform evaluations like the ones in Eq. (8). The code has been used to solve nonlinear structural response problems with up to one million elements and involving up to 36 random variable inputs. The AMV technique is elucidated with example applications in Wu (1994) and Wu, Millwater and Cruse (1990).

A class of stochastic finite element techniques that takes a different approach than the reliability-based techniques is the one introduced by Ghanem and Spanos (1991). [See also Ghanem and Brzakala (1996) and Ghanem (1999).] It is what might be termed a random field method. It casts the coefficient in the equation of equilibrium as separable into deterministic and random parts. In addition, the excitation may be stochastic. The parameters that underlie the coefficient are assumed to be random fields with known autocorrelation functions. The eigenproblem is solved for the autocorrelation function of the underlying parameter; then the stochastic coefficient is replaced with its truncated Karhunen-Loeve expansion. Next, the response variable (defined as a stochastic field over the system analyzed) is expressed as a truncated

Karhunen-Loeve expansion with unknown coefficients and unknown eigenvectors. The coefficients in this expression are typically non-Gaussian distributed, and they are expanded in a homogeneous chaos - a multivariate Hermite polynomial. Finally, the eigenvectors in the response expression are expanded using a finite element basis. Both sides of the resulting expression are multiplied by an arbitrary term in the homogeneous chaos, and expected values are taken. The resulting approximate governing equations are collected into a block matrix equation. The dimension of one block in the set of block equations is the same as the dimension of the equivalent deterministic problem. The block dimension of the equilibrium equations is a function of the number of terms retained in the truncated Karhunen-Loeve expansion approximation to the coefficient random field and the order of the homogeneous chaos used in the representation of the response coefficients.

The objective in solution of the block equations is to find the amplitudes of the finite element shape functions associated with terms in the homogeneous chaos expansion for the response. Once these are obtained, the random field expression for the system response can be expressed, and this can be used to obtain arbitrary measures of the system response. For example, moments of any function of the response can be computed; further, marginal and joint probability distributions of the response can be obtained.

Other papers dealing with stochastic finite elements are those by Contreras (1980); Manohar and Adhikari (1998); Elishakoff, Ren, and Shinozuka (1996); Saigal and Kaljevic (1996).

An important problem that arises in connection with the stochastic finite element formulations is that of specifying and identifying the random fields that are used to characterize system properties. Some papers that deal with this issue are Zeldin and Spanos (1996, 1998b); Hoshiya and Yoshida (1996, 1998); and Noda and Hoshiya (1998).

The general problem described here is very practical in the sense that both excitation and material parameters often need to be considered as random. For this reason the field of stochastic finite elements and the implementation of stochastic capabilities into existing codes merits great attention.

6.   EXPERIMENTAL PROBABILISTIC STRUCTURAL DYNAMICS

Most of random vibration analysis is intended to realistically characterize the behavior of structural systems excited by random inputs; however, the responses of actual systems are only known when they are measured during the application of physical environments (and even then only approximately). The effects caused by the use of simplifying assumptions in the process of numerical simulation of physical systems are seldom evaluated. Most important, because of the depth and breadth of our capability to analyze linear systems subjected to stationary environments, we often idealize real systems as linear and inputs as stationary. Practically all real systems are nonlinear to a small or great extent; therefore, the response characterization can only be approximate.

Real physical system responses are measured, then characterized, in two frameworks. First, when real structural systems are designed, then built (or, perhaps, placed) in the field, their responses can be measured when subjected to natural or man-made environments. Second, when systems are of the appropriate size, they can be experimentally excited in the laboratory. Most laboratory tests of physical systems performed today are done on electrodynamic or electrohydraulic shakers and subject the system tested to a quasi-stationary random vibration environment.

Quasi-stationary random vibration tests to be performed on a shaker are controlled today using digital, closed loop control systems. The objective of a random vibration test is to maintain the autospectral density of a single measure of motion on or near a test system within some preestablished limits for a preestablished length of time. The algorithm used for control was derived in Tebbs and Hunter (1974). Prior work leading to this development is summarized in Hunter and Helmuth (1968) and Otts and Hunter (1970).

Before digital, closed loop control algorithms were available for the control of random vibration tests, analog, quasi-closed loop control systems that were operated manually were used. The control scheme employed during the 1950s and 1960s (and in some laboratories, the 1970s and even 1980s) is described by Metzgar (1958). The system used manual equalization to achieve the desired test spectral density. The purpose of the manual equalizer setting was to account for (a) the electro-mechanical frequency response function between the drive signal and the control point and (b) the shape of the desired spectral density at the control point. Because the electro-mechanical system, consisting of the signal generation electronics, the transmission system, the power amplifier and shaker, and the test item, is usually nonlinear, it was usually very difficult to establish and maintain the desired control. The control algorithm developed for digital control of random vibration tests is an attempt to mimic the analog control system.

With reference to Figure 3, the digital control algorithm consists of the following operations. Starting at top left a digital computer generates finite duration realizations of a stationary, Gaussian random process. Generated realizations are output to a digital-to-analog converter. The analog signal may then be filtered and amplified and transmitted to a shaker system. The signal is first amplified then used to drive the shaker. Motion is generated in the shaker armature. The test item is attached to the shaker armature, perhaps via a fixture or through a table that permits the testing of a much larger system. Control of the test is sought on or near the test item - control that is gauged in terms of the autospectral density of motion. Motion at the control point is measured, usually in terms of absolute acceleration. The motion is measured with a transducer, the output of which is amplified and filtered, then transmitted to the control digital computer. Before processing in the computer, the signal must be converted from analog to digital. The digitized signal is read by the control system. The continuously arriving signal is used to estimate the running, or real-time, spectral density (see Wirsching, Paez, and Ortiz, 1995).

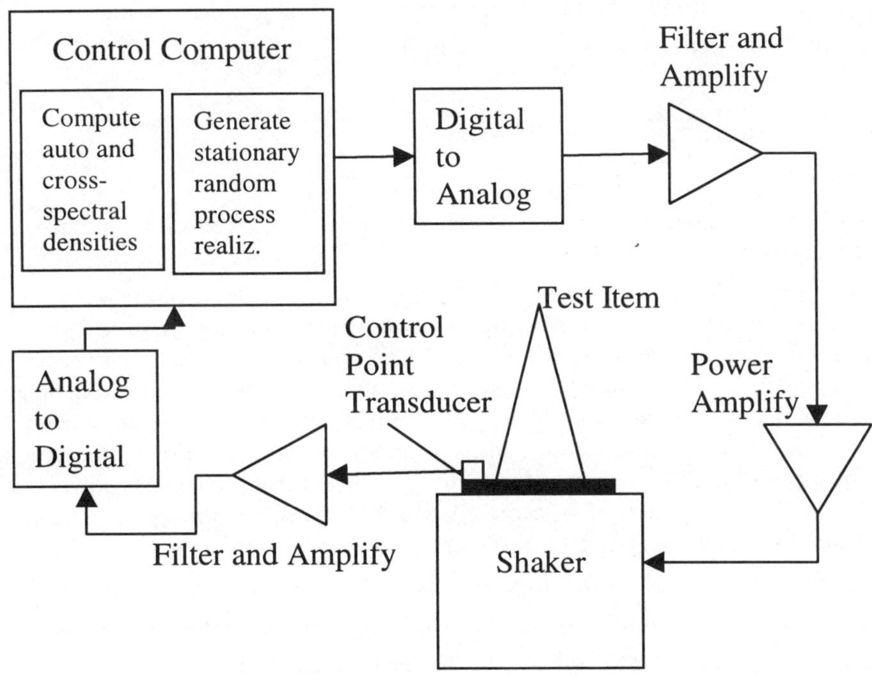

Figure 3. Elements and operations of the closed loop control system.

A critical element permitting realization of the digital control system is the ability to combine finite duration realizations of stationary random process into a continuous realization of stationary random process of arbitrary length. This is accomplished using a procedure called overlap processing, described in Gold and Rader (1969).

With the control system defined above, a random vibration test is performed following a two-step process. The first step is the system identification phase. In this phase a band-limited white noise drive signal is generated in the digital control computer and used to drive the shaker and the control point at a level that is low relative to the full level of the test to be performed. The drive signal and control point motion signals are stored and used to estimate the autospectral density function of the drive and the cross-spectral density function between the drive and the control. The latter is ratioed to the former to create an estimate of the electro-mechanical system frequency response function; denote this quantity $H(f_k), k = 0,...,n/2$.

The second phase involves the performance of the actual test. This is done in a sequence of steps where the excitation level is gradually increased until the full level of the test is realized. Denote the autospectral density desired at the control point with $G_{XX}(f_k), k = 0,...,n/2$; then the test is initiated by generating a drive signal from a Gaussian source with zero mean and spectral density

$$G_{DD}(f_k) = \alpha_0 \frac{G_{XX}(f_k)}{|H(f_k)|} \qquad k = 0,...,n/2. \tag{14}$$

This signal is meant to drive the system so that the motion at the control point has the spectral density $\alpha_0 G_{XX}(f_k), k = 0,...,n/2$. The constant α_0 is normally chosen so that the RMS motion is some number of decibels (db) below the full level of the test. Data are gathered at this level, and the drive autospectral density and the drive to control cross-spectral density estimates are initiated and then updated using real-time spectral density estimation. These are used to update the estimate of the electro-mechanical system frequency response function. When the test either equalizes at the current level - i.e., the estimated control autospectral density matches, within some limits, the target control autospectral density - or the test operator is satisfied that the test has equalized as well is it can, then the test level is increased, and the next step in the test is initiated. This is accomplished by changing the coefficient α_0 to α_1, where α_1 is some multiple of α_0. The running estimates of the drive autospectral density and the drive to control cross-spectral density are modified by the ratio α_1/α_0, and the test continues. When the motion finally reaches its full level, the test is allowed to continue for a preestablished duration or until the test operator aborts the test. Though there are many other details to the operation of a random vibration test, and there are many potential pitfalls, the foregoing description characterizes the fundamental ideas behind random vibration test operation.

The digital computers used today for random vibration control are many orders of magnitude faster than the ones used when digital control was first introduced in the early 1970s. Nevertheless, the algorithm used in most commercial control systems is substantially the same as the one proposed decades ago and described above. Because of limitations in the capabilities of the standard control system and because of hardware limitations, the typical random vibration test performed in the laboratory is controlled in a single axis. Because of the physical limitations associated with testing hardware on a shaker - i.e., one axis of controlled-input motion, test item attached to a very stiff armature, etc. - there are many situations where it is thought preferable to control not a single measure of motion in one axis and at one point, but rather a single measure of many motions at multiple points and, perhaps, in multiple directions. In these situations the test operator uses a control scheme known as average control or one known as extremal control (see Smallwood and Gregory, 1977). As their names imply, the former seeks to control the average autospectral density at a number of locations, and the latter seeks to control the maximal spectral density at a set of locations. These control schemes are used when knowledge of the autospectral density in an actual system is inaccurate, or when, because of physical testing limitations, a known autospectral density at a particular critical location will be surpassed when motion is controlled to a particular level at another point.

More than 99 percent of "state-of-the-art" random vibration testing is performed on a single shaker, using single-point control; therefore, it is fair to state that the limitations to this type of testing are real limitations to the actual state of the art. The fundamental limitations are that testing is performed in a single axis, and sometimes equalization cannot be achieved - i.e., motion with the desired autospectral density cannot be excited at the control point. The limitation to single-axis testing arises from two sources - economics and the difficulty of performing more realistic multi-axis/multi-shaker tests. The rationale underlying the first source is obvious as is its effect on testing realism. Few real random vibration environments are limited to motion that occurs in a single axis. During a single-axis controlled random vibration test, only one measure of motion is sought to be controlled, though motions in all axes occur. Sometimes, depending on the system under test, the off-axis motions have greater RMS value than the controlled motions. The off-axis motions that occur during performance of a single-axis random vibration test have autospectral densities that practically never match the autospectral densities of the system motions in the field. Further, the cross-spectral densities between motions realized during a single-axis test practically never match the cross-spectral densities between motions realized in the field.

The inability to generate motion at the control point with the desired autospectral density during a single-axis controlled random vibrations test arises from the nonlinearity of the electromechanical test system and the system under test. This occurs in, perhaps, half of all random vibration tests of complex systems. A frequent side effect of nonlinear response is the occurrence of harmonic distortion - i.e., the generation of response harmonics associated with strong motion response at a particular frequency, especially system resonances. These harmonics occur because during strong motion, system response excited by a simple harmonic signal is often not a simple sine, cosine, or combination of these. Because the algorithm used to control motion assumes that the system under test is linear, the response at the harmonic frequencies is misinterpreted and cannot always be controlled. In particular when the response at a higher harmonic has an autospectral density greater than the desired control autospectral density, then the motion at the control point will have an autospectral density that is out of tolerance on the high side, since the drive signal is not designed to diminish RMS motion at any frequency.

Clearly there are other limitations to random vibration testing in the laboratory. Among these are the force limits to any shaker. A large electrodynamic shaker might generate up to 50,000-lb force in the frequency range [5,5000] Hz; a large electrohydraulic actuator might produce the same force but in the frequency range [0,500] Hz. It would appear, of course, that if economics permit, many shakers can be used in parallel to overcome this limitation. Such appearance is not realized in real systems though. Multiple shakers tied together via a slip table or other fixture actually excite an elastic (or inelastic) system - the system that connects the shakers. The system has modes starting at a frequency that is a function of the size of the system and the acoustic velocity of the system materials (approximately the same for the materials used to construct armatures and fixtures). Large seismic simulations can have fundamental resonances below 100 Hz, and small component shakers can have fundamentals of 1000 to 2000 Hz. The control system must account for these in order to generate stable motion in the fixture/test article system. Separate control computers cannot, in general, be used to control the separate shakers; the system is sometimes unstable.

There are software plus hardware solutions to some of the problems mentioned above, although they are not widely applied, in practice, perhaps for economic reasons. One of these is the multi-axis/multi-shaker testing capability. This capability was developed by Smallwood (1982a, 1982b, 1999). It operates following the basic principles of Figure 3 except that multiple coherent drive signals are generated, these are separately conditioned, and separate power amplifiers are used to drive multiple shakers. The shakers connect to a test item or fixture at multiple points and perhaps in multiple directions. When N shakers are used to excite a system, then N^2 measures of system motion can, within certain constraints, be controlled (see Paez, Smallwood and Buksa, 1987). For example, in a three-axis test the autospectral density of motion in each of three axes can be controlled, as well as the real and imaginary parts of the cross-spectral densities between the pairs of motions; these are nine quantities. Of course, motions at all the control points are required, and the control computer must be capable of estimating the auto and cross-spectral densities of the control point motions.

The fundamental capabilities required to make the control algorithm work is the ability to generate multiple signals with arbitrary auto and cross-spectral densities and the ability to make the algorithm stable. The former capability is achieved via Cholesky or eigenvalue decomposition of the cross-spectral density matrix of the multiple drives. See Smallwood and Paez (1993) for details on the signal generation algorithm.

There are many less obvious shortcomings of standard laboratory random vibration tests. Among the most important is the high - effectively infinite - impedance of the shaker system. This means a shaker will exert as much force as is required to match the control autospectral density. Overtesting problems can arise in this connection, and force limiting must be imposed to achieve realistic tests. Scharton (1995), Chang and Scharton (1998), and Smallwood and Coleman (1993) treat this issue.

There are many other acitvities in probabilistic testing that merit our attention including the generation of non-Gaussian environments (see Smallwood, 1996); nonstationary random vibration and random shock testing (see Smallwood, 1973); and the control of nonlinear systems in random vibration testing.

7.   SUMMARY AND CONCLUSIONS

Some recent and earlier activities in the area of random vibration analysis and testing have been summarized in this contribution. A brief history was given, and linear random vibration was discussed. Some fundamental areas that merit continued investigation are: robust and convenient frameworks and algorithms for the analysis of nonstationary response of structures, general methods for the analysis of structural response to non-Gaussian excitations, improved techniques for the analysis of first passage probabilities of complex systems.

Some nonlinear models and computation of response measures for nonlinear systems were discussed. This field is wide open in the sense that there is an extremely wide variety of types of nonlinear behavior in real systems, and the modeling of almost all could stand improvement. Among many other things general models and techniques for nonlinear analysis are required; general techniques for analysis of large systems are required.

Large system analysis is normally performed today using finite element codes. There are many commercial and proprietary codes, most of which are limited to spectral density analysis - i.e., the computation of response spectral density, given input spectral density. The capabilities of these codes need to be broadened. Further, reliability-based codes that yield a more traditional random vibration response characterization need to be developed, and stochastic finite element codes need to be made practical.

Random vibration testing is the most practical of the areas discussed in this paper. There are many investigations that could improve the state of the art in testing. Some areas requiring development work are: random vibration control algorithms for nonlinear systems, procedures for nonstationary excitation identification and nonstationary testing, means for making multi-shaker/multi-axis testing more robust and economical, improved hardware and standard procedures for force controlled testing.

These are a few of the areas, among many others, that require the attention of investigators in the field of random vibrations.

280

ACKNOWLEDGEMENT
Sandia is a multiprogram laboratory operated by Sandia Corporation, a Lockheed Martin Company, for the United States Department of Energy under Contract DE-AC04-94AL85000.

REFERENCES
Augusti, G., Baratta, A., and Casciati, F., (1984), *Probability Methods in Structural Engineering*, Chapman and Hall, New York.

Bailer-Jones, C., MacKay, D., Withers, P., (1998), "A Recurrent Neural Network for Modeling Dynamical Systems," *Network: Comput. Neural Syst.*, V. 9, pp. 531-547.

Bendat, J., Piersol, A., (1982), "Spectral Analysis of Nonlinear Systems Involving Square-Law Operations," *Journal of Sound and Vibration*, V. 82, No. 2, p. 199.

Bendat, J., (1983), "Statistical Errors for Nonlinear System Measurements Involving Square-Law Operations," *Journal of Sound and Vibration*, V. 90, No. 2, p. 275.

Bendat, J., (1990), *Nonlinear System Analysis and Identification from Random Data*, John Wiley & Sons, New York.

Bendat, J., (1998), *Nonlinear System Techniques and Applications*, John Wiley & Sons, New York.

Bendat, J., Piersol, A., (1986), *Random Data: Analysis and Measurement Procedures*, 2nd Ed., Wiley-Interscience, New York.

Bolotin, V., (1984), *Random Vibration of Elastic Systems*, Martinus Nijhoff, The Hague, The Netherlands.

Caughey, T., (1963), "Derivation of the Fokker-Planck equation to Discrete Nonlinear Systems Subjected to White Random Excitation," Journal of the Acoustic Society of America, V. 35, No. 11, pp. 1683-1692.

Chance, J., Worden, K., Tomlinson, G., (1998), "Frequency Domain Analysis of NARX Neural Networks," *Journal of Sound and Vibration*, V. 213, No. 5, pp. 915-941.

Chang, K., Scharton, T., (1998), "Cassini Spacecraft Force Limited Vibration Testing," *Sound and Vibration*, pp. 16-20.

Coleman, J., (1959), "Reliability of Aircraft Structures in Resisting Chance failure," *Operations Research*, V. 7, No. 5, pp. 639-645.

Contreras, H., (1980), "The Stochastic Finite element Method," *Computes &Structures*, V. 12, pp. 341-348.

Crandall, S., (Ed.), (1958), *Random Vibration*, Technology Press of MIT and John Wiley and Sons, New York.

Crandall, S., (1958), "Statistical Properties of Response to Random Vibration," Chapter 4 in *Random Vibration*, S. Crandall, Ed. (1958).

Crandall, S., (Ed.), (1963), *Random Vibration*, MIT Press, Cambridge, MA.

Crandall, S., Mark, W., (1963), *Random Vibration in Mechanical Systems*, Academic, New York.

Dodds, C., Robson, J., (1975), "Partial coherence in Multivariate Random Processes," *J. Sound Vibrat.*, Vol. 42, pp. 243-247.

Doob, J., (1942), "The Brownian Movement and Stochastic Equations," *Annals of Mathematics*, Vol. 43, No. 2, pp. 351-369. Also reprinted in Wax (1954).

Drenick, R., (1970), "Model-Free Design of Aseismic Structures," *Journal of the Engineering Mechanics Division*, ASCE, V. 96, No. EM4, pp. 483-493.

Drenick, R., (1977), "The Critical Excitation of Nonlinear Systems," *Proceedings of the ASME Applied Mechanics Summer Conference*, ASME, New Haven, Connecticut.

Einstein, A., (1905), "On the Movement of Small Particles Suspended in a Stationary Liquid Demanded by the Molecular Kinetic Theory of Heat," *Annalen der Pyhsik*, V. 17, p. 549. Also, reprinted in Einstein (1956).

Einstein, A., (1956), *Investigations on the Theory of the Brownian Movement*, Dover Publications, New York, Edited by R. Furth.

Elishakoff, I., (1983), *Probabilistic Methods in the Theory of Structures*, Wiley, New York.

Elishakoff, I., Ren, Y., Shinozuka, M., (1996), "Variational Principles Developed for and Applied to Analysis of Stochastic Beams," *Journal of Engineering Mechanics*, V. 122, No. 6.

Engelhardt, C., (1999), "Random Vibration Analysis Using Statistically Equivalent Transient Analysis," *Proceedings of the International Modal Analysis Conference*, SEM, Kissimmee, Florida.

Eubank, S., Farmer, D., (1990), *An Introduction to Chaos and Randomness*, 1989 Lectures in Complex Systems, SFI Studies in the Sciences of Complexity, Lect. Vol. 11, Ed. Erica Jen, Addison Wesley.

Farmer, D., Sidorowich, (1988), "Exploiting Chaos to Predict the Future and Reduce Noise," *Evolution, Learning and Cognition*, World Scientific, Y. C. Lee, Ed.

Feng, Q., Pfeiffer, F., (1998), "Stochastic Model on a Rattling System," *Journal of Sound and Vibration*, V. 215.

Fokker, A., (1913), Dissertation, Leiden.

Ghanbari, M., Dunne, J., (1998), "An Experimentally Verified Non-Linear Damping Model for Large Amplitude Random Vibration of a Clamped-Clamped Beam," *Journal of Sound and Vibration*, V. 215, No. 2, pp. 343-379.

Ghanem, R., (1999), "Stochastic Finite Elements with Multiple Random Non-Gaussian Properties," *Journal of Engineering Mechanics*, V. 125, No. 1.

Ghanem, R., Spanos, P., (1991), *Stochastic Finite Elements: A Spectral Approach*, Springer-Verlag, New York.

Ghanem, R., Brzakala, W., (1996), "Stochastic Finite-element Analysis of Soil Layers with Random Interface," *Journal of Engineering Mechanics*, V. 122, No. 4.

Gnedenko, B., (1997), *Theory of Probability, 6th Ed.*, Gordon and Breach Science Publishers, UK.

Gold, B., Rader, C., (1969), *Digital Processing of Signals*, McGraw-Hill, New York.

Gregory, D., Paez, T., (1990), "Use of Chaotic and Random Vibrations to Generate High Frequency Test Inputs – Part I, The System," *Proceedings of the 36th Annual IES Technical Meeting*, Institute of Environmental Sciences, New Orleans, Louisiana, pp. 96-102.

Hoshiya, M., Yoshida, I., (1996), "Identification of Conditional Stochastic Gaussian Field," *Journal of Engineering Mechanics*, V. 122, No. 2.

Hoshiya, M., Yoshida, I., (1998), "Process Noise and Optimum Observation in Conditional Stochastic Fields," *Journal of Engineering Mechanics*, V. 124, No. 12.

Hunter, N. F., (1997) State Analysis of Nonlinear Systems Using Local Canonical Variate Analysis, Thirtieth Annual Hawaii Conference on System Sciences, 1997.

Hunter, N., (1998), "Bilinear System Character from Nonlinear Time Series Analysis," *Proceedings of the International Modal Analysis Conference*, Kissimmee, Florida.

Hunter, N., Helmuth, J., (1968), "Control Stabilization for Multiple Shaker Tests," *Shock and Vibration Bulletin.*

Hunter, N., Theiler, J., (1992), "Characterization of Nonlinear Input-Output Systems Using Time Series Analysis," *Proceedings of the first Experimental Chaos Conference*, Wiley.

Hunter, N., Paez, T., (1987), "Experimental Identification of Nonlinear Structural Models," *Proceedings of the International Modal Analysis Conference*, IES, Orlando, Florida.

Ibrahim, R., (1985), *Parametric Random Vibration*, Wiley, New York.

Iyengar, R., Roy, D., (1996), "Conditional Linearization in Nonlinear Random Vibration," *Journal of the Engineering Mechanics Division*, V. 122, No. 3, p. 197.

James, H., Nichols, N., Phillips, R., (Eds.) (1947), *Theory of Servomechanisms*, Radiation Laboratory Series, Vol. 25, MIT, McGraw-Hill, New York.

Jing, H-S., Young, M., (1990), Random Response of a Single-Degree-of-Freedom Vibro-Impact System with Clearance," *Earthquake Engineering and Structural Dynamics*, V. 19, pp. 789-798.

Karhunen, K., (1947), "Uber Lineare Methoden in der Wahrschienlichkeitsrechnung," *Amer. Acad. Sci., Fennicade, Ser. A., I*, Vol. 37, pp. 3-79. (Translation: RAND Corporation, Santa Monica, California, Rep. T-131, Aug. 1960)

Koopmans, L., Qualls, C., Yao, J., (1973), "An Upper Bond on the Failure Probability for Linear Structures," *Journal of Applied Mechanics*, ASME.

Laning, J., Battin, R., (1956), *Random Processes in Automatic Control*, McGraw-Hill, New York.

Larimore, W., (1983), "System Identification, Reduced Order Filtering and Modeling Via Canonical Variate Analysis," *Proceedings of the 1983 American Control Conference*, H. S. Rao and P. Dorato, Eds., 1982, pp. 445-451.

Lin, H., Yim, S., (1996a), "Nonlinear Rocking Motions. I: Chaos Under Noisy Periodic Excitations," *Journal of Engineering Mechanics*, ASME, V. 122, No. 8.

Lin, H., Yim, S., (1996b), "Nonlinear Rocking Motions. II: Overturning Under Random Excitations," *Journal of Engineering Mechanics*, ASME, V. 122, No. 8.

Lin, Y., (1967), *Probabilistic Theory of Structural Dynamics*, McGraw-Hill, New York. Republished in 1976 by Krieger, Huntington, New York.

Loeve, M., (1948), "Fonctiones Aleatores du Second Ordre," supplement to P. Levy, *Processus Stochastic et Mouvement Brownien*, Paris, Gauthier Villars.

Madsen, H., Krenk, S., Lind, N., (1986), *Methods of Structural Safety*, Prentice-Hall, Englewood Cliffs, NJ.

Manohar, C., Adhikari, S., (1998), "Statistics of Vibration Energy Flow in Randomly Parametered Trusses," *Journal of Sound and Vibration*, V. 217.

Marmarelis, P., Marmarelis, V., (1978), *Analysis of Physiological Systems: The White Noise Approach*, Plenum Press, New York.

Masri, S., Smyth, A., Traina, M., (1998) "Probabilistic Representation and Transmission of Nonstationary Processes in Multi-Degree-of-Freedom Systems," *Journal of Applied Mechanics*, ASME, Vol, 65, June, pp. 398-409.

Metzgar, K., (1958), "The Basis for the Design of Simulation Equipment," Chapter 10 in *Random Vibration*, S. Crandall, ed.

NESSUS (Reference Manual), (1996), Version 2.3, Southwest Research Institute, San Antonio, Texas.

Newland, D., (1984), *Random Vibrations and Spectral Analysis*, Longman, New York.

Nigam, N., (1983), *Introduction to Random Vibrations*, MIT Press, Cambridge, MA.

Noda, S., Hoshiya, M., "Kriging of Lognormal Stochastic Field," *Journal of Engineering Mechanics*, V. 124, No. 11.

Otts, J., Hunter, N., (1970), "Shock Reproduction on Shakers," *Instrumentation Society of America Transactions*.

Paez, T., Gregory, D., (1990), "Use of Chaotic and Random Vibrations to Generate High Frequency Test Inputs – Part II, Chaotic Vibrations," *Proceedings of the 36th Annual IES Technical Meeting*, Institute of Environmental Sciences, New Orleans, Louisiana, pp. 103-111.

Paez, T., Hunter, N., (1997), "Dynamical System Modeling Via Signal Reduction and Neural Network *Modeling,*" *Proceedings of the 68th Shock and Vibration Symposium*, SAVIAC, Baltimeor, Maryland.

Paez, T., Smallwood, D., Buksa, E., (1987), "Random Control at n^2 Points Using n Shakers," *Proceedings of the Institute of Environmental Sciences*, IES, pp. 271-275.

Pandey, M., Ariaratnam, S., (1996), "Crossing Rate analysis of NonGaussian Response of Linear Systems," *Journal of Engineering Mechanics*, V. 122, No. 6.

Parssinen, M., (1998), "Hertzian Contact Vibrations Under Random External Excitation and Surface Roughness," *Journal of Sound and Vibration*, V. 214, No. 4, pp. 779-783.

Phillips, R., (1947), "Statistical Properties of Time Variable Data," Chapter 6 in James, Nichols and Phillips (1947).

Planck, M., (1927), *Berl. Ber.*, p. 324.

Rayleigh, Lord, (1880), "On the Resultant of a Large Number of Vibrations of the Same Pitch and of Arbitrary Phase," *Philosophical Magazine*, V. 10, pp. 73-78.

Rayleigh, Lord, (1919a), "On the Problem of Random Vibrations, and of Random flights in One, Two, or Three Dimensions," *Philosophical Magazine*, V. 37, pp. 321-347.

Rayleigh, Lord, (1919b), "On the Resultant of a Number of Unit Vibrations, Whose Phases Are at Random Over a Range Not Limited to an Integral Number of Periods," *Philosophical Magazine*, V. 37, pp. 498-515.

Red-Horse, J., Paez, T., (1998), "Uncertainty Evaluation in Dynamic System Response," *Proceedings of the 16th International Modal Analysis Conference*, SEM, Santa Barbara, California.

Rice, S., (1944, 1945), "Mathematical Analysis of Random Noise," *Bell System Technical Journal*, V. 23, pp. 282-332, V. 24, pp. 46-156. Reprinted in Wax (1954).

Richards, C., Singh, R., (1998), "Identification of Multi-Degree-of-Freedom Non-Linear Systems Under Random Excitations by the "Reverse Path" Spectral Method," *Journal of Sound and Vibration*, V. 213, pp. 675-708.

Roberts, J., Spanos, P., (1990), *Random Vibration and Statistical Linearization*, Wiley, New York.

Robson, J., (1964), *An Introduction to Random Vibration*, Elsevier, New York.

Rojwithya, C., (1980), "Peak Response of Randomly Excited Multi-Degree-of-Freedom Structures," PhD Dissertation, The University of New Mexico.

Rosenblatt, M. (1952), "Remarks on a Multivariate transformation," *Annals of Mathematical Statistics*, **23**, 3, pp. 470-472.

Roy, V., Spanos, P., (1989), "Wiener-Hermite Functional Representation of Nonlinear Stochastic Systems," *Structural Safety*, V. 6, pp. 187-202.

Ruelle, D., (1991), *Chance and Chaos*, Princeton University Press, Princeton, New Jersey.

Saigal, S., Kaljevic, I., (1996), "Stochastic BEM – Random Excitation and Time-Domain Analysis," *Journal of Engineering Mechanics*, V. 122, No. 4.

Scharton, T., (1995), "Vibration-Test Force Limits Derived from Frequency-Shift Method," *AIAA Journal of Spacecraft and Rockets*, V. 32, No. 2, pp. 312-316.

Schetzen, M., (1980), *The Volterra and Wiener Theories of Nonlinear Systems*, Wiley, New York.

Schroeder, M., (1991), *Fractals, Chaos, Power Laws*, W.H.Freeman and Dompany, New York.

Schueller, G., Shinozuka, M., (1987), (Eds.), (1987), *Stochastic Methods in Structural Dynamics*, Martinus Nijhoff, Boston.

Schuster, A., (1906), "The Periodogram and Its Optical Analogy," *Proceedings of the Royal Society*, V. 77, pp. 136-140.

Shinozuka, M., (1970), "Maximum Structural Response to Seismic Excitations," *Journal of the Engineering Mechanics Division*, ASCE, V. 96, No. EM5, pp. 729-738.

Smallwood, D., (1973), "A Transient Vibration Test Technique Using Least favorable responses," *Shock and Vibration Bulletin*, No. 43, Part I, pp. 151-164.

Smallwood, D., (1982), "Random Vibration Testing of a Single Test Item with a Multiple Input Control System," *Proceedings of the IES Annual Meeting*, IES.

Smallwood, D., (1982), "Random Vibration Control System for Testing a Single Test Item with Multiple Inputs," *Advances in Dynamic Analysis and Testing*, SAE Publication SP-529, Paper No. 821482.

Smallwood, D., (1996), "Generation of Partially Coherent Stationary Time Histories with non-Gaussian Distributions," *Proceedings of the 67th Shock and Vibration Symposium*, Vol. 1, pp. 489-498.

Smallwood, D., (1999), "Multiple Shaker Random Vibration Control - An Update," *Proceedings of the IEST Speint Meeting*, IEST, Los Angeles.

Smallwood, D., Coleman, R., (1993), "Force Measurements During vibration Testing," *64th Shock and Vibration Symposium*, SAVIAC.

Smallwood, D., Gregory, D., (1977), "Bias errors in Random Vibration Extremal Control Strategy," *Shock and Vibration Bulletin*, No. 50, Part II.

Smallwood, D., Paez, T., (1993), A Frequency Domain Method for the Generation of Partially Coherent Random Signals," *Shock and Vibration*, Vol. 1, No. 1, pp. 45-53.

Smoluchowski, M., (1916), *Phys. Zeits.*, V. 17, p. 557.

Soong, T., Grigoriu, M., (1993), *Random Vibration of Mechanical and Structural Systems*, Prentice-Hall, Englewood Cliffs, NJ.

Taylor, L., Flanagan, D., (1987), "PRONTO 3D: A Three-Dimensional Transient Solid Dynamics Program," Sandia Report SAND87-1912, Albuquerque, New Mexico.

Tebbs, J., Hunter, N., (1974), "Digitally Controlled Random Vibration Tests on a Sigma V Computer," *Proceedings of the Institute of Environmental Sciences Meeting*, pp. 36-43.

Uhlenbeck, G., Ornstein, L., (1930), "On the Theory of the Brownian Motion," *Physical Review*, V. 36, pp. 823-841.

Urbina, A., Hunter, N., Paez, T., (1998), "Characterization of Nonlinear Dynamic Systems Using Artificial Neural Networks," *Proceedings of the 69th Shock and Vibration Symposium*, SAVIAC, St. Paul, Minnesota.

Volterra, V., (1959), *Theory of Functionals and of Integral and Integro-Differential Equations*, Dover Publications, New York.

Wang, M., Uhlenbeck, G., (1945), "On the Theory of Brownian Motion II," *Reviews of Modern Physics*, V. 17, Nos. 2 and 3, pp. 323-342.

Wax, N. (ed.), (1954), *Selected Papers on Noise and Stochastic Processes*, Dover Publications, New York.

Wiener, N., (1923), "Differential Space," *J. Math, Phys.*, V. 2, pp. 131-174.

Wiener, N., (1930), "Generalized Harmonic Analysis," *Acta Mathematica*, V. 55, No. 118.

Wiener, N., (1942), "Response of a Nonlinear Device to Noise," Report No. 129, Radiation Laboratory, MIT, Cambridge, Massachusetts.

Wiener, N., (1958), *Nonlinear Problems in Random Theory*, Wiley, New York.

Wirsching, P., Paez, T., Ortiz, K., (1995), *Random Vibrations: Theory and Practice*, Wiley, New York.

Wu, Y. -T., (1994), "Computational Methods for Efficient Structural Reliability and Reliability Sensitivity Analysis," *AIAA Journal*, Vol. 32, No. 8, pp. 1717-1723.

Wu, Y. -T., Millwater, H., Cruse, T., (1990), "An Advanced Probabilistic Structural Analysis Method for Implicit Performance Functions," *AIAA Journal*, Vol. 28, No. 9.

Wu, Y. –T., Wirsching, P., (1987a), "Demonstration of a New Fast Probability Integration Method for Reliability Analysis," *Journal of Engineering for Industry*, V. 109.

Wu, Y. -T., Wirsching, P. H. (1987b), "A New Algorithm for Structural Reliability Estimation," *Journal of the Engineering Mechanics Division*, ASCE, **113**, 9, pp. 1319-1334.

Yang, C., (1986), *Random Vibration of Structures*, Wiley, New York.

Zeldin, B., Spanos, P., (1996), "Random Field Representation and Synthesis Using Wavelet Bases," *Journal of Applied Mechanics*, V. 63.

Zeldin, B., Spanos, P., (1998a), "Spectral Identification of Nonlinear Structural Systems," *Journal of Engineering Mechanics*, V. 124, No. 7.

Zeldin, B., Spanos, P., (1998b), "On Random Field Discretization in Stochastic Finite Elements," *Journal of Applied Mechanics*, V. 65.

Zhu, W., Lei, Y., (1997), "Equivalent Nonlinear System Method for Stochastically Excited and Dissipated Integrable Hamiltonian Systems," *Journal of Applied Mechanics*, V. 64.

SIGNAL PROCESSING IN VIBRATION ANALYSIS

R. B. Randall
School of Mechanical and Manufacturing Engineering
The University of New South Wales
Sydney 2052, Australia

1.   INTRODUCTION

Signal processing in vibration analysis can be divided into two main categories: single channel or "**signal analysis**", and multiple channel, mainly used for "**system analysis**". Vibration signals result from the action of forces on structures, and are a combination of the characteristics of the forcing function and the structural dynamic properties of the structure. Where one forcing function dominates, in so-called SIMO (single input, multiple output) situations, and where linearity applies as in a very large number of cases where stresses are below yield, the forcing function and structural response effect (transfer function) between the points of force application and response are convolved in the time domain, multiplied in the frequency domain and additive on taking logs of the frequency functions. This simplifying effect is one of the major reasons for the widespread use of **frequency analysis** (Fourier analysis) in processing vibration signals. A consequential property of linear systems is that when excited at one frequency, they respond only at that frequency, which is another reason for using frequency analysis. Even when the system is slightly nonlinear, this often just means that the response becomes distorted and higher harmonics (multiples) of the forcing frequency are generated, which are very easily related to the fundamental frequency. A further very important reason for the widespread use of frequency analysis is the ease with which it can now be carried out since the introduction of the so-called "**fast Fourier transform**" (FFT) algorithm in 1965 [1.1] which is a very efficient way of calculating the "**discrete Fourier transform**" (DFT).

Signal analysis is often carried out in cases where the forcing functions cannot be measured, but changes in the response signals are assumed to be dominated either by the forcing function or the structural response, based on other criteria. An example is the condition monitoring or diagnostics of operating machines. Even though frequency analysis using the FFT is the most commonly applied technique, other signal analysis procedures have advantages in certain circumstances as discussed below. For short signals, better frequency resolution can be obtained using parametric analysis techniques such as the **maximum entropy method** [1.2]. Since basic Fourier analysis assumes infinitely long time records, information is lost about time localisation of events. Analysis in both time and frequency can be achieved by moving a finite time window along a signal and calculating the frequency spectrum for each position (so-called **short time Fourier transform** - STFT). Because of the property of the DFT that the frequency resolution is the reciprocal of the length of record transformed, the better the time resolution the poorer the frequency resolution and vice versa. Techniques have been developed to improve the combined resolution in time-frequency analysis, based on the **Wigner-Ville distribution** [1.3, 1.4]. Another time-frequency analysis technique which has become increasingly used is "**wavelet analysis**" [1.5], where signals are decomposed in terms of a family of wavelets (of which there are many types) where the basic wavelet can be both translated and dilated in time. The dilation typically occurs in octave (2:1) steps so that the "scale", which represents frequency, is naturally a logarithmic axis. This means that the time resolution of a wavelet analysis is much better at high scales or frequencies. The same applies to **octave-based analysis** with 1/n-octave filters such as the 1/3-octave filters which have long been used in the analysis of acoustic signals. This is because the human ear tends to interpret equal changes on a logarithmic scale (eg octaves) as equal intervals, and the same applies to the response of the human body to vibrations, where standards recommend analysis in 1/3-octave bands. Before the advent of the FFT algorithm, most frequency analysis was done with analogue filters, measuring the power of time signals transmitted by a series of band-pass filters each covering part of the frequency range to be analysed. Although most narrow band analysis is now done digitally using the FFT, **digital filters** working in the time domain in a similar manner to their analogue counterparts have advantages in certain situations, in particular for constant percentage bandwidth (1/n-octave) analysis over a wide frequency range on a logarithmic scale.

Some effects are more easily interpreted in the time domain rather than the frequency domain, and there are a number of techniques to enhance such effects. An example is **autocorrelation analysis** which gives a measure of how well a signal correlates with delayed versions of itself. This can be useful for enhancing periodicity and echoes, for example. A related time domain function is the **cepstrum**, which in general is better for detecting echoes, and which has two other main applications in vibration analysis:

- Detecting families of equally spaced components in a spectrum such as harmonics and sidebands
- Helping to separate forcing and transfer functions in response signals as they are additive in the cepstrum.

A very important time domain technique is **synchronous averaging**, where signal sections are averaged together synchronously with a timing signal, for example a once-per-rev tacho signal, to enhance events occurring at that repetition rate. The best known example is the application to gears, where the signals from a pair of gears in mesh can be separated from each other (and background noise) by averaging synchronously with each gear in turn. In signals from rotating machines synchronous averaging must often be combined with so-called "**order analysis**" to remove the effects of speed variation by sampling the signals synchronously with the speed of the machine. This is now most efficiently done by **digital interpolation** (resampling) techniques. Apart from eliminating the effects of small speed fluctuations, this process can also be used to study how the vibrations at the various harmonic orders vary over a wide speed range such as during a machine run-up or coast-down. Synchronous averaging is most useful when signals are exactly periodic (or synchronous); however, if they are only approximately periodic but their autocorrelation function is periodic, then their **cyclostationarity** and **spectral correlation** properties may be of interest.

Other effects in vibration signals are often due to modulation, either amplitude or phase modulation, so that the signals obtained by **demodulation** are of more interest than the raw signals. An example where **amplitude demodulation** is of interest is in "envelope analysis" of signals from faulty rolling element bearings, where the series of impulse responses due to impacts with the fault contain the diagnostic information in the envelope signal, obtained by amplitude demodulation. As regards **phase demodulation**, this is useful in the measurement of torsional vibration of machines, which is a phase modulation if expressed in terms of shaft angular displacement, and a frequency modulation if expressed in terms of angular velocity. Demodulation is readily carried out using **Hilbert transform techniques**, which have a close relationship with Fourier analysis.

In multiple channel analysis, one or more forces are applied (and measured) by the user, and when combined with the corresponding response measurements from all over a structure can be used to determine the dynamic properties of the structure. Once again the most commonly applied techniques are based on the FFT process to extract **frequency response functions** (FRFs) from which a complete scaled modal model can be determined. FRFs are only strictly valid for linear systems and a measure of the degree of linearity of the relationship between the forcing function and response is given by the **coherence**. For calculation of FRFs and coherence it is necessary to measure cross spectra between pairs of signals, the **cross spectrum** being the Fourier transform of the **cross correlation function**, similar to the autocorrelation function but relating one signal to a delayed version of the other. The inverse Fourier transform of an FRF is the corresponding **impulse response function**, which represents the system dynamic properties in the time domain. In some situations, such as when the vibration signals represent "free decay" after all forcing functions have been removed, then other techniques can be applied to determine the modal properties of a structure, such as the "**Ibrahim time domain**" (ITD) method [1.6] and **Prony** method. The **complex cepstrum** and **differential cepstrum** have also been found useful for determining modal properties of structures based on response measurements only, because of the previously mentioned property of the cepstrum that forcing and transfer function components in response signals are additive.

In what follows, the above highlighted techniques are explained in somewhat more detail, after a discussion of some basic theoretical concepts. However, the analysis of multiple channel signals for modal analysis is only treated sketchily as it is the subject of other specialist papers.

2.   BASIC THEORY

2.1   The Fourier Transform

2.1.1 Fourier Series

The basic concept of Fourier analysis is to express signals as a summation of sinusoidal components, and with few exceptions virtually all signals can be decomposed in this way. Fourier's original analysis (now called Fourier Series) was applied to finite length signals (or periodic signals, as the resulting solution is periodic with the finite length as period). In vibration analysis it is used almost exclusively for periodic signals, as produced by a machine rotating at constant speed. Thus, for any periodic signal $g(t)$ of period T for which $g(t) = g(t+T)$ it can be shown that:

$$g(t) = \frac{a_0}{2} + \sum_{k=1}^{\infty} a_k \cos(k\omega_0 t) + \sum_{k=1}^{\infty} b_k \sin(k\omega_0 t)$$

(2.1)

where ω_0 is the fundamental angular frequency in rad/s ($= 2\pi / T$) and where the coefficients of the cosine and sine terms can be obtained by correlating them with $g(t)$, as follows:

$$a_k = \frac{2}{T} \int_{-T/2}^{T/2} g(t)\cos(k\omega_0 t)dt$$

(2.2)

$$b_k = \frac{2}{T} \int_{-T/2}^{T/2} g(t)\sin(k\omega_0 t)dt$$

(2.3)

Thus, the total component at frequency $\omega_k (= k\omega_0)$ is given by:

$$a_k \cos(\omega_k t) + b_k \sin(\omega_k t) \qquad \text{which can alternatively be written as:}$$

$$C_k \cos(\omega_k t + \phi_k)$$

(2.4)

where $\quad C_k = \sqrt{a_k^2 + b_k^2} \qquad$ and $\qquad \phi_k = \tan^{-1}(b_k / a_k)$

Expression (2.4) can again be represented as:

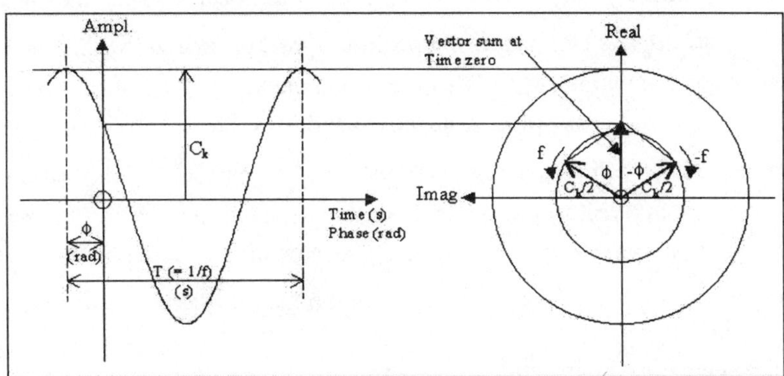

Figure 2.1. Representation of a sinusoid as a sum of two rotating vectors

$$(2.5)$$

$$\frac{C_k}{2}[\exp\{j(\omega_k t + \Phi_k)\} + \exp\{-j(\omega_k t + \Phi_k)\}]$$

which can be interpreted as two rotating vectors, each of length $C_k/2$, one rotating at angular frequency ω_k with initial phase ϕ_k and the other rotating at angular frequency $A_k \cos(k\omega_0 t + \phi_k)$ with initial phase $-\phi_k$, as illustrated in Fig.2.1.

Using this interpretation of Fourier analysis as representing $g(t)$ as a sum of rotating vectors leads to the alternative version of Equ.(2.1) as:

$$g(t) = \sum_{k=-\infty}^{\infty} A_k \exp(j\omega_k t) \qquad (2.6)$$

where the coefficients A_k are now complex and incorporate the phase shift in the form

$$A_k = \frac{C_k}{2} \exp(j\phi_k) \qquad (2.7)$$

The equation for calculating the coefficients A_k (equivalent to Equ.(2.2, 2.3)) now becomes:

$$A_k = \frac{1}{T} \int_{-T/2}^{T/2} g(t) \exp(-j\omega_k t)\, dt \qquad (2.8)$$

292

This has the simple physical interpretation that multiplication by $\exp(-j\omega_k t)$ subtracts angular frequency ω_k from each component, meaning that the one originally rotating at ω_k is stopped in the position it had at time zero (this then being extracted by the integral) while all other components still rotate at some other multiple of ω_k (either positive or negative) and thus integrate to zero over the periodic time. Thus each frequency component A_k represents the position (and value) of the rotating vector at time zero, so that to obtain its position at any other time t it is necessary to cause it to rotate at angular frequency ω_k by multiplying by $\exp(j\omega_k t)$. Summing then over all frequency components gives Equ.(2.6).

Before leaving this interpretation of Fourier series analysis as a sum of rotating vectors each of amplitude half that of the corresponding sinusoid (ie $C_k/2$), but having a two-sided spectrum in that each positive frequency component is accompanied by its complex conjugate at negative frequency, it can be seen that the same result can be achieved by retaining the positive frequency components only, but doubling their length to C_k and then taking the projection of each vector on the real axis. This is illustrated in Fig. 2.2. A signal with a one-sided spectrum like this is complex (and known as an **analytic signal**) but it will later be shown that the projection on the imaginary axis is the Hilbert transform of the real part. This applies equally to the vector sum of a number of components as to each individual component.

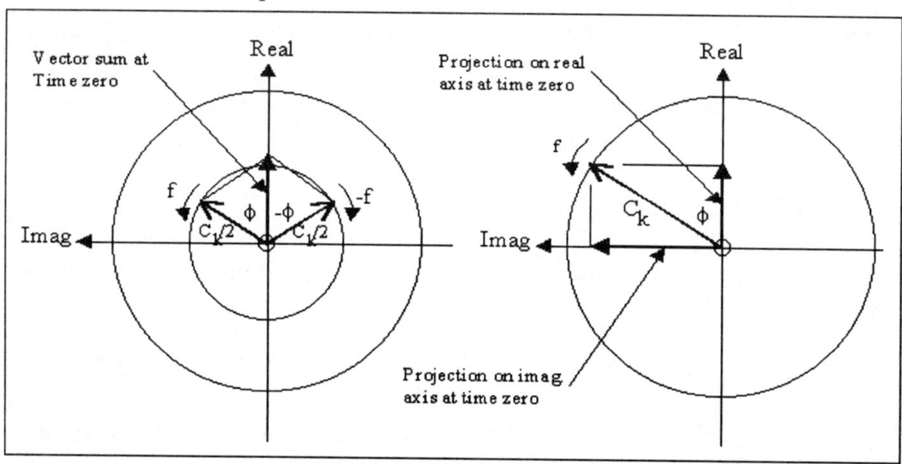

Figure 2.2. Equivalence of vector sum and projection on the real axis

2.1.2 Fourier integral transform

Signals other than periodic can also be expressed as a sum of complex exponentials, in particular transients to which the so-called Fourier transform applies. The Fourier transform can be derived from the Fourier series by allowing the periodic time to tend to infinity and at the same time removing the division by T because transients have finite energy rather than finite power. Equations (2.8) and (2.6) then become:

$$G(f) = \int_{-\infty}^{\infty} g(t)\ \exp(-j2\pi ft)\ dt \tag{2.9}$$

$$g(t) = \int_{-\infty}^{\infty} G(f)\ \exp(j2\pi ft)\ df \tag{2.10}$$

respectively, where angular frequency ω_k in rad/s has been replaced by the continuous frequency function f expressed in Hz. Equations (2.9) and (2.10) are known as the forward and inverse Fourier (integral) transforms, respectively.

At this point it is convenient to highlight the connection between the Fourier transform of Equ.(2.9) and the Laplace transform, defined by:

$$G(s) = \int_{0}^{\infty} g(t)\ \exp(-st)\ dt \tag{2.11}$$

where s is a complex variable which can be represented as $\sigma + j\omega$ in terms of its real and imaginary parts. Thus, for impulse response functions, which are necessarily causal (ie they do not exist for negative time) their Fourier transform, known as the frequency response function (FRF) is equal to their Laplace transform (transfer function) for the special case that s is restricted to the imaginary axis ($s = j\omega = j2\pi f$).

Note the symmetry between the forward and inverse transforms of Equ.(2.9, 2.10), the only difference being the sign of the exponent. This means that, most often, results which apply to the forward transform also apply to the inverse transform, for example the convolution theorem to be discussed below. This is particularly the case for real, even functions, for which it makes no difference if time or frequency run forwards or backwards.

2.1.3 Sampled time signals

All signals which are to be processed digitally have to be digitised or discretely sampled. As will be seen in Fig.2.3(c) [2.1], this is the inverse case of the Fourier series (Fig.2.3(b)), where the spectrum is discretely sampled, and the symmetry of the Fourier transform means that the spectrum of a sampled time signal is periodic. The corresponding versions of the forward and inverse transforms are:

$$G(f) = \sum_{n=-\infty}^{\infty} g(t_n) \exp(-j2\pi f\, t_n) \tag{2.12}$$

$$g(t_n) = \frac{1}{f_s} \int_{-f_s/2}^{f_s/2} G(f) \exp(j 2\pi f\, t_n)\, df \tag{2.13}$$

where $t_n = n\, \Delta t = n/f_s$

2.1.4 The discrete Fourier transform (DFT)

The sampled time signals in 2.1.3 are in principle of infinite length, but when the record length is finite, this leads to the same situation as with the Fourier series in that the spectrum is discrete and the time record implicitly periodic. As seen in Fig.2.3(d), this leads to a combination of the cases of Fig.2.3(b, c) so that both the time record and frequency spectrum are discretely sampled and periodic. The continuous infinite integrals of the Fourier transform become finite sums, usually expressed as:

$$G(k) = 1/N \sum_{n=0}^{N-1} g(n) \exp(-j 2\pi kn/N) \tag{2.14}$$

$$g(n) = \sum_{k=0}^{N-1} G(k) \exp(j 2\pi kn/N) \tag{2.15}$$

This version corresponds most closely to the Fourier series in that the forward transform is divided by the length of record N to give correctly scaled Fourier series components. If the DFT is used with other types of signals, eg transients or stationary random signals, the scaling must be adjusted accordingly as discussed below. Note that with the very popular signal processing package Matlab®, the division by N is done in the inverse transform, which requires scaling in every case, as even though the forward transform is then closer to the Fourier integral, it still must be multiplied by the discrete equivalent of dt.

Note also that for convenience, the time and frequency zero positions have been shifted from the centre of the record to the beginning, but because of the implicit periodicity this just means that the second half of each record represents the negative axes (see Fig.2.3(d)).

The forward DFT operation can be understood as the matrix multiplication:

$$\mathbf{G_k} = \frac{1}{N} \mathbf{W_{kn}} \mathbf{g_n} \tag{2.16}$$

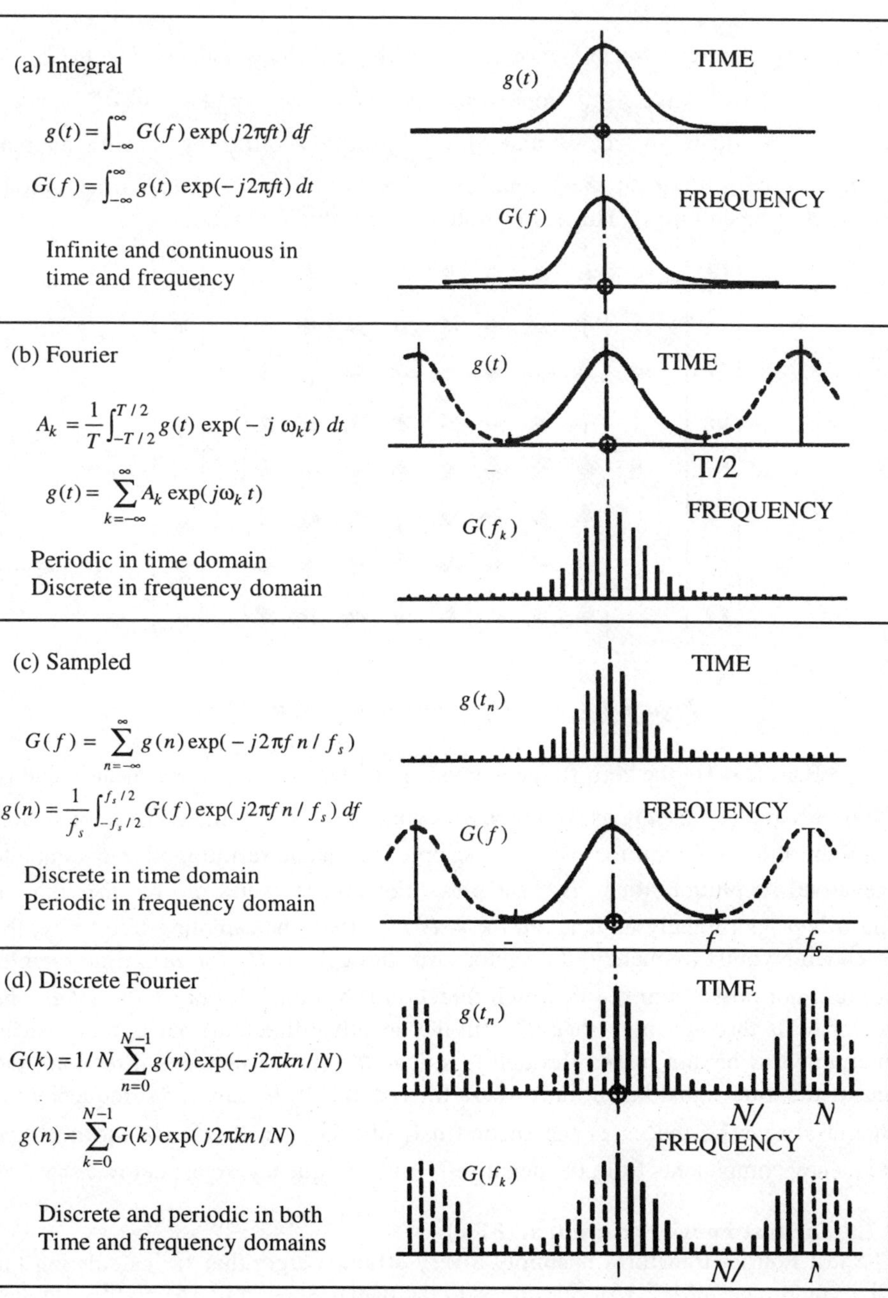

Figure 2.3. Various forms of the Fourier transform [2.1](a) Fourier integral transform (b) Fourier series (c) Sampled functions (d) Discrete Fourier transform

296

where $\mathbf{G_k}$ represents the vector of N frequency components, the $G(k)$ of Equ.(2.14, 2.15), while $\mathbf{g_n}$ represents the N time samples $g(n)$. $\mathbf{W_{kn}}$ represents a square matrix of unit vectors $\exp(-j2\pi kn/N)$ with angular orientation depending on the frequency index k (the rows) and time sample index n (the columns). This is illustrated graphically in Fig.2.4.

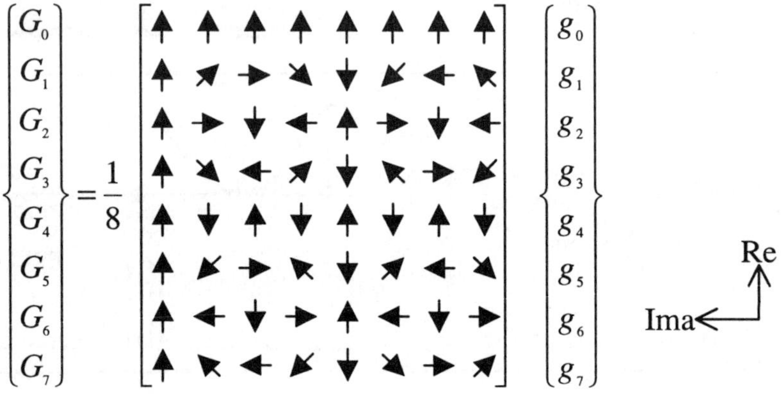

$$\begin{Bmatrix} G_0 \\ G_1 \\ G_2 \\ G_3 \\ G_4 \\ G_5 \\ G_6 \\ G_7 \end{Bmatrix} = \frac{1}{8} \begin{bmatrix} \cdots \end{bmatrix} \begin{Bmatrix} g_0 \\ g_1 \\ g_2 \\ g_3 \\ g_4 \\ g_5 \\ g_6 \\ g_7 \end{Bmatrix} \qquad \begin{array}{c} \text{Re} \\ \uparrow \\ \text{Ima} \leftarrow \end{array}$$

Figure 2.4. Matrix representation of the DFT

For $k = 0$ the zero frequency value $G(0)$ is simply the mean value of the time samples $g(n)$ as would be expected. For $k = 1$ the unit vector rotates - $1/N$ th of a revolution for each time sample increment, resulting in one complete (negative) revolution after N samples. For higher values of k the rotation speed is proportionally higher. For $k = N/2$ (half the sampling frequency, the so-called Nyquist frequency) the vector turns through $-\pi$ for each time sample, but it is not possible to see in which direction it has turned. For $k > N/2$ the vector turns through more than π (in the negative direction) but is more easily interpreted as having turned through less than π (in the opposite direction) and thus if the time signal has been lowpass filtered at half the sampling frequency (as should always be the case) the second half of $\mathbf{G_k}$ will contain the negative frequency components from the negative Nyquist frequency to just below zero.

2.1.5 The fast Fourier transform (FFT)

The fast Fourier transform is simply a very efficient algorithm for calculating the DFT equations (2.14, 2.15). Starting with the matrix version (2.16), in the simplest form (the so-called radix 2 algorithm) the FFT is based on N being a power of 2, and factorises the matrix $\mathbf{W_{kn}}$ into $\log_2 N$ matrices each with the property that multiplication by them only requires N complex operations as compared with the

N^2 operations required for direct multiplication by $\mathbf{W_{kn}}$. Thus the total number of complex operations is reduced from N^2 to $N \log_2 N$, a saving by a factor of more than 100 for the typical case where $N = 1024 \quad (= 2^{10})$. Other savings can be made in special cases, and a similar but not so effective gain can be made for factorisation other than in powers of 2, but the main point is that the properties of the FFT are those of the DFT.

2.2 The Convolution Theorem

This very important theorem states that a Fourier transform in either direction converts a multiplication into a convolution and vice versa. One example already mentioned is that the convolution between an input force and the impulse response of a structure results in the frequency domain in a product of their respective Fourier transforms to form the Fourier transform of the response vibration. An example of the inverse is that multiplication of a time signal by a windowing function results in the frequency domain in a convolution of their respective Fourier transforms.

The convolution operation represents a moving average or swept filtration of one function by the other reversed, as expressed by the Duhamel integral:

$$x(t) = \int_{-\infty}^{\infty} g(\tau)h(t - \tau)\, d\tau \qquad \text{represented as}$$

$$x(t) = g(t) * h(t) \tag{2.17}$$

When one of the functions is a Dirac delta function (or impulse function), with or without scaling and physical dimensions, the convolution operation is very simple and involves shifting the origin of the convolving function to the position of the delta function and multiplication by the scaling factor. Note that the scaling factor can be complex and thus involve a phase shift as well as a gain factor. In what follows, considerable use will be made graphically of these convolution relationships.

2.2.1 Fourier Series from the Fourier transform

As a first example it can be seen that any periodic signal can be generated by convolving the transient representing one period with a train of delta functions with spacing T, the periodic time. It can be shown that the Fourier (Series) transform of the latter is likewise a train of delta functions with spacing $1/T$ but also value $1/T$, so that the Fourier series is obtained by multiplying the Fourier transform of the transient by the train of delta functions (thus sampling it at the harmonics of the fundamental frequency) and scaling the samples by the factor $1/T$. Note that at the same time this corrects the physical dimensions of the result so that the Fourier series components have the same units as the original signal.

Figure 2.5 shows an example of how the Fourier series of a full-wave rectified cosine can be obtained from the Fourier transform of a single half-cosine pulse. To obtain the Fourier series of a half-wave rectified cosine it is simply necessary to sample the spectrum at intervals of $1/2T$ and multiply by $1/2T$.

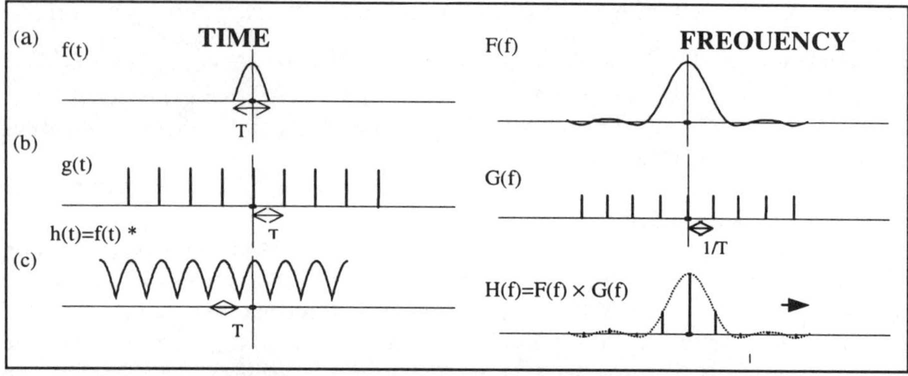

Figure 2.5. Determining a Fourier series from a Fourier transform
(a) half-cosine pulse and its Fourier transform
(b) train of delta functions and its spectrum
(c) full-wave rectified cosine and its Fourier series spectrum

2.2.2 Multiplication in the time domain

The case taken for illustration is that of the response of a single-degree-of-freedom (SDOF) system, which in the time domain is an exponentially damped sinewave represented by

$$\exp(-\sigma t)\sin(2\pi ft) \tag{2.18}$$

The spectrum of the sinewave can be represented by two delta functions, each of amplitude 1/2, the one at $+f_0$ having a phase angle of $-\pi/2$ and the one at $-f_0$ having a phase angle of $+\pi/2$. The Fourier transform of the exponential function is:

$$\frac{1}{\sigma + j2\pi f} \tag{2.19}$$

which changes in phase from $+\pi/2$ at $-\infty$ through zero at zero frequency to $-\pi/2$ at $+\infty$. Figure 2.6 shows how the multiplication of these two functions in the time domain results in a convolution of their respective Fourier transforms, and giving the well-known frequency response function with a peak at the resonance frequency (where the phase is turned through $-\pi/2$), zero phase at zero frequency and $-\pi$ at $+\infty$.

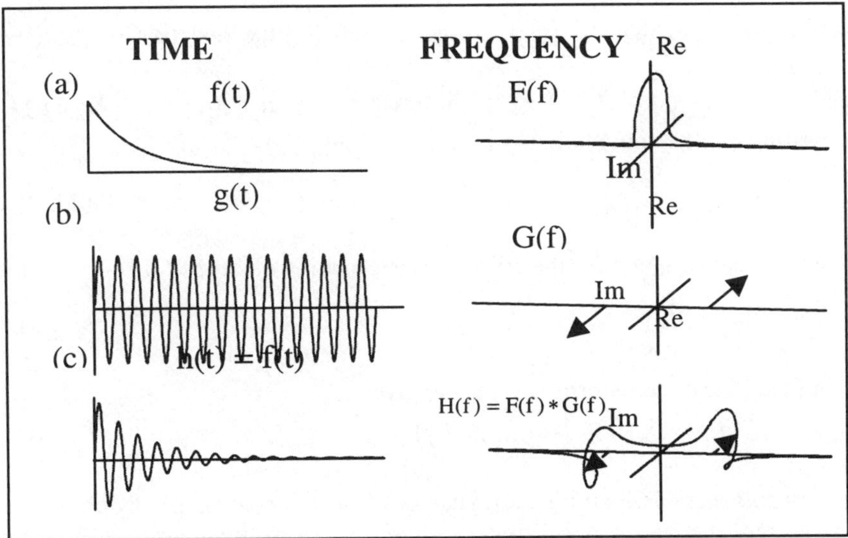

Figure 2.6. Spectrum of an exponentially damped sinewave.
In the spectrum of (c) the arrows depict the resonance peak

2.3 Hilbert transform relationships

The Hilbert transform can be said to be the relationship between the real and imaginary parts of the Fourier transform of a one-sided function. For example, any impulse response function is causal and thus one-sided in the time domain, and this means that the real and imaginary parts of the corresponding frequency function (eg that shown in Fig.2.6) are related by a Hilbert transform. That there should be a relationship becomes evident when it is considered that a causal function is made up of even and odd components which are identical for positive time, and thus cancel for negative time. Thus:

$$x(t) = x_e(t) + x_o(t) \qquad (2.20)$$

and

$$x_e(t) = x_o(t) \times \mathrm{sgn}(t) \qquad (2.21)$$

where sgn(t) is the sign function. Since the even part of a time function transforms to the real part of its Fourier transform, and the odd part to the imaginary part, by applying the convolution theorem to Equ.(2.21) it can be seen that:

$$X_e(f) = X_o(f) * \Im\{\mathrm{sgn}(t)\} \qquad (2.22)$$

The Fourier transform of the sign function is the imaginary hyperbolic function $\dfrac{1}{j\pi f}$ so that the final expression relating the real part of the Fourier transform ($X_R(f) = X_e(f)$) to the imaginary part ($X_I(f) = X_o(f)/j$), and writing out the convolution in full, is given by:

$$X_R(f) = \frac{1}{\pi} \int_{-\infty}^{\infty} X_I(\phi) \cdot \frac{1}{(f - \phi)}\, d\phi \qquad (2.23)$$

The equivalent equation for the Hilbert transformation of a time function $x(t)$ is:

$$\tilde{x}(t) = \frac{1}{\pi} \int_{-\infty}^{\infty} x(\tau) \frac{1}{(t - \tau)}\, d\tau \qquad (2.24)$$

Taking the Fourier transform of Equ.(24) gives:

$$\tilde{X}(f) = X(f) \cdot \left(- j\,\mathrm{sgn}(f)\right) \qquad (2.25)$$

which shows that a Hilbert transform can be achieved more simply by transforming into the frequency domain, shifting the phase of positive frequency components by $-\pi/2$ and of negative frequency components by $+\pi/2$, and then transforming back to the time domain.

A one-sided frequency spectrum can similarly be divided into conjugate even and conjugate odd components which transform by the inverse Fourier transform to real and imaginary time signals, respectively, which are related by a Hilbert transform. The sum of these two components is known as an "analytic signal", which can be formed from a given real time signal by adding j times its Hilbert transform. Alternatively, it can be obtained more simply by transforming the real time signal into the frequency domain, obtaining the equivalent one-sided spectrum by multiplying by $2H(f)$, where $H(f)$ is the Heaviside or unit step function, and transforming back to the time domain. This is also a very efficient way of performing a Hilbert transform. Note from Fig.2.2 that the Hilbert transform of a cosine is obviously a sine function.

As a corollary, it is worth pointing out that when working with a signal processing package such as Matlab®, modifying frequency spectra and then transforming back to the time domain, it is not necessary to adjust all negative frequency components in the same way (but complex conjugate) as the positive frequency components; it is much simpler to multiply the positive frequency components by 2 (but not the zero frequency component), set the second half of the spectrum (the negative frequency components) to zero, perform an inverse transform to an analytic signal and simply take the real part (this last operation is usually necessary even when working with 2-sided spectra, as the program does not know that the answer is supposed to be real, and will usually calculate a very small imaginary part). This process is illustrated in Fig.2.7.

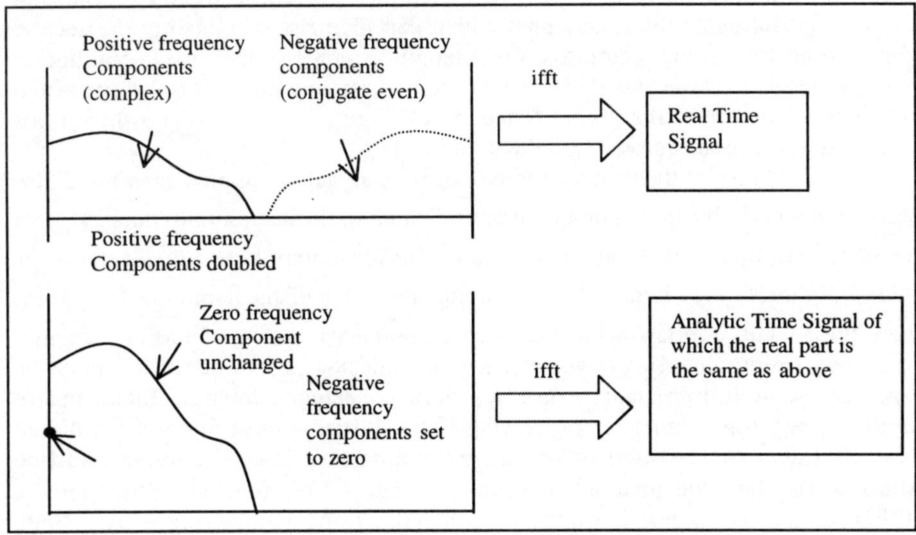

Figure 2.7. Manipulation of the positive frequency spectrum to obtain a real time signal

3.   PRACTICAL FFT ANALYSIS

3.1   Pitfalls

The so-called pitfalls of the FFT are all properties of the DFT and result from the three stages in passing from the Fourier integral transform to the DFT. The first step is digitisation of the time signal which can give rise to **aliasing**; the second step is truncation of the record to a finite length, which can give rise to **leakage** or **window effects**, while the third results from discretely sampling the spectrum, which can give rise to the **picket fence effect**. Figure 3.1 shows these three steps graphically, using the convolution theorem [2.1].

In Fig.3.1(a-c) the infinite continuous time signal is sampled as in Fig.2.3(c) producing a periodic spectrum with a period equal to the sampling frequency f_s . It can be seen that if the original signal contains any components outside the range $\pm f_N$, where f_N is the "Nyquist frequency" or half the sampling frequency, then these will overlap with the true components giving "aliasing" (higher frequencies represented as lower ones). Once aliasing is introduced it cannot be removed, so it is important to use appropriate analogue lowpass filters before digitising any time signal for processing. After initial correct digitisation, digital lowpass filters can be used to permit resampling at a lower sampling rate (the Matlab® function "decimation" achieves this, but not the function "resampling"). In Fig.3.1(d-e) the signal is truncated to length T by multiplying it by a finite (rectangular) window. The spectrum is thus convolved with the Fourier transform of the window, which acts as a filter characteristic. Energy at a single frequency is spread into adjacent frequencies in the form of this characteristic, hence the term "leakage". Finally, in Fig.3.1(f-g) the continuous spectrum is discretely sampled in the frequency domain, which corresponds in the time domain to convolution with a train of delta functions of spacing T , making the time signal periodic. The spectrum is not necessarily sampled at peaks; hence the term "picket fence effect"; it is as though the spectrum is viewed through the slits in a picket fence.

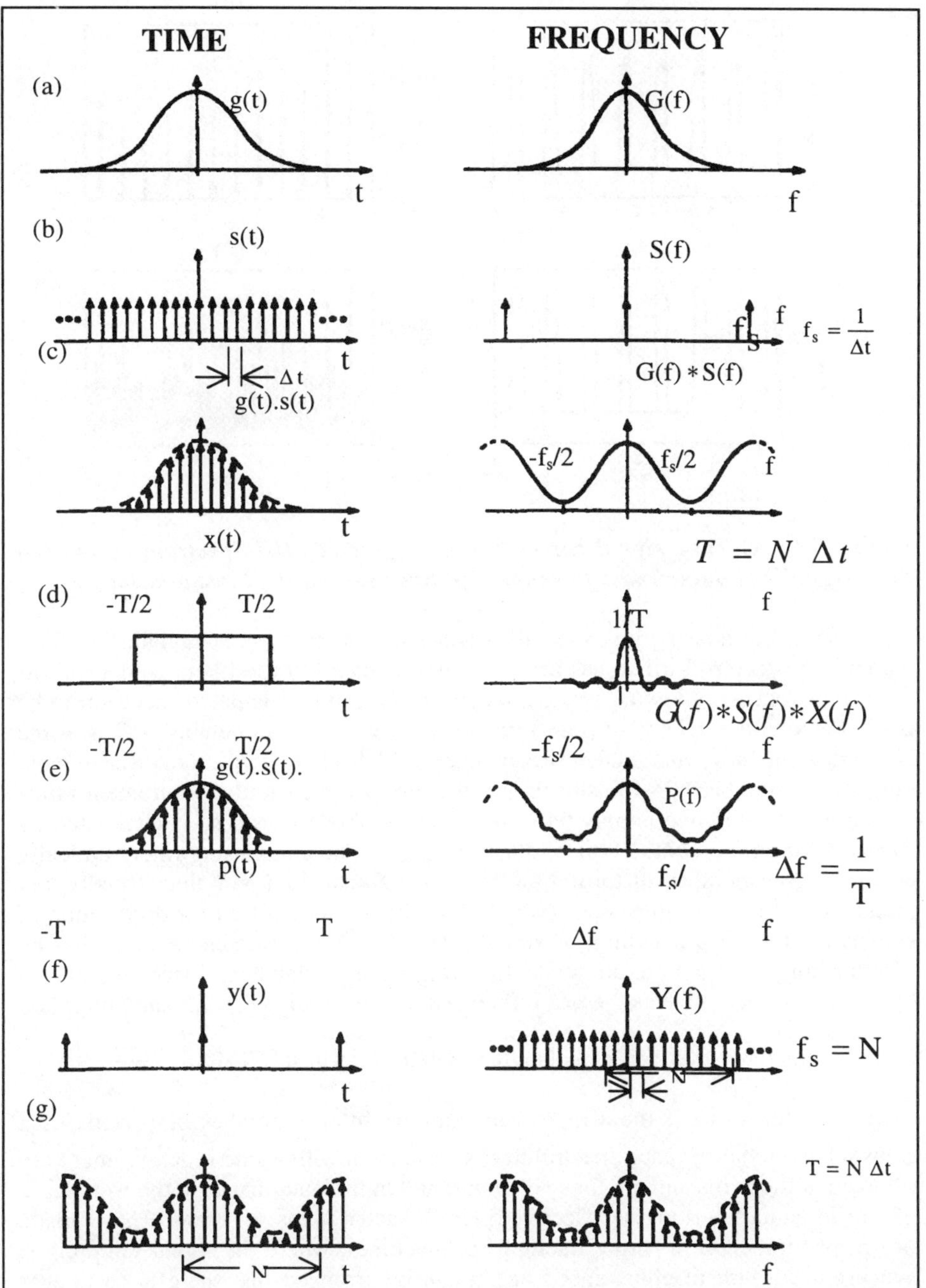

Figure 3.1. Three steps in passing from the FT to the DFT [2.1]
(a-c) Time sampling (d-e) Truncation (f-g) Frequency sampling

304

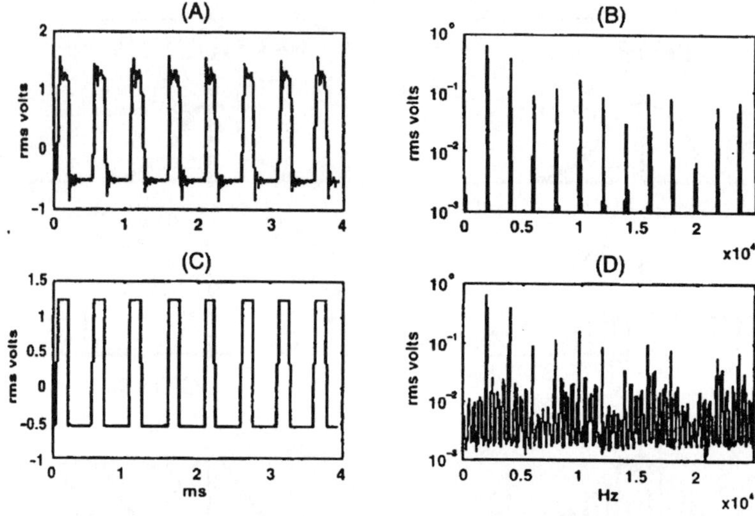

Figure 3.2. (A) Time signal correctly lowpas filtered (B) Spectrum of (A) (C)
Time signal without lowpass filtration (D) Spectrum of (C) (note aliasing)

To avoid aliasing it is virtually always necessary to use an antialiasing filter with a very steep roll-off. It has become fairly standard to use filters with a roll-off of 120 dB/octave, allowing approximately 80% of the calculated spectrum to be used. Thus with a 1K (1024 point) transform, spectrum line number 512 is at the Nyquist frequency, and higher frequencies fold back towards the measurement range. Line number 624 folds back into the top of the desired measurement range (line number 400), and is only 64% of an octave above it, and so is attenuated by 77 dB, taking it below the typical dynamic range. The antialiasing filters typically result in considerable distortion of the time signal, and are thus usually not included in digital oscilloscopes (which thus should not be used for digitisation of signals for further processing). Figure 3.2 (from [3.1]) illustrates how antialiasing filtering corrects the spectrum while distorting the time signal, and vice versa.

The effects of leakage are influenced by the final spectrum sampling, and Fig.3.3 illustrates that for a rectangular window (whose FT is a $\dfrac{\sin(x)}{x}$ or $\text{sinc}(x)$ function) if the window contains an integer number of periods of a sinusoid, even though each spectral line is associated with a sinc function, these are sampled at the zeros and are thus not apparent. On the other hand, in the worst case of a residual half period, the effective filter characteristic is very poor. Use is made of this phenomenon in "order tracking" of machines, where the signal sampling is synchronised with machine speed and it can be arranged that there is an integer number of periods of (all harmonics of) the rotational frequency within the record length, in which case a rectangular window can be used. Otherwise, for continuous signals it is usually necessary to choose a data window other than rectangular to achieve a better filter characteristic. This is discussed below.

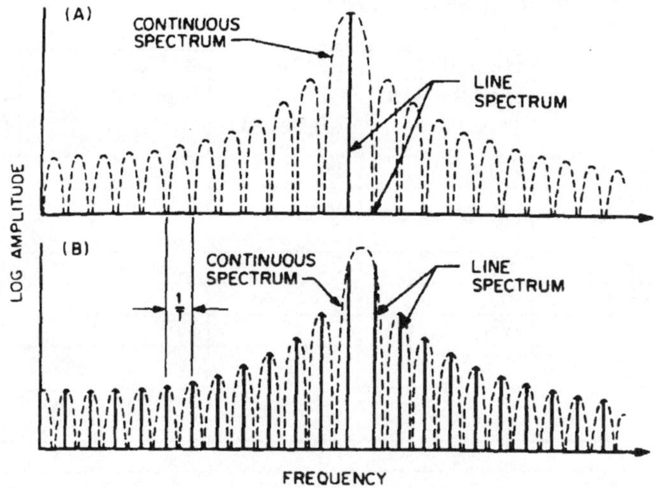

Figure 3.3. How frequency sampling affects the apparent filter characteristic of a window (A) Integer number of periods in record (B) Extra half period in record

Because of the picket fence effect, spectral functions are not necessarily sampled at their peaks and the "picket fence error" is the difference between the true value and the value of the maximum spectral line. For rectangular weighting this can be as much as 3.9 dB, and most other windows have a reduced value.

3.2   Data windows

3.2.1   Continuous signals

For continuous signals, a major function of the window is to reduce the effect of the discontinuity which usually arises when a random section of signal is made periodic. Practically, it means minimising the sidelobes in the filter characteristic, both the highest and the remaining ones (by maximising their rate of falloff). To improve enhancement of discrete frequency components with respect to broadband noise it is desirable to minimise the noise bandwidth of the characteristic, but on the other hand, attention must also be paid to minimising the picket fence effect. Table 3.1 gives a comparison of the properties of the most common windows applied to stationary signals, and Figure 3.4 (from [3.1])compares their worst case filter characteristics. The Hanning window, which can be considered as one period of a sine squared function, is a good general purpose window, with picket fence effect limited to 1.4 dB, noise bandwidth 1.5 (times Δf the line spacing) and desirable characteristics with respect to overlap averaging, to be discussed below. The best window with respect to separating adjacent components of widely

Table 3.1. Comparison of window properties

Window type	Highest sidelobe (dB)	Sidelobe roll-off (dB/decade)	Noise bandwidth $(\times \Delta f)$	Max. picket fence error, (dB)
Rectangular	-13.4	-20	1.00	3.9
Hanning	-32	-60	1.50	1.4
Hamming	-43	-20	1.36	1.8
Kaiser-Bessel	-69	-20	1.80	1.0
Truncated Gaussian	-69	-20	1.90	0.9
Flattop	-93	0	3.70	<0.1

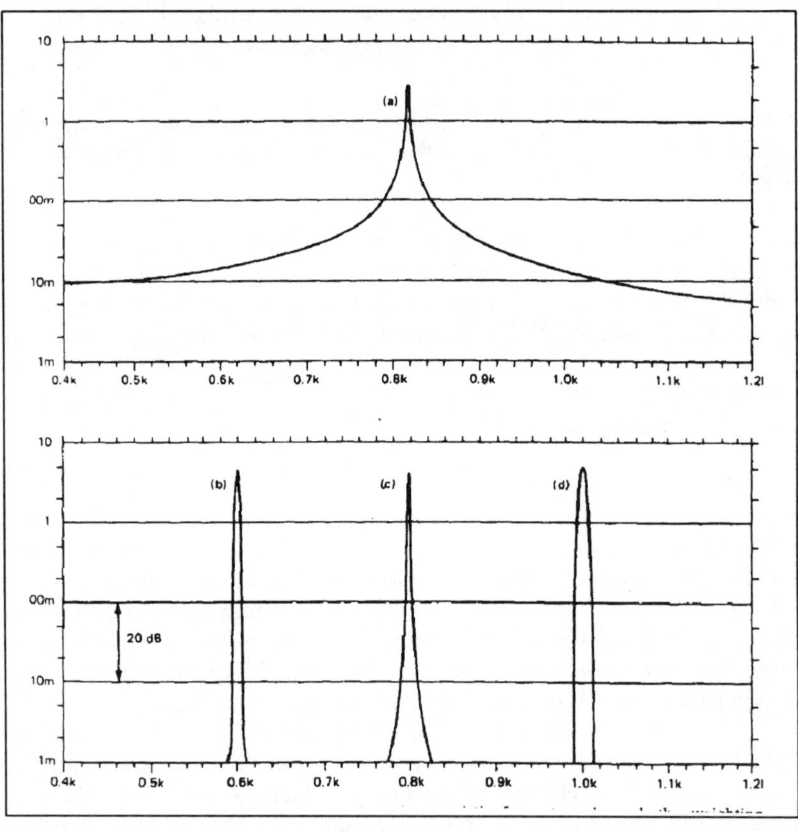

Figure 3.4. Worst case filter characteristics for some common window functions (a) Rectangular (b) Kaiser-Bessel (c) Hanning (d) Flattop

differing levels is probably the Kaiser-Bessel, but the same can usually be achieved by simply analysing with more resolution (zoom analysis or a larger transform). The flattop window is specifically designed to minimise the picket fence effect, and is thus usually the best choice when calibrating measurements with a calibration signal whose frequency can fall anywhere between two analysis lines. It can also be useful when analysing a signal dominated by one or more families of harmonics, since as long as they are resolved (keeping in mind that the noise bandwidth is $3.7 \, \Delta f$) there is no need to compensate the indicated values of the various harmonics. Figure 3.5 shows how compensation can be made for both picket fence error and frequency error when using a Hanning window. Provided frequencies are stable along the record length, the difference in dB between the two highest samples around a frequency peak (ΔdB) determines the errors. As mentioned above, the Hanning window has the desirable property that the effective weighting in overlap averaging (see later) can be made completely uniform with an overlap of $2/3$, $3/4$, etc. With 50% overlap, the weighting varies by $2{:}1$, but this is not serious with stationary signals. When finding the averaged spectrum of a long transient signal (longer than the transform size), a uniform weighting is preferable.

3.2.2 Transient signals

Analysis of transient signals is common in impact measurements for modal analysis. The force signal is always short, and a rectangular window is suitable, although this may be tailored to just more than the length of the force pulse in order to exclude noise. For the response signal it must be ensured that it has died away (ie by 50 to 60 dB) by the end of the record. This can be achieved by extending the record length (ie by zoom or a larger transform size), but where this is restricted by other constraints it is common to apply an exponential

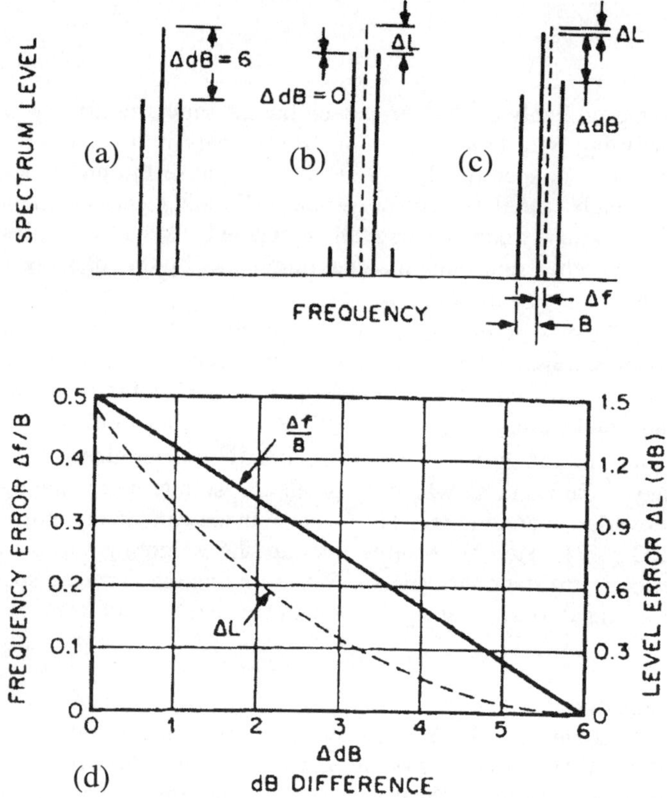

Figure 3.5 Compensation for the picket fence effect with a Hanning window.
ΔdB = *difference between two highest spectrum samples.*
Δ L = *picket fence error.* Δf = *frequency error (a), (b), (c) Minimum, maximum and intermediate error cases (d) Error nomogram*

window, starting just before the response signal, which attenuates the signal sufficiently. This is equivalent to applying additional damping, which is known very precisely, and so can be subtracted from the resulting measurements.

3.2.3 Application in the frequency domain

Since multiplication by a window function corresponds in the frequency domain to a convolution with its FT, this is sometimes the most efficient way to apply them. Examples of where this is advantageous are firstly where the FT of the window is very simple, such as with the Hanning function, secondly where only a part of the spectrum is required as with zoom spectra, and thirdly where several different windows can be applied to exactly the same FT. The basic principle can be explained using the Hanning window, which when repeated periodically (as

happens implicitly with the DFT) can be represented as $\sin^2 \theta$ or $\dfrac{1}{2} - \dfrac{1}{2}\cos 2\theta$

which has the convolution coefficients $\left[-\dfrac{1}{4}\,,\,\dfrac{1}{2}\,,\,-\dfrac{1}{4}\right]$ (keeping in mind that the frequency corresponds to one period along the record length, and thus to one line spacing. Convolution with such a simple function is often more efficient than direct multiplication in the time domain. Note that the coefficients as stated are for a window with maximum value one, and are usually modified to scale the result (see below under "Scaling").

3.3 Zoom Analysis

The basic DFT transform of Equ.(2.14) extends in frequency from zero to the Nyquist frequency and has a resolution equal to the sampling frequency f_s divided by the number of samples N. Sometimes it is desired to analyse in more detail in a limited part of the frequency range, in which case use can be made of so-called "zoom analysis". Since resolution $\Delta f = {f_s}/{N}$, the two ways to improve it are:

1) Increase the length of record N. Some analysers include this option in the form of "non-destructive zoom", which make use of an algorithm to perform a transform of size $m \times N$ by combining the results of m undersampled transforms of size N. This was useful when hardware restrictions limited the size of transform which could be performed, but in modern analysers, and in signal processing packages such as Matlab® there is virtually no restriction on transform size, and so zoom can be achieved by performing a large transform and then viewing only part of the result.

2) Reduce the sampling frequency f_s. This can be done if the centre of the desired zoom band is shifted to zero frequency so that the zoom band around the centre frequency can be isolated by a lowpass filtration. The highest frequency is then half the zoom band and the sampling frequency can be reduced accordingly without aliasing problems. This process is illustrated in Fig.3.6. The lowpass filtering and resampling process is usually done in octave (2:1) steps, as a digital filter will always remove the highest octave, relative to the sampling frequency, and halving the sampling frequency simply means discarding every second sample. This type of zoom is normally done in real-time by a specialised hardware processor, the advantage being that the sampling rate is reduced before signals have to be stored. It is discussed below that the zoom process is a useful precursor to demodulation, even where the further processing is to be done in a computer. Note that the time signal output from the zoom processor is complex, as the corresponding spectrum is not conjugate even.

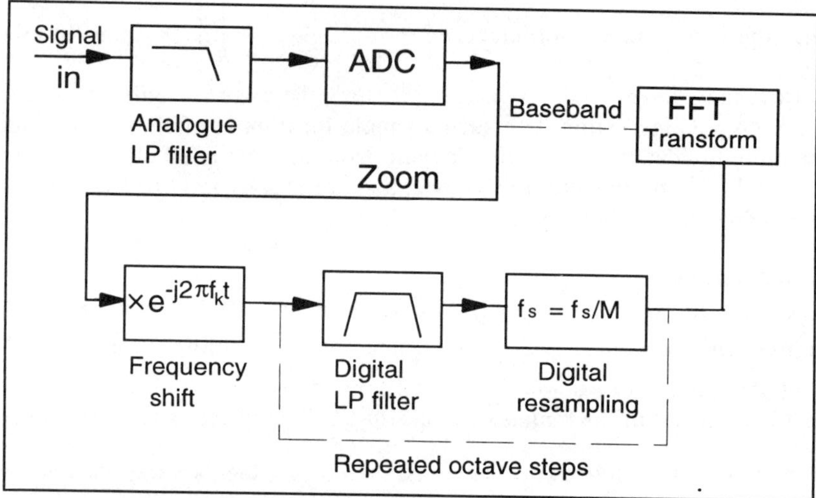

Figure 3.6. Schematic diagram of FFT zoom process

3.4   Spectrum Averaging

The need for averaging of FFT spectra is determined by whether the signal contains random components or not. Averaging should always be done in terms of signal power (ie amplitude squared) as it is this which is conserved independent of phase. The DFT spectra of discrete frequency components always have the same amplitude, and therefore little is achieved by averaging the squared amplitudes, although this can be useful for clarifying which components are discrete frequency and which are random. Where analysis is being done for diagnostic purposes (eg a zoom spectrum to measure a single frequency or family of sidebands very accurately) it is preferable not to average, as in practice machine speeds vary slightly with time and averaging results in a smearing of frequencies and in particular disguises frequency spacings.

When meaningful spectra are to be obtained from random signals, which in mechanical signals are typically caused by fluid flow (turbulence, cavitation) road roughness etc, it is necessary to average a number of power spectrum estimates. The number of averages required is determined by the desired accuracy, as the standard deviation of the result (for gaussian signals) is given by [2.1]:

$$\varepsilon = \frac{1}{2\sqrt{n}} \tag{3.1}$$

where n is the number of independent averages. Thus, for $n = 16$, $\varepsilon = 12.5\%$ or 1 dB, meaning that there is a 68% probability that the result will be within ± 1 dB, 95% probability that it will be within ± 2 dB and 99.7% probability that it will be within ± 3 dB. To halve the error it is necessary to make four times as many averages etc.

With a rectangular window, "independent" means non-overlapping, but with other windows such as Hanning, advantage can be gained by overlapping, as information is lost near the two ends where the weighting is near zero. In fact, very little is lost statistically by overlapping 50%, and so this is recommended for stationary random signals, as twice as many effective averages can be obtained from a given length of signal. As mentioned above, the overall weighting is not uniform in that case, as at the point where the successive Hanning windows overlap, their amplitude weighting is $\frac{1}{2}$ and thus their power weighting $\frac{1}{4}$, meaning that the final weighting of that part of the signal in the overlap average is $\frac{1}{2}$ and the total power weighting along the signal varies between $\frac{1}{2}$ and unity. This gives no problem for stationary signals, but to extract all information from a given length of record, in particular if it is non-stationary, it is advisable to overlap by a factor of at least $\frac{2}{3}$ although with typical FFT record lengths in powers of 2 it is often simpler to overlap by $\frac{3}{4}$. In the latter case the effective number of averages to insert in Equ.(3.1) is half the actual number. The Matlab® function PSD performs overlap averaging (as does CSD for multiple signals).

3.5   Scaling

3.5.1   Discrete frequency signals

As stated above, the DFT operations of Equ.(2.14, 2.15) result in a correctly scaled Fourier series spectrum for the forward transform and exact reconstitution of a (periodic) time signal for the inverse transform. Note that this means that the resulting value A_k at the positive frequency ω_k has an amplitude half that of the corresponding sinewave (C_k). To obtain the equivalent RMS (root mean square) value, $\left| A_k \right|$ must be multiplied by $\sqrt{2}$ to take account of the power in the negative frequency component. The RMS value is also equal to $C_k / \sqrt{2}$. If the signal contains discrete frequency components which do not have an integer number of periods along the record length, then a window function such as Hanning will generally be used to reduce leakage. However, if the window is scaled to a maximum value of unity, the average "power" (ie mean square value) of the signal will obviously be reduced, and it is necessary to compensate for this. From section 3.2.3 it will be seen that the convolution coefficients for a Hanning window scaled to a maximum value of 2 are $\left[-\frac{1}{2}, 1, -\frac{1}{2} \right]$ meaning that a single component in one line would be replaced by three components of which the central one would have the same value (ie the peak value would be scaled correctly). This is usually the best scaling to use for discrete frequency components as it means that the maximum value around a spectral peak can be read off directly after correction for picket fence error (eg using Fig. 3.5). For windows other than Hanning, the same effect can be achieved by scaling the window such that its central convolution coefficient is unity. Note that because of the extraneous sidebands introduced, the total "power" in the spectrum has been increased as discussed below.

3.5.2 Parseval's theorem

At this stage it is convenient to introduce Parseval's theorem which in broad terms states that the total power (or energy) in a signal can be obtained by integrating over all time or all frequency, and in both domains is related to amplitude squared. For a stationary signal with finite power, the frequency spectrum will either contain discrete frequency components whose amplitude squared directly represents the power at each frequency, or for random signals the squared amplitude spectrum is continuously distributed over frequency and represents "power spectral density" (PSD) which has to be integrated over a finite bandwidth to give finite power. In both cases the equivalent "power" in the time domain is the mean square value, obtained by integrating the instantaneous squared value (instantaneous power) over a sufficiently long time and dividing by that time. For transient signals with finite "energy" (integral of "power" over time) the squared amplitude of its Fourier transform represents "energy spectral density" (ESD) which when integrated over all frequency gives the same total energy as integrating the instantaneous power of the signal over all time.

Henceforth, the terms power and energy will be used without inverted commas to represent signal amplitude squared and its time integral, respectively, as these are generally related to physical power and energy by an impedance or admittance function (eg electrical power = $I^2 R = V^2/R$ in terms of current I, voltage V and resistance R). Thus for a signal with units U (where U represents m, ms^{-1}, ms^{-2}, g, N etc) power has units U^2, energy has units U^2s, PSD has units U^2/Hz (= U^2s), ESD has units U^2s/Hz (= U^2s^2).

3.5.3 Stationary random signals

Each signal record transformed will be treated by the DFT algorithm as a periodic signal, but the power in each spectral line can be assumed to represent the integral of the PSD over the frequency band of width Δf (= $1/T$), and thus the average PSD is obtained by multiplying the squared amplitude by T. The required averaging over a number of records does not change this scaling. How well the average PSD represents the actual PSD depends on the width of peaks (and valleys) in the spectrum. The width of such peaks is typically determined by the damping associated with a structural resonance excited by the broadband random signal, and the 3dB bandwidth is given by twice the value of σ (of Equ.(2.19)) expressed in Hz. The PSD will be sufficiently accurate if the 3dB bandwidth is a minimum of five analysis lines.

If a window such as Hanning has been used to reduce leakage, and if it is scaled so as to read the peak value of discrete frequency components (as recommended in 3.5.1) the calculated PSD value will have to be divided by the "noise bandwidth" indicated in Table 1 to compensate for the extra power given by the spectral sidebands. This noise bandwidth is the sum of the squares of the convolution coefficients (for Hanning it is $0.5^2 + 1^2 + 0.5^2 = 1.5$). When integrating over several frequency lines (eg to convert a constant bandwidth spectrum to constant percentage bandwidth) the total power in each integrated band must be divided by the noise bandwidth of the window because of the extra power associated with each line; this is the same as integrating the PSD over the required bandwidth.

Note that discrete frequency components cannot be represented on a PSD scale as they are concentrated in an infinitely narrow bandwidth and thus have infinite PSD. Note also that their power is independent of the analysis bandwidth Δf used to analyse them, whereas the power of spectral lines of random signals varies directly with the analysis bandwidth. Zoom analysis is sometimes used to make discrete frequency components stand out from random background noise.

3.5.4 Transient signals

Transient signals are also treated as being one period of a periodic signal, so not only does the power in a spectral line have to be converted to an average spectral density by dividing by Δf, but also the average power must be converted to energy per period by a further multiplication by T, altogether a multiplication by T^2 to obtain a result scaled as ESD. Generally, transient signals will be shorter than the transform length and thus a rectangular window will be used, and if the signal has decayed to near zero at the end of the record (eg following the recommendations of 3.2.2) the signal bandwidth will be sufficiently greater than the analysis bandwidth for the average ESD to represent the true ESD. The extra damping given by an exponential window will genuinely give a reduction in signal energy.

4.   OTHER SPECTRAL ANALYSIS TECHNIQUES

4.1   Parametric Spectral Analysis

With Fourier analysis, as is evident from Fig.3.1(d), the spectral resolution is determined by the Fourier transform of the time window which limits temporal resolution. For a window of length T the spectral resolution is of the order of $1/T$, and thus the better the time localisation the poorer the frequency localisation, and vice versa. This is one expression of Heisenberg's uncertainty principle, and is because no assumption is made about the behaviour of the time function outside the window (effectively it is set to zero, which is extremely improbable).

With parametric spectral analysis [4.1, 4.2], better spectral resolution can be obtained for short records, basically because it assumes that the behaviour of the function outside the window is most similar to its behaviour inside the window. This is valid for sinusoidal or near sinusoidal signals. With parametric analysis, the signal is modelled as the output of a physical system described by a limited number of parameters when excited by a unit white noise input. Thus the frequency response of the system represents the signal spectrum. Generally, the improvement in spectral resolution is accompanied by a deterioration in amplitude accuracy.

314

4.1.1 MA models

Perhaps the easiest case to understand is where the system is modelled as an FIR (finite impulse response) filter, in which case the output is the (digital) convolution of the input signal with the finite length impulse response of the filter, as expressed by the equation:

$$y_i = \sum_{k=0}^{M} b_k x_{i-k} \tag{4.1}$$

where x_i represents the input signal, y_i represents the output signal, and the b_k represent the convolution weights or samples of the impulse response. Equ.(4.1) is a digitised, finite length version of the convolution equation (2.17) and is effectively a "moving average", giving rise to the term MA model. Applying a Z-transform to Equ.(4.1), which is the equivalent of a Laplace transform for discrete time signals, the convolution becomes the product:

$$Y(z) = \sum_{k=0}^{M} b_k z^{-k} \, X(z) = B(z) \, X(z) \tag{4.2}$$

from which the transfer function can be seen to be:

$$B(z) = \sum_{k=0}^{M} b_k z^{-k} = b_0 \prod_{k=1}^{M} \left(1 - z^{-1} z_k\right) \tag{4.3}$$

which has no poles and is thus an "all-zero" model.

This type of model is obviously most efficient when the effective length of the impulse response is short, meaning that it is highly damped and thus without sharp spectral peaks.

4.1.2 AR models

AR or "autoregressive" models are more efficient where there are sharp spectral peaks, and thus the required transfer function has poles. This is the case with IIR (infinite impulse response) filters where outputs are generated recursively from the previous outputs and the current input.

The relationship between input and output signals can be expressed as:

$$y_i = -\sum_{k=1}^{N} a_k \, y_{i-k} + x_i \tag{4.4}$$

where in principle $N \to \infty$ but can be truncated when the terms become sufficiently small. After Z-transformation this gives:

$$Y(z)A(z) = X(z) \tag{4.5}$$

from which the transfer function can be seen to be:

$$1/A(z) = 1 \bigg/ \sum_{k=0}^{N} a_k z^{-k} = 1 \bigg/ \prod_{k=1}^{N} \left(1 - z^{-1} p_k\right) \qquad (4.6)$$

which has no zeros and is an all-pole model.

There are a number of techniques which result in such an AR model, one of which is the "maximum entropy" method mentioned above. In this, the coefficients are found by maximising the entropy (disorder) of the signal, while ensuring that the autocorrelation function is determined by the signal within the window. This really means that the signal outside the window will be most similar to the signal within the window because of the elimination of any biasing effect. Other AR techniques include "linear prediction" used in speech analysis, and statistical "autoregression" from which it takes its name.

Figure 4.1 shows the results of applying maximum entropy analysis to a very short record of envelope signal from a bearing with an inner race fault [4.3]. The record length comprised only 1.29 revolutions of the shaft speed which determined the spacing of modulation sidebands in the envelope spectrum. The maximum entropy spectrum of Fig.4.1(d) appears to give very good resolution of the sidebands, but Fig.4.1(c) shows that Fourier analysis can give almost as much information provided a sufficient degree of spectrum interpolation is used. The spectrum interpolation was achieved by padding the data record with zeros to 7 times its original length. Note that the maximum entropy spectrum had to be represented on a logarithmic amplitude scale because of the much wider range of amplitude values than for the Fourier analysis cases.

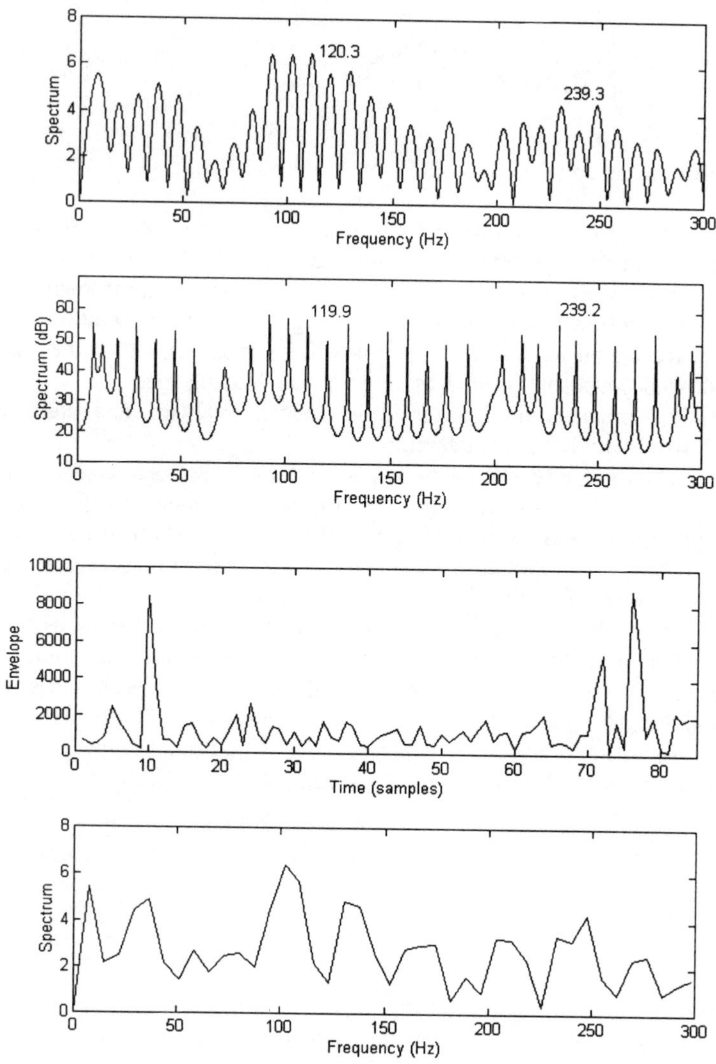

Figure 4.1 *Envelope spectra of short time recording (1.29 revolutions) just spanning the first two sections where the inner race fault passes the loading zone. [4.3] (a) Envelope signal after bandpass filtration in frequency range 2.7 ~ 3.3 kHz, (b) Envelope spectrum without interpolation, (c) Envelope spectrum with interpolation, (d) Maximum entropy envelope spectrum*

AR models have been found useful for characterising a signal with a minimum number of parameters, eg to use as a feature vector for input to a neural network to distinguish different cases. Where signals are nonstationary, the concept of "evolutive" AR models [4.4] may be useful. In this case the AR parameters are updated by moving a finite length window along the record. In [4.4] an evolutive AR model gave a better separation of faults in reciprocating machine vibration signals.

4.1.3 ARMA models

Where the spectrum contains sharp peaks and valleys, it may be preferable to model it with an ARMA model which contains both poles and zeros. The relationship between input and output signals is then:

$$y_i = \sum_{k=0}^{M} b_k x_{i-k} - \sum_{k=1}^{N} a_k y_{i-k} \qquad (4.7)$$

for which the transfer function is:

$$B(z)/A(z) \qquad (4.8)$$

with both poles and zeros. In general it is much more onerous in terms of computational effort to fit an ARMA model than either an AR or MA model, although this may be compensated by a reduction in the number of parameters required to obtain a good fit. Note that software to fit these models is available in signal processing packages such as the Matlab® System Identification Toolbox.

4.1.4. Other models

For vibration measurements made when the structure is in "free decay" after being excited, it is natural to fit a number of complex exponentials corresponding to the impulse responses of the various modes. One method which achieves this is the ITD method [1.6] referred to in the Introduction while another is the Prony method described in [4.1]. These should give a result similar to an ARMA model, in that the transform of a sum of complex exponentials results in a ratio of polynomials with poles and zeros.

4.2 Digital Filters

Section 4.1 discussed one application of digital filters, with both infinite and finite impulse responses, to model a system which would produce a measured signal when excited by a white noise source. Digital filters can also be used to frequency analyse a signal in the same way as an analogue filter, ie by measuring the power transmitted by a series of filters covering the required frequency band. Where analysis in constant bandwidths is required, this is most efficiently done using the FFT algorithm, but when constant percentage bandwidth is required it is most efficiently done using a series of 1/n-octave digital filters, the different octaves being covered by repetitively halving the sample rate. There are two ways in which the properties of a digital filter are varied:

1)   By varying the filter coefficients (either FIR or IIR)
2)   By varying the sample rate, as for given coefficients the frequency characteristic is proportional to the sampling frequency.

Thus for 1/3-octave analysis, for example, three bandpass filters are required (ie three sets of coefficients) covering the octave from 33% to 66% of the Nyquist frequency. The signal is sampled at a rate corresponding to the highest octave (ie the upper cutoff frequency of the upper 1/3-octave at 66% of the Nyquist frequency) and is passed through the three 1/3-octave filters and a digital lowpass filter cutting off at 40% of the current Nyquist frequency. The sample rate of the lowpass filtered signal can then be halved (by discarding every second sample) after which passage through the same digital filters will produce the filter outputs corresponding to the next lower octave in actual frequency. By repeating this process, as many octaves can be covered as desired. Note that by being able to perform the filtering calculations twice as fast as required for the upper octave, it is possible to process any number of octaves in real-time, as each time the sample rate is halved only half as many samples have to be processed in a given time. If the time taken to process the upper octave is taken to be unity the time taken to

process all octaves is $\left[1 + \frac{1}{2} + \frac{1}{4} + \frac{1}{8} + \cdots \right] = 2$.

This process is best done in real-time by specialised hardware, but can be done as post-processing using a signal processing package. However, in general the record to be post-processed would have to be extremely long if the analysis is to cover, say, three decades or more, with a reasonable amount of averaging. For example, for a 1/12-octave (6% bandwidth) analysis over three decades (eg 20 Hz - 20kHz) of a random signal for which a BT product (equivalent to n in Equ.(3.1)) of 16 is required, the filter output would have to be averaged over $16 \times 80 = 1280$ samples of the lowest filter sample rate, corresponding to $640,000$ samples of the original record.

The power of each filter output is measured by squaring it, and either calculating the average over a defined time period, or smoothing it with an "exponential averager", a first order lowpass filter (which can also be applied as a digital filter in real-time to give a running average). The equivalent linearly weighted averaging time is twice the time constant of the first order filter.

4.3 Order Tracking

In analysing rotating machine vibrations it is often desired to have an x-axis based on harmonics or "orders" of shaft speed. This can be to avoid smearing due to speed fluctuations or can be to see how the strength of the various harmonics changes over a greater speed range, for example as they pass through various resonances. If a constant amplitude signal which is synchronous with the rotation of a shaft, for example, is sampled a fixed number of times per revolution, the digital samples are indistinguishable from those of a sinusoid, and thus give a line spectrum, whereas if normal temporal sampling is used the spectrum spreads over a range corresponding to the variation in shaft speed. Thus, for order analysis it is necessary to generate a sampling signal from a tacho signal synchronous with shaft speed. It is sometimes possible to use a shaft encoder mounted on the shaft in question to provide a sampling signal, but more often the latter has to be generated electronically. Formerly, this was done using a phase-locked loop to track the tacho signal and then generate a specified number of sampling pulses per period of the tracked frequency. However, an analogue phase-locked loop has a finite response time and cannot necessarily keep up with random speed fluctuations such as occur with an internal combustion engine from cycle to cycle. The best method is to digitally resample each record based on the corresponding period of the tacho signal. This can be done in a number of ways, based on digital interpolation. One way is simply to increase the sample rate of each section by say a factor of 10, and then select the nearest sample to the theoretical interpolated position. Increasing the sample rate by an integer factor can be achieved by inserting the appropriate number of zeros in between each actual sample, and then applying a digital lowpass filter to limit the frequency range to the original maximum, thus smoothing the curve (it will also require rescaling proportional to the resampling factor). Resampling by a factor of 4 is illustrated in Fig.4.2. The same result can be achieved using the FFT by padding the spectrum with zeros in the centre (ie around the Nyquist frequency) and then inverse transforming the increased (2-sided) spectrum to the same increased number of time samples. Note that the record length in seconds is the reciprocal of the frequency line spacing in Hz which is not affected by the zero padding. This latter procedure can also be used to resample a record consisting of an integer number of samples to another (though greater) integer number, and is the basis of the Matlab® function INTERPFT. In general, more accurate interpolation, not limited to a ratio of integer numbers, can be achieved by fitting a curve to a group of samples (eg two for a linear curve, three for a quadratic etc) and then calculating the value of the polynomial at the interpolated positions. The accuracy of the interpolation can be judged by considering that the interpolation in the time domain corresponds to a multiplication in the frequency domain by a filter characteristic which aliases back into the measurement range. For example, choosing the nearest sample value is the same as convolving the original samples with a rectangular function of width equal to the sample spacing, while a linear interpolation corresponds to a convolution with a triangular function of base width twice the sample spacing (the convolution of the rectangular function with itself). In the former case the filter characteristic is a sinc(x) function with zeros at multiples of the sampling frequency (so that all

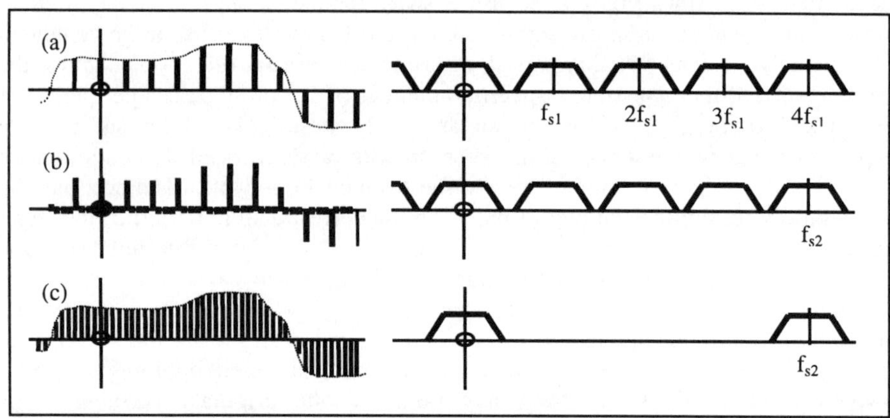

Figure · 4.2.   Digital resampling with four times higher sampling frequency(a)Signal sampled at f_{s1} and its spectrum (b) Addition of zeros which changes sampling frequency to f_{s2} (c) Lowpass filtration and rescaling

sidelobes fold back into the measurement range), while in the latter case it is the square of the sinc(x) function, which has much smaller sidelobes. In practice, cubic interpolation involving two samples on each side of a central one gives good results without excessive computational effort [4.5].

Quite apart from errors introduced by the interpolation, when resampling at a lower frequency (for example as a machine speed reduces), it is necessary to ensure that the signal is adequately lowpass filtered to prevent aliasing. Digital filtering can be useful here as the cutoff frequency varies directly with the sampling frequency, but the initial analogue lowpass filtration must be such that aliasing components do not enter the measurement range. Digital oversampling can solve this problem, as from Fig.4.2(c) it can be seen that the sampling frequency can be reduced by a large factor before overlap occurs. In fact, the factor is 5.9 in this case because a tracking digital filter cutting off at a particular shaft order proportionally reduces the useful band in terms of frequency.

Figure 4.3 illustrates the use of tracking to avoid smearing in the spectrum of the vibration signal from a gearbox in a variable speed mining shovel. The discrete frequency components in the spectrum after tracking come mainly from gear-related components which were removed using synchronous averaging as described in the next section.

5.   TIME DOMAIN ANALYSIS

5.1   Time Synchronous Averaging

Synchronous averaging is useful to extract that part of a signal which is periodic with the same period as a trigger signal, eg a once-per-rev tacho signal from a shaft in a rotating machine. In practice it is done by averaging together a series of signal segments each corresponding to one period of the synchronising signal. Thus:

$$y_a(t) = 1/N \sum_{n=0}^{N-1} y(t + nT) \qquad\qquad (3.5)$$

This can be modelled as the convolution of $y(t)$ with a train of N delta functions displaced by integer multiples of the periodic time T, which corresponds in the frequency domain to a multiplication by the Fourier transform of this signal, which can be shown to be given by the expression [4.1, 5.1]:

$$C(f) = 1/N \, \sin(N\pi Tf)/\sin(\pi Tf) \qquad\qquad (3.6)$$

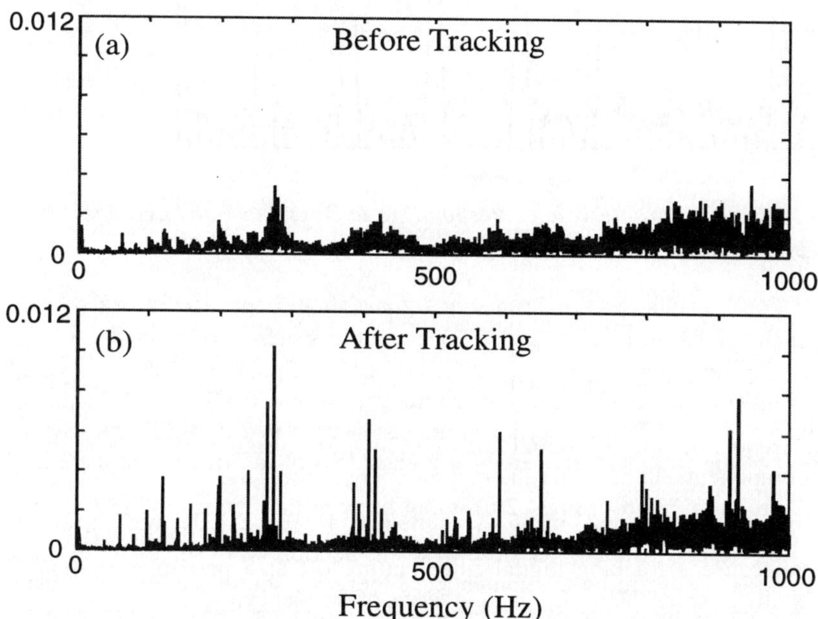

Figure 4.3. Use of tracking to avoid smearing of shaft speed related components

The filter characteristic corresponding to this expression is shown in Fig.(5.1) for the case where $N = 8$, and is seen to be a comb filter selecting the harmonics of the periodic frequency. The greater the value of N the more selective the filter, and the greater the rejection of non harmonic components. The noise bandwidth of the filter is $1/N$, meaning that the improvement in signal/noise ratio is $10 \, \log_{10}N$ dB for additive random noise. For masking by discrete frequency signals, it should be noted that the characteristic has zeros which move with the number of averages, so it is often possible to choose a number of averages which completely eliminates a particular masking frequency. The above characteristic is for an infinitely long time signal $y(t)$, and in Ref.[5.1] it is

322

shown that for the practical situation of a finite length of signal with finite sampling frequency, it is possible to calculate an optimum number of averages to completely remove a discrete masking signal, in particular when the frequency is related by a rational fraction to the synchronous frequency. This is always the case for different shafts in gearboxes.

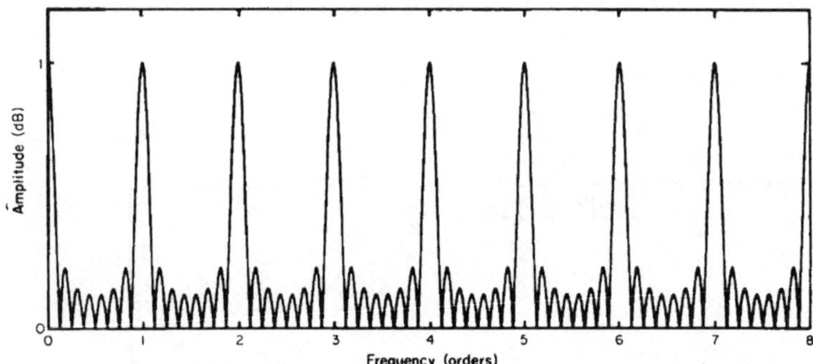

Figure 5.1. Filter characteristic corresponding to 8 synchronous averages (from [5.1]

For good results the synchronising signals should correspond exactly with samples of the signal to be averaged, as one sample spacing corresponds to 360° of phase of the sampling frequency, and thus to 144° of phase at 40% of it which is a typical maximum signal frequency. Moreover, even a 0.1% speed fluctuation would cause a jitter of the same order of the last sample in a (typical) 1K record, with respect to the first, and thus an even greater loss of information at the end of the record, after averaging.

Sampling the signal using a sampling frequency derived from the synchronising (tacho) signal, as described in Section 4.3, solves both these problems and is always to be recommended. Figure 5.2 shows the results of using synchronous averaging on the data of Fig.4.3. The order tracked data was arranged to have an integer number of samples per period of the low speed gear, which allowed determination of the harmonics of this gear speed by synchronous averaging. The spectrum of this signal is shown in Fig.5.2(a). After a periodic repetition of this signal was subtracted from the overall tracked signal (Fig.4.3(b)) the data was resampled to have an integer number of samples per period of the high speed gear, after which its harmonics could be determined in the same way (Fig.5.2(b)). Finally, after subtraction of this periodic signal from the data, the remaining signal was dominated by the effects of an inner race bearing fault (Fig.5.2(c)).

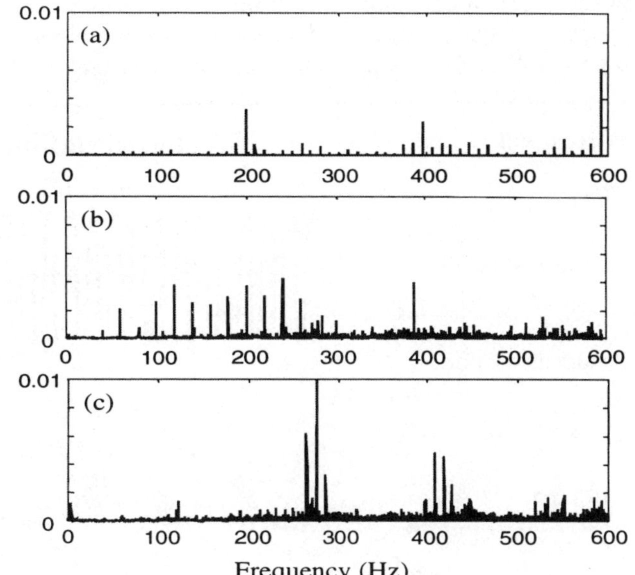

Figure 5.2. Application of synchronous averaging to data of Fig.4.3.
(a) Spectrum synchronous with low speed gear (b) Spectrum synchronous with high speed gear (c) Spectrum dominated by bearing fault after effects of two gears removed

5.2   Autocorrelation Analysis

The autocorrelation function measures how well a signal correlates with delayed versions of itself. The equation for calculating it for a transient is:

$$R_{xx}(\tau) = \int_{-\infty}^{\infty} x(t)x(t+\tau)dt \qquad (5.3)$$

while for a stationary function (for which the above integral would be infinite) it is:

$$R_{xx}(\tau) = \lim_{T\to\infty} \frac{1}{T} \int_{-T/2}^{T/2} x(t)x(t+\tau)dt \;=\; E[x(t)x(t+\tau)]$$

$$(5.4)$$

For $\tau = 0$, Equ.(5.3) gives the total energy, and Equ.(5.4) the mean square value or power, and it is common to normalise the autocorrelation function to a maximum value of unity by dividing through by this value (the correlation cannot be better than the perfect correlation with itself). The Wiener-Khinchin relationship states that the autocorrelation function is the inverse Fourier transform of the power (amplitude squared) spectrum, even for a stationary random function. This can easily be understood for a transient (with power spectrum replaced by energy spectrum) by comparing Equ.(5.3) with (2.17) where it can be seen that the

324

autocorrelation represents a convolution with the same function reversed in time, so that its spectrum is the product of the original spectrum with its complex conjugate, and thus is the squared amplitude or energy spectrum.

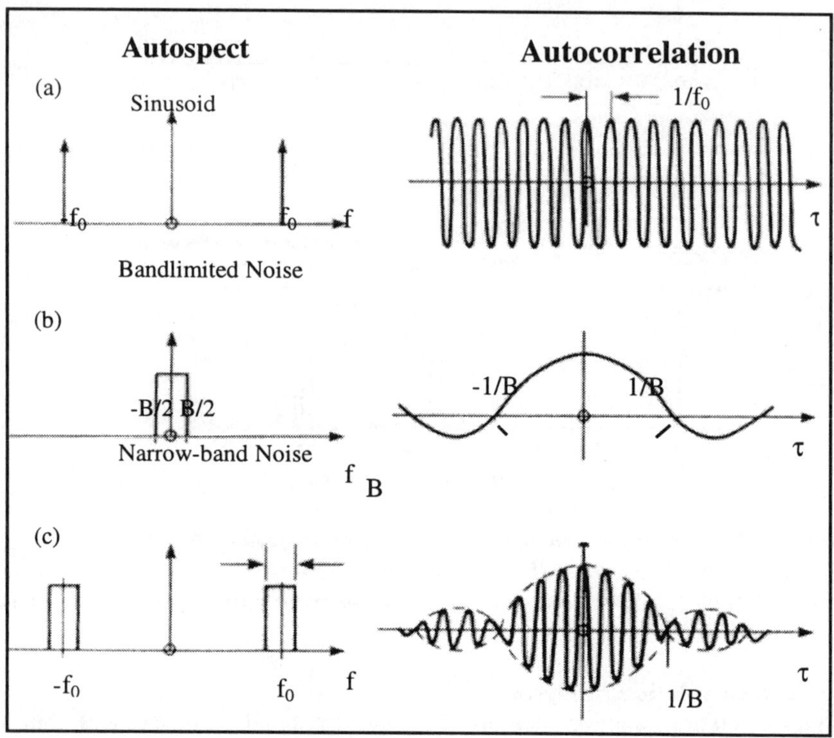

Figure 5.3. Autocorrelation vs autospectrum for three signals. Note that spectrum of (c) is the convolution of (a) and (b)

Figure 5.3 uses this relationship to derive the autocorrelation for a sinusoid and a bandlimited noise, and then a narrow-band noise as its spectrum can be generated as the convolution of the first two. This shows that the narrower the bandwidth of a noise signal, whether or not it is shifted from zero frequency, the more its autocorrelation is spread out in time. On the other hand, for a white noise, whose spectrum extends uniformly to infinity, its autocorrelation is concentrated at zero time lag.

The autocorrelation function can be used to enhance periodicity, as it separates this from broadband noise, and converts all sinusoids to cosines. This means that periodic functions become more pulse-like and thus apparent, since the phase of all harmonics is aligned once per period. The autocorrelation function can also be used to detect echos, but this is limited by the abovementioned spreading out of narrow-band functions, since even a perfect echo will just reproduce the autocorrelation at a lag corresponding to the echo delay time. As seen below, the cepstrum is a better echo detector, as in principle the echo delay is given by a delta function, independent of the bandwidth.

Figure 5.4 shows a case where masking noise was reduced considerably in an envelope spectrum for a bearing fault, by clipping the autocovariance function near zero time lag. The autocovariance is the autocorrelation performed after removing the mean value, which gives considerable distortion in an all-positive envelope signal. A further squaring of the autospectrum, corresponding to a further autocorrelation operation on the autocorrelation function, gives even better rejection of the noise but at the same time enhances large discrete peaks at the expense of small ones.

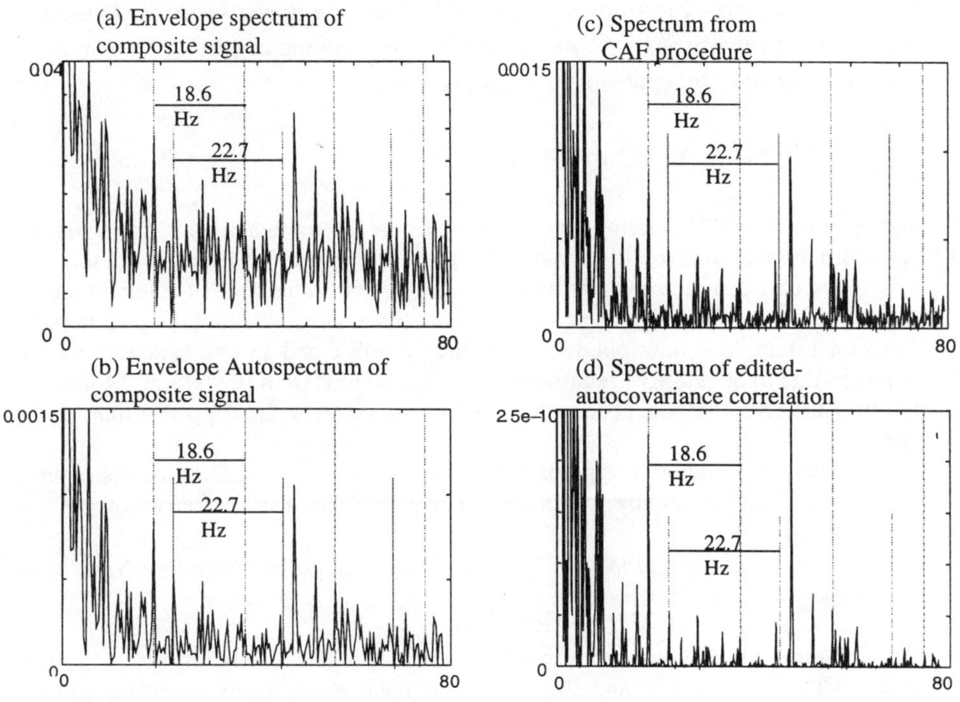

Figure 5.4. Use of CAF (clipped autocovariance function) (c) and its squared spectrum (d) to reduce noise effects

5.3 Cepstrum Analysis

The cepstrum is usually calculated from the spectrum, and often used to extract information about a spectrum, but will be treated here as a time domain technique as its x-axis is time, and it is closely related to the autocorrelation function just discussed. The most general definition of the cepstrum is as the inverse Fourier transform of a log spectrum, or:

$$C(\tau) = \mathfrak{J}^{-1}\left[\log(X(f))\right] \qquad (5.5)$$

326

where

$$X(f) = \Im[x(t)] = A(f)\exp(j\phi(f))$$
(5.6)

in terms of its amplitude and phase, so that:

$$\log(X(f)) = \ln(A(f)) + j\phi(f)$$
(5.7)

When $X(f)$ is complex as in this case, the cepstrum of Equation (5.5) is known as the "complex cepstrum", although since $\ln(A(f))$ is even and $\phi(f)$ is odd, the complex cepstrum is real-valued. When the power spectrum is used to replace the spectrum $X(f)$ in Equation (1), the resulting cepstrum, known as the "power cepstrum" or "real cepstrum", is given by:

$$C_{xx}(\tau) = \Im^{-1}[2\ln(A(f))]$$
(5.8)

and is thus a scaled version of the complex cepstrum where the phase of the spectrum is set to zero. It can also be seen from Equ.(5.5) and (5.8) that the difference between the cepstrum and the autocorrelation function is given by taking the log of the amplitude squared spectrum. Since forcing function and transfer function effects are multiplied in both the complex and power spectra, the log converts this to an additive relationship, which remains in the cepstrum. Moreover, the two additive components are often better separated in the cepstrum than in the spectrum.

There is yet another type of cepstrum known as the "differential cepstrum", which is defined as minus the inverse Z-transform of the derivative of the log spectrum, or:

$$C_d(n) = -Z^{-1}\left[z\frac{d}{dz}(\log\{X(z)\})\right]$$
(5.9)

and can be directly calculated from the time signal as:

$$C_d(n) = \Im^{-1}\left[\frac{\Im\{n\,x(n)\}}{\Im\{x(n)\}}\right]$$
(5.10)

where the Fourier transforms are to be interpreted as evaluating the Z-transform on the unit circle. One advantage is that it is not then necessary to unwrap the phase of the spectrum to a continuous function as is the case with the complex cepstrum.

Where the frequency spectrum $X(f)$ in Equ.(5.5) is a frequency response function (FRF) which can be represented in the Z-plane by a gain factor K and the zeros and poles inside the unit circle, a_i and c_i, respectively, and the zeros and poles outside the unit circle, $1/b_i$ and $1/d_i$, respectively, (where $|a_i|, |b_i|, |c_i|, |d_i| < 1$) then it can be shown that the complex cepstrum is given by the analytical formulae:

$$C(n) = \ln(K) \qquad\qquad , n = 0$$

$$C(n) = -\sum_i \frac{a_i^{\,n}}{n} + \sum_i \frac{c_i^{\,n}}{n} \qquad , n > 0 \qquad\qquad (5.11)$$

$$C(n) = \sum_i \frac{b_i^{\,-n}}{n} - \sum_i \frac{d_i^{\,-n}}{n} \qquad , n < 0$$

in terms of the time index n (known as quefrency). When grouped in complex conjugate pairs, these terms represent exponentially damped cosines, further weighted by the hyperbolic function $1/n$. For the differential cepstrum, the differentiation in the frequency domain results in a multiplication by n in the quefrency domain, giving exponentially damped cosines which are mathematically similar to modal impulse responses. For minimum phase FRFs, there are no poles or zeros outside the unit circle, and the cepstrum becomes causal (positive quefrency only), so that the real and imaginary parts of its Fourier transform (the log amplitude and phase of the spectrum, respectively) are related by a Hilbert transform. This means that the phase does not have to be measured or unwrapped, as the complex cepstrum can be obtained from the real cepstrum by multiplying by 2 times the Heaviside function H(n).

The cepstrum was originally proposed by Bogert et al [5.2] to detect echos, as it can be shown that an echo manifests itself in both the log amplitude and phase spectra as a periodic function and thus transforms to a series of delta functions (known as "rahmonics"), with a spacing equal to the echo delay time, in the cepstrum. Figure 5.5 illustrates this for a signal with two evenly spaced echos, and shows how the latter may be removed, even when they overlap the original function. This leads to one set of applications for the cepstrum.

328

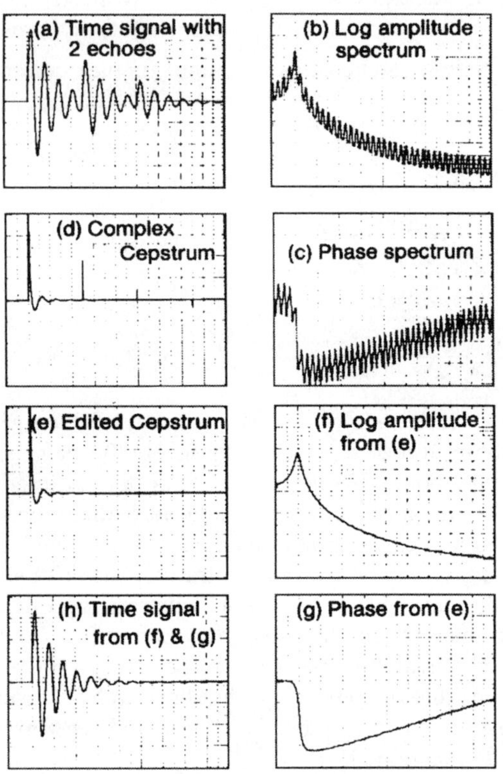

Figure 5.5. Use of complex cepstrum to remove echos

Taking the logarithm of the spectrum often enhances the pattern of equally spaced spectral components, such as harmonics and sidebands, in for example signals from gearboxes, where on a linear scale only the very largest would be visible. The cepstrum condenses such patterns into a small number of rahmonics and makes for easier comparison [5.3]. Recently it has been shown that the two sets of rahmonics produced by two gears in mesh are complementary in that the sum of the values of the first rahmonic in each series is a constant equal to 0.5 in the absence of noise, or somewhat less with noise [5.4]. If the condition of one gear deteriorates its component will increase, but is accompanied by a corresponding decrease in the other. This leads to a very general and sensitive indicator of change which is only dependent on the signal/noise ratio. This research has also led to a technique to detect and locate spalled gear teeth [5.5].

Because the forcing function and transfer function are additive in the response cepstrum, at least in SIMO situations, it is possible to separate them in response measurements if the log spectrum of the excitation is smooth and flat, so that the corresponding cepstrum is short. In [5.6] it was shown how the poles and zeros of the FRF could be extracted from the response cepstrum, in a windowed range of quefrency with negligible force effect. To use this information to reproduce the FRF requires an overall scaling factor, and an equalisation curve to compensate for the effects of unmeasured out-of-band modes. In [5.6] it was shown that these could be determined from an initial measurement, or a reasonably accurate finite element model, even when actual natural frequencies were different by up to 10%. Figure 5.6 illustrates this for two cases, one where a number of FRFs for a beam were determined by combining the poles and zeros from the response cepstrum with a scaling and equalisation function from an FEM model, and another where this was further extended by tracking the changes due to milling a slot in the middle of the beam. In [5.6] the measurements were made using impulsive excitation with very little background noise, but the method has recently been extended [5.7] to a case where broadband random excitation was used in the presence of other excitations. The response to the dominant excitation could be extracted by singular value decomposition provided it was larger by a factor of four than the next largest force.

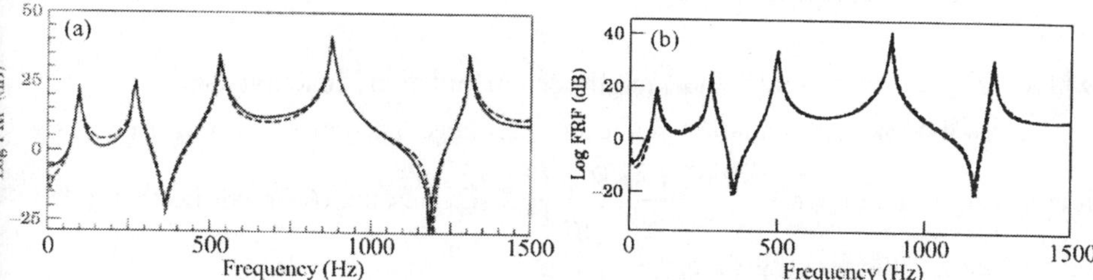

Figure 5.6. FRFs regenerated from response cepstra (full) compared with direct measurement (dashed) for impact measurements on a beam. (a) Using an FEM model for scaling and equalisation (b) After milling a slot in the middle of the beam (note reduction in frequency of odd-numbered modes)

6. OTHER SIGNAL ANALYSIS TECHNIQUES

6.1 Demodulation

Modulation occurs when an otherwise sinusoidal signal, a so-called carrier signal, has its amplitude or frequency made to vary with time. In the first case it is known as amplitude modulation, and in the second it can be considered as a frequency or phase modulation. Phase modulation is the deviation in phase (angular displacement) from the linearly increasing phase of the carrier, while frequency modulation is the difference in instantaneous frequency (angular velocity) from the constant carrier frequency. Thus, frequency modulation is the derivative of phase modulation. A direct mechanical example of phase/frequency modulation is shaft torsional vibration, which when expressed in terms of shaft angle is a phase modulation, and when in terms of shaft speed is a frequency modulation. There is no modulation term for the angular acceleration obtained by further differentiation.

A mechanical example of amplitude modulation is the variation in vibration amplitude at the meshing frequency in a gearbox, as the increase in tooth deflection with load gives an increasing departure from ideal involute profiles, and often tooth load varies periodically with the rotation of the gears.

The signals produced by faults in rolling element bearings are a series of high frequency bursts as resonance frequencies are excited by near periodic impacts. The diagnostic information is contained in the repetition frequency, not in the resonance frequencies excited, but spectra obtained by direct Fourier analysis are dominated by the latter, and the important information is disguised by smearing of the high order harmonics. Such signals can be modelled as an amplitude modulation of a carrier signal at the resonance frequency by a near periodic series of exponential pulses (though in general there will also be a jump in phase at the start of each new pulse). In so-called "envelope analysis" the signal envelope is extracted by amplitude demodulation, and frequency analysed to reveal the repetition frequencies even when these have a small random fluctuation.

Thus, a generally modulated signal can be represented by:

$$A_m(t) \cos\left(2\pi f_c t + \phi_m(t)\right) \tag{6.1}$$

where $A_m(t)$ represents the amplitude modulation function and $\phi_m(t)$ represents the phase modulation function in radians. The corresponding frequency modulating function (in Hz) is $\dfrac{1}{2\pi} \dfrac{d\phi_m(t)}{dt}$. Expression (6.1) will be seen to be the real part of the rotating vector:

$$A_m(t) \exp\left\{ j\left(2\pi f_c t + \phi_m(t)\right)\right\} \tag{6.2}$$

whose modulus is the amplitude modulating function and whose phase is the phase modulating function plus the linear carrier component. Thus, if it is desired to demodulate a real signal such as (6.1), it is desirable to find the corresponding imaginary part so as to form the complex expression (6.2). Provided the fluctuating part of (6.2),

$$A_m(t) \exp\left\{ j\phi_m(t)\right\} \tag{6.3}$$

has a half bandwidth less than the carrier frequency f_c , the spectrum of (6.2) will be one-sided, and (6.2) will be an analytic function. In this case the required imaginary part is the Hilbert transform of the real part, and the methods of Section 2.3 can be used. As the spectra of the two parts are convolved, the total bandwidth is less than the sum of the individual bandwidths. The bandwidth of the amplitude part is directly that of $A_m(t)$, and even though that of $\exp\left\{ j\phi_m(t)\right\}$ is theoretically infinite, if the maximum phase deviation is less than 1 radian, the effective bandwidth (within the dynamic range) is less than twice that of $\phi_m(t)$.

Note that a zoom processor can be used directly both to extract that part of the spectrum to be demodulated, and to remove the carrier component by zooming at the carrier frequency. Generally, the zoom process results at the same time in a considerable reduction in the sampling rate to be more compatible with the bandwidth of the modulating functions. The modulus of the complex output from the zoom processor is the amplitude modulating function, while the argument is the phase modulating function. This may have to be unwrapped to a continuous phase function (ie eliminating jumps over 2π), but in general this is not a problem for well-behaved functions. Demodulating a larger bandwidth decreases the time step, and thus phase jump, between samples and may facilitate unwrapping.

As illustrated in Figure 6.1, the same thing can be achieved using FFT transforms, although the first one will have to be large to accommodate the high carrier frequency while being long enough to contain sufficient periods of the lower modulating frequencies. Where phase demodulation is required, the centre of the demodulation band will have to be shifted to zero frequency (and negative frequency components shifted to the other end of the frequency record). However, for amplitude demodulation the result is unaffected by the frequency shift, and it is more convenient to shift the left hand end of the band to zero frequency, and pad the negative frequency side with zeros, thus maintaining an analytic signal. In either case there should be at least as many contiguous zeros in the spectrum as components, since the modulus is the square root of the amplitude squared, and the latter corresponds to the convolution of the spectrum with its complex conjugate reversed end-for-end. The zeros prevent extraneous wrap-around errors.

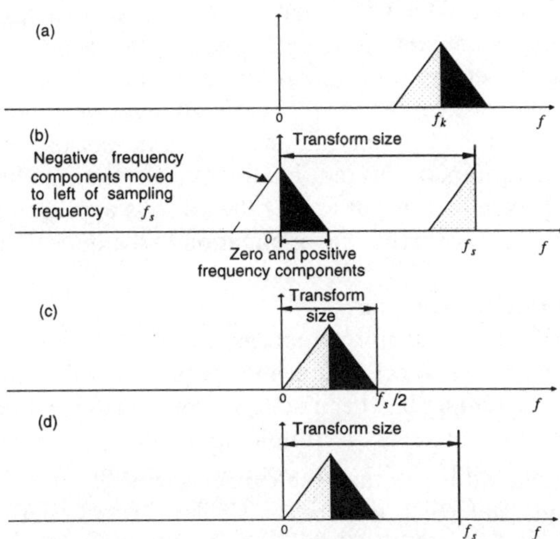

Figure 6.1 Block shift procedure for selecting frequency band for demodulation (a) Original spectrum of one-sided bandpass section (b) Frequency shift by f_k (centre of passband) (c) Frequency shift by amount corresponding to lower passband limit (half size transform) (d) Frequency shift by amount corresponding to lower passband limit (full size transform)

332

6.2 Cyclostationarity and Spectral Correlation

Many machine generated signals are almost but not completely periodic, even after compensation for small speed changes by order tracking. One example are the vibration signals from rolling element bearing faults, where even though theoretical fault impact frequencies can be calculated on the basis of pure rolling contact, the actual frequencies fluctuate a small amount because of random slip. Another is the vibration signal from an internal combustion engine, as even though some events such as piston motion and valve operation are completely periodic, the combustion pressure signal varies randomly from cycle to cycle. In such cases synchronous averaging cannot be used, at least without losing the variable part of the signal. However, such signals are often cyclostationary [6.1] meaning that their second order statistics, such as autocorrelation functions, are periodic.

In the definition of the autocorrelation function of Equ.(5.4), it was assumed to be a function of delay time τ only, as the function was assumed stationary with statistical properties independent of t. However, in the more general case of non-stationary signals, the second order statistical moment, the autocorrelation function, varies with both t and τ, and it is possible to make two-dimensional Fourier transforms with respect to each. By the Wiener-Khinchin relationship, that with respect to t gives autospectra against normal frequency, while that with respect to τ gives the spectral correlation against "cyclic frequency", normally represented by α. For stationary signals, there is a peak only for $\alpha = 0$, but for cyclostationary signals there will be peaks also for frequency shifts where spectrum bands correlate with each other. As an example, where two different carrier frequencies are amplitude modulated by the same narrow-band noise signal, there will be a perfect correlation for a frequency shift corresponding to the separation of the two carrier frequencies, giving a very narrow peak in the cyclic frequency direction, even though the width of peaks in the normal frequency direction is determined by the bandwidth of the modulating noise function. Ref.[6.1] is an excellent discussion of the various applications of cyclostationarity, while [6.2] and [6.3] describe the application to the detection of gear faults.

6.3 Time-Frequency Analysis

In theory the Fourier transform requires integration over all time, but we are all aware that the ear can detect changes in frequency with time (what is music?), and so an analysis technique has been sought which matches the ear's ability to follow changing frequency patterns. A simple approach is to move a short time window along the record and obtain the Fourier spectrum as a function of time shift. However, the uncertainty principle means that the frequency resolution is the reciprocal of the effective time window length, and this does not seem to accord with the ear's appreciation of a tonal quality of a note even if it lasts for a short time. However, the so-called STFT (short time Fourier transform) is sometimes useful for tracking changes in frequency with time, even with the restriction of resolution. It is described by the formula:

$$S(f,\tau) = \int_{-\infty}^{\infty} x(t)\, w(t-\tau) \exp(-j2\pi t)\, dt \qquad (6.4)$$

where $w(t)$ is a window which is moved along the record. Normally, the amplitude squared $|S(f;\tau)|^2$ is displayed on a time vs frequency diagram, in which case it is sometimes known as a spectrogram. The window could be of finite length such as a Hanning window, or theoretically infinite such as a gaussian window, but in practice of course it must be truncated.

6.3.1 The Wigner-Ville distribution

The Wigner-Ville distribution (WVD) appears to violate the uncertainty principle in appearing to give better resolution than the STFT, but suffers from interference components between the actual components. The original Wigner distribution [1.3] was modified by Ville [1.4] who proposed the analysis of the corresponding analytic signal so as to eliminate interference between positive and negative frequency components. The WVD is one of the so-called "Cohen's class" of time-frequency distributions [6.4], most of which have been proposed to improve on the WVD in some way. Even the STFT falls into this class. Cohen's class may be represented by the formula:

$$C_x(t, f, \phi) = \Im\{R(t;\tau)\} \qquad (6.5)$$

where $R(t;\tau)$ is a weighted autocorrelation-like function defined by:

$$R(t;\tau) = \int_{-\infty}^{\infty} x\left(u+\frac{\tau}{2}\right) x^*\left(u-\frac{\tau}{2}\right) \phi((t-u),\tau)\, du \qquad (6.6)$$

and $\phi(u,\tau)$ is a kernel function used to smooth the WVD (with $\phi = 1$ the WVD is obtained). The "pseudo Wigner-Ville distribution" is a finite windowed version of the WVD and the "smoothed pseudo Wigner-Ville distribution" suppresses interference in both the time and frequency directions. Figure 6.2 compares the WVD and the smoothed pseudo-WVD against the STFT for a vibration signal from a portion of a diesel engine cycle, and shows that at least in this case the smoothing gives a simultaneous resolution in both directions that is better than the STFT while still suppressing the major interference components. In Ref.[6.5] the proposal is made to use various smoothing techniques to locate the interference components, and then remove them from the unsmoothed WVD to retain optimum resolution.

334

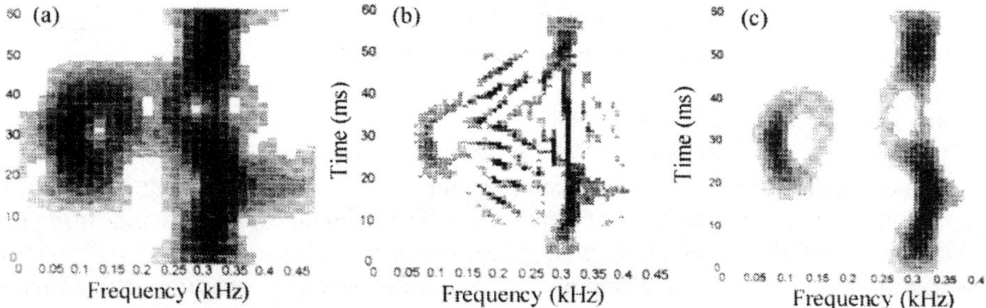

Figure 6.2.Comparison of time-frequency distributions for a diesel engine vibration signal (a) STFT (b) Wigner-Ville distribution (c) Smoothed pseudo Wigner-Ville

6.3.2 Wavelet analysis

Another approach to time-frequency analysis is to decompose the signal in terms of a family of "wavelets" which have a fixed shape, but can be shifted and dilated in time. The formula for the wavelet transform is:

$$W(a;b) = \frac{1}{\sqrt{a}} \int_{-\infty}^{\infty} x(t)\,\psi*\!\left(\frac{(t-b)}{a}\right) dt \qquad (6.7)$$

where $\psi(t)$ is the mother wavelet, translated by b and dilated by factor a. Since this is a convolution, the wavelets can be considered as a set of impulse responses of filters, which because of the dilation factor have constant percentage bandwidth properties. In principle, they are not very different from $1/n$-octave filters, but with zero phase shift because the mother wavelet is normally centered on zero time. The dilation factor a is known as scale, but represents log frequency, as for constant percentage bandwidth filters. Wavelets give a better time localisation at high frequencies, and for that reason can be useful for detecting local events in a signal. Many authors have described their use for detecting local faults in gears and bearings (eg [6.6, 6.7]).

6.3.3 Other time-frequency techniques

A number of other time-frequency techniques have been proposed, including the evolutive AR methods mentioned in Section 4.1.2, and time-frequency "worms" [6.8]. It is still an area where considerable development is going on. One problem with time-frequency representations is the problem with interpreting the results. Even if a trained eye can see the desired effect, quite advanced image processing is sometimes required to extract the information.

7. MULTIPLE CHANNEL ANALYSIS

For system analysis, it is normally necessary to measure both input and output signals at the same time, and process them in pairs to obtain transfer function information, such as FRFs. With multiple inputs and/or outputs, it can be necessary to process more than two signals at a time, by matrix methods, but that will not be considered in detail here, as it is covered in other specialist papers.

7.1   The Cross Spectrum

The cross spectrum between two signals is obtained by multiplying the spectrum of one by the complex conjugate of the spectrum of the other (the latter being considered as the input, since the phase of the result has its phase as datum). As with the autospectrum, to which it reduces if the two signals are identical, for random signals the individual estimates must be averaged to obtain meaningful results. Thus:

$$G_{AB}(f) = \mathrm{E}\big[G_A{}^*(f) \cdot G_B(f)\big] \qquad (7.1)$$

where $\mathrm{E}[.]$ is the statistical expectation, or averaging, operation.

By the Wiener-Khinchin relationship, the cross spectrum inverse transforms to the cross correlation function:

$$g_{AB}(t) = \mathrm{E}\big[a(t)\, b(t + \tau)\big] \qquad (7.2)$$

which reverts to the autocorrelation function (Equ.(5.4)) when the signals are identical.

7.2   The Frequency Response Function (FRF)

In the frequency domain the FRF is basically the ratio of output over input, or:

$$H_{AB}(f) = \frac{G_B(f)}{G_A(f)} \qquad (7.3)$$

but in the presence of noise a better estimate can be obtained as described in other specialist papers. If noise is primarily located in the output signal (over which the experimentalist has less control), the procedure is modified by multiplying numerator and denominator by the complex conjugate of the input spectrum, and averaging, to give:

$$H_1(f) = \frac{\mathrm{E}\big[G_B(f)\, G_A{}^*(f)\big]}{\mathrm{E}\big[G_A(f)\, G_A{}^*(f)\big]} = \frac{G_{AB}(f)}{G_{AA}(f)} \qquad (7.4)$$

since averaging reduces the effect of noise on the cross spectrum. If noise is primarily located in the input signal, a better estimate is given by:

$$H_2(f) = \frac{E\left[G_B(f)\,G_B{}^*(f)\right]}{E\left[G_A(f)\,G_B{}^*(f)\right]} = \frac{G_{BB}(f)}{G_{BA}(f)} \qquad (7.5)$$

The system impulse response function can be obtained by inverse transformation of the FRF.

7.3 Coherence

The coherence gives a measure of the degree of linear relationship between two signals as a function of frequency. It is calculated by the formula:

$$\gamma^2(f) = \frac{\left|G_{AB}(f)\right|^2}{G_{AA}\,G_{BB}} \qquad (7.6)$$

This has values between zero and one depending on the degree of linearity. For a single estimate, it will always be one, since $\left|G_{AB}\right| = \left|G_A\right| \cdot \left|G_B\right|$ and in the absence of noise and nonlinearity each estimate of G_{AB} will have the same amplitude and orientation so that the average value will not change. However, in the presence of noise and/or nonlinearity the various estimates will change in length and particularly phase, so that the average obtained by vector summation will be less than if they were aligned. If there is no linear relationship, the various estimates will have random orientation and the average will tend to zero.

It is possible for signals to have partial coherence [7.1], but to determine this requires multiple channel processing and is beyond the scope of this contribution.

7.4 Separation of Sources

This is an area of signal processing which is receiving considerable attention and will presumably result in far-reaching consequences in the next millennium. So far there has been little application to mechanical problems, but if the type of result which has already been achieved in, for example, communications can be translated to structural dynamics, there is a strong likelihood that it will be possible to determine modal properties of structures from response measurements only. Such results obtained in service would give the values which actually apply in each situation. In Section 5.3 a case was mentioned where it was possible using singular value decomposition to separate out the response to the largest forcing function, provided it was at least four times larger than the next largest. This is a limitation given by "second order statistics". However, in Refs.[7.2-7.4] it is demonstrated that it should be possible to separate responses to different sources, even of comparable strength, if their higher order statistics are different.

References

1.1 Cooley, JW & Tukey JW, (1965) "An Algorithm for the Machine Calculation of Complex Fourier Series", *Math. Of Comp.*, **19**(90), pp297-301.

1.2 Burg, JP, (1975), *Maximum Entropy Spectral Analysis*, PhD Dissertation, Stanford University, Stanford, CA, USA.

1.3 Wigner, EP, (1932) "On the Quantum Correction for Thermodynamic Equilibrium." *Physical Review*, **40**, pp749-759.

1.4 Ville, J, (1948) "Théorie et Applications de la Notion de Signal Analytique," *Cables et Transmission*, **2 A**, pp61-74.

1.5 Newland, DE, (1993) *An Introduction to Random Vibrations, Spectral and Wavelet Analysis.* 3rd Ed., Longmans Scientific, Harlow, UK

1.6 Ibrahim, SR (1984) "A Modal Identification Algorithm for Higher accuracy requirements". *AIAA Proc. 25th Structures, Structural Dynamics & Materials Conf.*Paper 84-0928, pp117-122.

2.1 Randall, RB (1987), *Frequency Analysis*, 3rd ed, Bruel & Kjaer, Naerum, Denmark.

3.1 Randall, RB (1996) "Vibration Measurement Instrumentation", Ch. 13, and "Spectrum Analyzers and their Use", Ch.14 *in Shock and Vibration Handbook*, (Ed. CM Harris), McGraw-Hill, NY.

4.1 Braun, S & Hammond JK (1986) "Parametric Methods" in *Mechanical Signature Analysis*, (Ed. S. Braun), Academic Press, London.

4.2 Robinson, EA (1982) "A Historical Perspective of Spectrum Estimation", *Proc.IEEE* **70**(9), 885-907.

4.3 Gao,Y, Randall RB & Ford, R (1998) "Estimation of Envelope Spectra using Maximum Entropy Spectral Analysis and Spectral Interpolation", *Int. J. of Comadem*, 1(3), pp15-22.

4.4 Bardou,O & Sidahmed, M (1994) "Early Detection of Leakages in the Exhaust and Discharge Systems of Reciprocating Machines by Vibration Analysis" *Mech. Systems & Signal Processing*, **8**(5), pp551-570

4.5 McFadden, PD (1986) "Interpolation Techniques for the Time Domain Averaging of Vibration Data with Application to Helicopter Gearbox Monitoring", *DSTO-ARL Report No.AR-004-488*, Comm. of Australia.

5.1 McFadden, PD (1987), "A Revised Model for the Extraction of Periodic Waveforms by Time Domain Averaging", *Mech. Systems & Signal Processing*, **1**(1), pp83-95.

5.2 Bogert BP et al (1963) "The Quefrency Analysis of Time Series for Echoes: Cepstrum, Pseudo-autocovariance, Cross-cepstrum and Saphe Cracking", in *Proc. Symp. Time Series Analysis*, (Ed. M Rosenblatt) Wiley, NY, pp209-243.

5.3 Randall,RB (1982/3) "The Application of Cepstrum Analysis in Gear Diagnostics" *Maintenance Management Int.* **3**, pp183-208.

5.4   El Badoui M, Guillet, F & Danière, J (1998) "Energy Cepstrum for the Diagnostic of Complex Gears" (in French). *3rd Int. Conf. on Acoustical and Vibratory Surveillance Methods and Diagnostic Techniques*, CETIM, Senlis, France, pp291-300.

5.5   Cahouet V, El Badaoui M, Velex P, Guillet F & Danière J (1999) "Simulations Numériques et Détection d'Avaries sur les Dentures d'Engrenages Cylindriques", *4th World Congress on Gears and Power Transmissions*, CNIT, Paris La Défense, 16-18 March.

5.6   Gao, Y & Randall, RB (1996) "Determination of Frequency Response Functions from Response Measurements."Part I: "Extraction of Poles and Zeros from Response Measurements" *Mechanical Systems and Signal Processing*, **10**(3), May 1996, pp293-317.Part II: "Regeneration of Frequency Response Functions from Poles and Zeros", pp319-340.

5.7   Randall, R B, Gao, Y & Swevers, J (1998) "Updating Modal Models from Response Measurements". *ISMA23 Conference*, KUL, Leuven, Belgium, pp1153-1160.

6.1   Gardner, WA (Ed), (1994) *Cyclostationarity in Communications and Signal Processing*, IEEE Press, NY.

6.2   Capdessus, C, Lacoume, JL & Sidahmed, M (1994) "Cyclostationarity: a New Signal Processing Concept for Vibration Analysis and Diagnostic" *1994 Int. Gearing Conference*, Newcastle, UK, Mechanical Engineering Publications, London.

6.3   Dalpiaz, G, Rivola, A & Rubini, R (1998) "Gear Fault Monitoring: Comparison of Vibration Analysis Techniques", *3rd Int. Conf. on Acoustical and Vibratory Surveillance Methods and Diagnostic Techniques*, CETIM, Senlis, France, pp623-637.

6.4   Cohen, L (1995) *Time-frequency Analysis*, Prentice-Hall, NJ.

6.5   Chiollaz, M & Favre, B, (1993) "Engine Noise Characterisation with Wigner-Ville Time-Frequency Analysis", *Mech. Systems & Signal Processing*, **7**(5), pp375-400.

6.6   Lin, ST & McFadden, PD (1995) "Vibration Analysis of Gearboxes by the Linear Wavelet Transform." *Second Int. Conf. on Gearbox Noise, Vibration and Diagnostics* IMechE, London

6.7   Staszewski, WJ & Tomlinson, GR (1994) "Application of the Wavelet Transform to Fault Detection in a Spur Gear." *Mech. Systems & Signal Processing* **8**(3) pp289-307, May.

6.8   Léonard, F (1998) "Object Transform and Time-frequency Worms" (in French), *3rd Int. Conf. on Acoustical and Vibratory Surveillance Methods and Diagnostic Techniques*, CETIM, Senlis, France, pp313-322.

7.1   Bendat JS & Piersol, AG (1993) *Engineering Applications of Correlation and Spectral Analysis*. 2nd Ed., Wiley, NY.

7.2   Capdeville, C, Servière, Ch & Lacoume, JL (1995) "Blind Separation of Sources in the Frequency Domain" *Proc. ICASSP95*, pp2080-2083.

7.3   Capdeville, C, Servière, Ch & Lacoume, JL (1996) "Blind Separation of Wide-Band Sources: Application to Rotating Machine Sources", *EUSIPCO*, pp.2085-2088

7.4   Gaeta, M, Harroy, F & Lacoume, JL (1992) "Utilisation de Statistiques d'Ordre Supérieur pour la Séparation de Sources Vibrantes", *1st Int. Conf. on Acoustical and Vibratory Surveillance Methods and Diagnostic Techniques*, CETIM, Senlis, France, pp523-533.

Structural Dynamics Measurements

Mark H. Richardson
Vibrant Technology, Inc.
Jamestown, CA 95327

1.   INTRODUCTION

In this contribution, the term "structural dynamics measurements" will more specifically mean the measurement of the vibration of mechanical structures and machinery. Because this topic is so broad in scope, modal analysis and signal processing are also discussed, but other papers at this conference are specifically devoted to those topics and cover them in more detail.

1.1   Why Vibration Measurements?

Why are vibration measurements important? Because vibration contributes to a variety of undesirable behavior in machinery and structures. A machine or structure,

- •· may be uncomfortable to ride in,
- •· is too difficult to control,
- •· makes too much noise,
- •· doesn't maintain tolerances,
- •· wears out too fast,
- •· fatigues prematurely,
- •· breaks unexpectedly.

1.2   Types of Vibration

All structural vibration can be characterized as a combination of forced and resonant vibration. No vibration can occur at all unless forces are applied to the structure. However, resonant vibration can still occur after the forces have been removed. Resonant vibration is also conveniently characterized in terms of the modes of vibration of a structure.

1.3   Resonant Vibration

A structure's modal parameters (resonant frequency, damping, and mode shape) can be estimated from certain kinds of structural dynamics measurements. If excited, modes (or resonances), can act like "mechanical amplifiers". Modes can cause excessive vibration responses that are orders of magnitude greater than responses due to static loading.

1.4 Key Issues in Structural Dynamics Testing

Since dynamic behavior can be unpredictable due to the excitation of structural resonances, the most important question to be answered from structural dynamics testing is,

- •· Is a **structural mode** being excited?"

In addition, several other questions need to be answered,

- •· What are the **excitation forces**, and where are they coming from?
- •· Is the system **stationary** (not changing over time)?
- •· Is the system **linear**?

Modes are only defined for linear, stationary mechanical systems. However, most real structures may exhibit non-stationary (or non-steady state) and non-linear dynamic behavior. When testing for the modes of a structure, these issues and others must be taken into account.

1.5 Spectral Analysis

Probably the most convenient way to analyze a vibration signal is to obtain its frequency content, or frequency spectrum. There are at least two good reasons for this,

1. Excitation forces often provide sinusoidal excitation at specific frequencies. This is especially true for rotating equipment. These forces are manifested as peaks in a frequency spectrum.
2. Resonances are also manifested as peaks in a frequency spectrum.

Prior to the late 1960s, all structural dynamic testing was done with analog instrumentation. Sine wave generators were used to artificially excite structures, one frequency at a time. Oscilloscopes were used to look at the signals. Analog filters were used to suppress either the low or high frequency content of the signals. Special analog filters that changed with the frequency of excitation, known as swept filters, were used to obtain the structural response, one frequency at a time. This response to each sinusoidal excitation frequency is a frequency spectrum.

In the 1960s, commercial spectrum analyzers were marketed that utilized swept analog filters, and constructed the frequency spectrum of structural vibrations, one frequency at a time.

2. THE FFT ANALYZER

The Fourier transform is a mathematical procedure that was invented by a Frenchman named Jean-Baptiste-Joseph Fourier in the early 1800s. The Fourier transform yields the frequency spectrum of a time domain function. It is defined for continuous (or analog) functions.

The discovery of the Fast Fourier Transform (FFT) algorithm in the late 1960s opened up a whole new area of signal processing using a digital computer [1]. The FFT computes a discretized (sampled) version of the frequency spectrum of a sampled time signal. This discretized, finite length spectrum is called a Discrete Fourier Transform (DFT). Following its discovery, the FFT was implemented in a new kind of spectrum analyzer called an FFT, or Fourier analyzer.

Present day FFT analyzers can compute a DFT in milliseconds, whereas it used to take hours using standard computational procedures. From a DFT, FFT analyzers can calculate a variety of other frequency domain functions, including Auto Power Spectra (APSs), Cross Power Spectra (XPSs), Frequency Response Functions (FRFs), Coherences, etc.

3. RULES OF DIGITAL MEASUREMENT

There are three key equations that govern the use of the DFT. The first one describes the sampled signal in the time domain, the second describes the sampled spectrum in the frequency domain, and the third is Shannon's Sampling Theorem, also called the Nyquist sampling rate.

3.1 Time Waveform

The DFT assumes that the sampled time waveform contains N uniformly spaced waveform samples, with an increment of (Δt) seconds between samples. (The most common FFT algorithms restrict N to being a power of 2, although this is not necessary.) The total time period of sampling (also called the sampling window), starts at ($t = 0$) and ends at ($t = T$). Therefore,

$$T = (\Delta t)\, N \quad \textbf{(seconds)}$$

3.2 Frequency Waveform

The DFT assumes that the digital frequency spectrum contains N/2 uniformly spaced samples of complex valued data, with frequency resolution (Δf) between samples. The frequency spectrum is defined for the frequency range ($f = 0$) to ($f = $ Fmax). Therefore,

$$\textbf{Fmax} = (\Delta f)\,(N/2) \quad \textbf{(Hertz)}$$

3.3 Nyquist Sampling

Shannon's Sampling Theorem says that a frequency spectrum can only contain unique frequencies in a range from ($f=0$) up to a maximum frequency ($f = $ Fmax) equal to one half the sampling rate of the time domain signal. Therefore,

$$\textbf{Fmax} = (1/2)\,(1/\Delta t) \quad \textbf{(Hertz)}$$

3.4   Fundamental Rule

The three equations above can be used to derive the most fundamental rule of digital spectrum based testing,

$$\Delta f = (1/T)$$

This equation says *"To Improve Frequency Resolution, You Have to Wait"*. In other words, the frequency resolution obtainable in a digital spectrum depends on the time domain sampling window length (T), not the sampling rate. Stated differently, to get better frequency resolution, you have to sample over a longer time period.

4.   ZOOM MEASUREMENTS

A popular digital signal processing technique that is implemented in most FFT analyzers is the Zoom transform, or Zoom measurement. A Zoom transform is essentially a digital filtering operation that takes place after the time waveform has been sampled. It involves re-sampling, frequency shifting, and low pass filtering of the sampled data to yield a DFT with increased frequency resolution, but over a smaller frequency band.

The Zoom transform is very useful for obtaining better frequency resolution without having to perform an FFT on a very large number of samples. From a practical standpoint, the Zoom transform is much faster than using a base band FFT (starting at zero frequency) with more samples to obtain more frequency resolution.

As an example, in order to obtain 1 milli-Hz of resolution in the vicinity of 100 Hz, a base band (0 to Fmax) FFT would have to transform at least 262,144 samples. This would yield a base band spectrum between 0 and 132 Hz.

$$132 \text{ Hz} = (0.001 \text{ Hz}) (262,144 / 2)$$

Even though the Zoom transform starts with the same 262,144 time samples, the Zoom band can be centered around 100 Hz, and, assuming that a 1 Hz bandwidth is sufficient, the FFT only needs to transform 2048 samples,

$$1 \text{ Hz} = (0.001 \text{ Hz}) (2048 / 2)$$

5.   DIGITAL MEASUREMENT DIFFICULTIES

The rules above are basically all that is required to make digital measurements. However, there are two remaining difficulties associated with the use of the FFT. They are called aliasing and leakage.

5.1   Aliasing

Aliasing of a signal occurs when it is sampled at less than twice the highest frequency in the spectrum of the signal. When aliasing occurs, the parts of the signal at frequencies above the sampling frequency add to the part at lower frequencies, thus giving an incorrect spectrum.

All modern FFT analyzers guarantee that aliasing will not occur by passing the analog signals through anti-aliasing filters before they are sampled. An anti-aliasing filter band limits (low pass filters) the signal so that it contains no frequencies higher than the sampling frequency. Since all filters have a roll off frequency band, the cutoff frequency of the anti-aliasing filters is typically set to 40% of the sampling frequency. Therefore, 80% of a DFT frequency band (0 to Fmax) is considered to be alias-free.

5.2   Leakage

The FFT assumes that the signal to be transforming is periodic in the transform window. (The transform window is the samples used by the FFT). To be periodic in the transform window, the waveform must have no discontinuities at its beginning or end, if it were repeated outside the window.

Signals that are always periodic in the transform window are,

1. Signals that are completely contained within the transform window.
2. Cyclic signals that complete an integer number of cycles within the transform window.

Many common types of signals (such as random signals), may not be periodic in the transform window. If a time signal is not periodic in the transform window, when it is transformed to the frequency domain, a smearing of its spectrum will occur. This is called leakage. Leakage distorts the spectrum and makes it inaccurate.

5.3   Minimizing the Effects of Leakage

If a signal is non-periodic in its sampling window, it will have leakage in its spectrum. In this case, leakage can never be eliminated but it can be minimized. To minimize the effects of leakage, specially shaped windows are applied to the time waveforms after they are sampled, but before they are transformed using the FFT.

- **Hanning Window**: the Hanning window is effective for minimizing the effects of leakage in the spectra of broadband signals, such as random signals.
- **Flat Top Window**: the Flat Top (Potter P301) window is effective for minimizing the effects of leakage in the spectra of narrow band signals, such as sinusoidal signals.
- **Exponential Window**: this window is effective for minimizing the effects of leakage in impulse responses that don't damp out within the sampling window.

346

6.   SPECTRUM AVERAGING

Spectrum averaging in an option in most modern FFT analyzers. It is done with a spectrum averaging loop, as shown in Figure 1. Spectrum averaging is used to remove the effects of,

1. extraneous random noise.
2. randomly excited non-linearities.

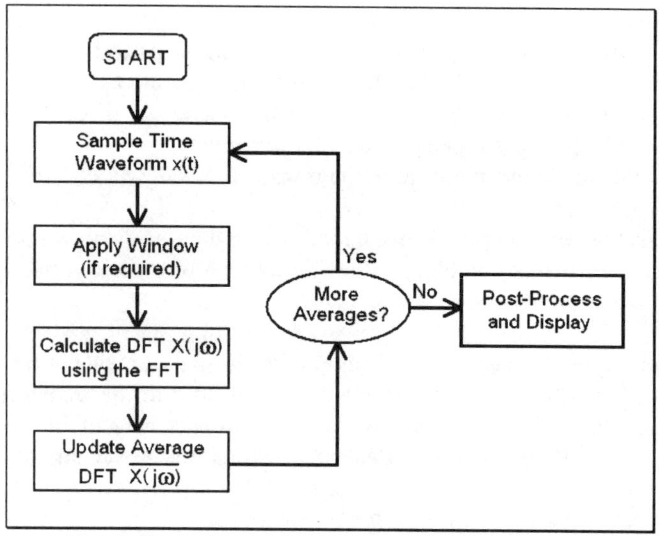

Figure 1. Spectrum Averaging Loop

In a spectrum averaging loop, multiple spectral estimates of the same signal are averaged together to yield a final estimate of the spectrum. Different types of averaging can be used, but the most common type (called stable averaging), involves summing all of the estimates together and dividing by the number of estimates.

The FFT is a linear, one-to-one transformation. That means that it uniquely transforms the vibration signal from a linear dynamic system into its correct digital spectrum, and vise versa. If a signal contains any additive Gaussian random noise or randomly excited non-linear behavior, these portions of the signal are transformed into spectral components that appear randomly in the spectrum.

6.1   Removing Random Noise & Non-Linearities

By summing together (averaging) multiple spectral estimates of the same signal, the linear spectral components will add up (re-enforce one another), while the random noise and non-linear components will sum toward zero, thus removing them from the resultant average spectrum.

In order to remove random noise and non-linearities while retaining the spectral components of the linear dynamics, we must guarantee that the magnitudes

& phases of the linear portion of all spectral estimates are the same. This depends on how the data are sampled in each sampling window.

6.2 Single Channel Versus Multi-Channel Measurements

FFT Analyzers can be classified into two categories, single channel and multi-channel. Each channel can process a unique signal. Single channel analyzers are the most popular because they cost less, but they also have limited measurement capabilities. The distinguishing feature of a multi-channel analyzer is that all channels are simultaneously sampled. Simultaneously sampled signals contain the correct magnitudes & phases relative to one another, since they are all sampled at the same moment in time.

In addition, the signals from all channels are filtering and processed so that they are all altered in the same way, within acceptable tolerances. (This requirement is a key reason why multi-channel systems are more costly.)

To help lower their cost, some multi-channel analysers use a multiplexer in front of a single ADC (analog-to-digital converter), instead of simultaneously sampling all signals with one ADC per channel. A multiplexer is essentially a very fast switch that switches from one channel to the next. Since the use of a multiplexer results in a time delay between the sampling of successive channels, each channel must be treated as if it were acquired with a single channel analyzer.

Any two simultaneously sampled signals can be used to form a Cross Power Spectrum (XPS), a fundamental multi-channel channel measurement function.

Using spectrum averaging, a single channel analyzer can remove noise and non-linearities from a spectrum if the measurement process is a repeatable process. A multi-channel analyzer only requires a less restrictive steady state (stationary) process.

6.3 Repeatable Process

In a repeatable measurement process, data acquisition must occur so that exactly the same time waveform is obtained in the sampling window, every time one is acquired. Figure 2 depicts a repeatable process. For a repeatable process, the magnitude & phase of each sampled signal are assumed to be unchanging throughout the measurement process.

Using spectrum averaging, a single channel analyzer can remove noise and non-linearities from a spectrum if the measurement process is repeatable. A repeatable process guarantees the same results as simultaneous sampling. That is, it guarantees that multiple signals will have the correct magnitudes & phases relative to one another, whether they are acquired one at a time or simultaneously. Therefore, if a repeatable measurement process can be achieved, multiple channels of data can be acquired one at a time if necessary.

348

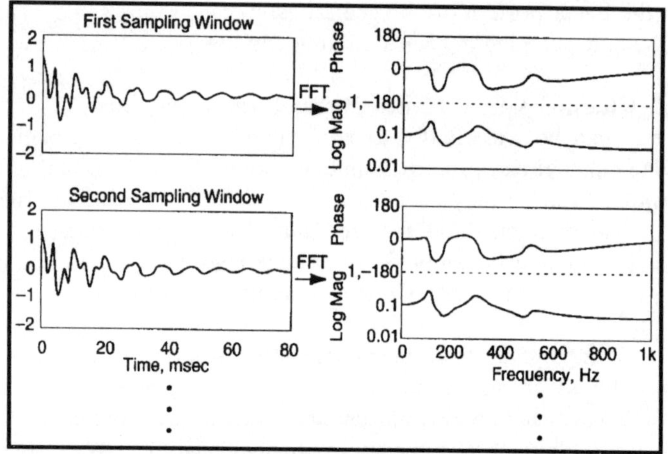

Figure 2. Repeatable Measurement Process.

To insure a repeatable process, an external trigger is usually required to capture a repeatable event in the sampling window. In machinery applications, the trigger is usually obtained as a tachometer signal from a rotating shaft.

With a repeatable process, time domain averaging can also be done to remove random noise and random non-linearities. This is also called synchronous averaging. Unfortunately, a repeatable measurement process cannot be achieved in many test situations.

6.4   Steady State Process

A steady state measurement process can often be achieved in situations where a repeatable process cannot. A steady state process is achieved when the Auto Power Spectrum (APS) of a signal does not change from measurement to measurement. (An APS is merely the magnitude squared of an FFT, or linear spectrum.) Figure 3 shows a steady state process. Notice that the time waveform can be different in each sampling window, but its APS does not change. No special triggering is required for steady state measurement.

7.   TRI-SPECTRUM AVERAGING

The measurement capability of most multi-channel FFT analyzers is built around a tri-spectrum averaging loop, as shown in Figure 4. This loop assumes that two or more time domain signals are simultaneously sampled. Three spectral estimates, an Auto Power Spectrum (APS) for each channel, and the Cross Power Spectrum (XPS) between the two channels, are calculated in the tri-spectrum averaging loop. After the loop has completed, a variety of other cross channel measurements (including the FRF), are calculated from these three basic spectral estimates.

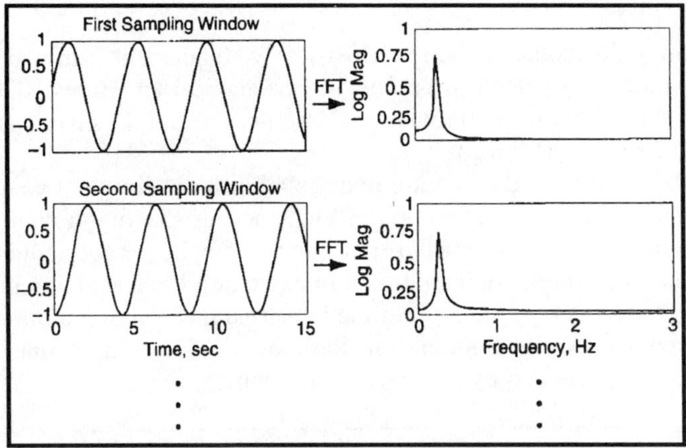

Figure 3. Steady State Measurement Process.

In a multi-channel analyzer, tri-spectrum averaging can be applied to as many signal pairs as desired. Tri-spectrum averaging will remove random noise and randomly excited non-linearities from signals taken during a steady state measurement process. This is particularly useful for measuring FRFs.

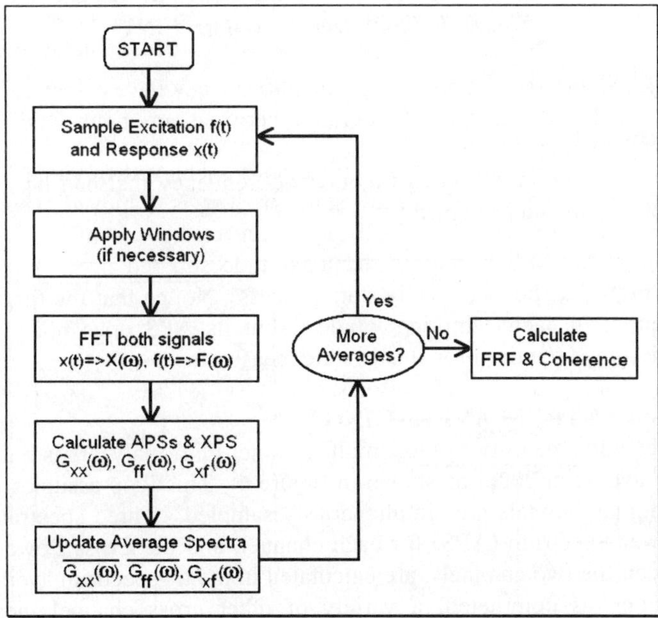

Figure 4. Tri-Spectrum Averaging Loop

8.   THE FRF

The Frequency Response Function (FRF) is a fundamental measurement that isolates the inherent dynamic properties of mechanical structures. Experimental modal parameters (resonant frequency, damping, and mode shape) are obtained from a set of FRF measurements.

The FRF describes the input-output relationship between two points (and directions) on a structure as a function of frequency, as shown in Figure 5. That is, the FRF is a measure of how much displacement, velocity, or acceleration response a structure has at an output point, per unit of excitation force at an input point.

The FRF is defined as the ratio of the Fourier transform of a motion output (or response) divided by the Fourier transform of the force input that caused the output. This is represented by the diagram in Figure 5.

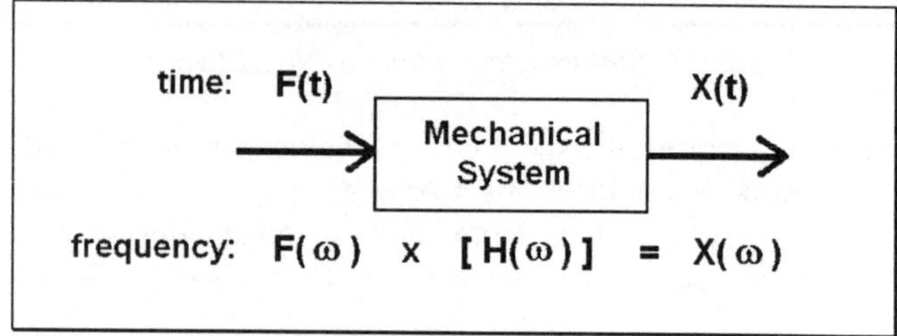

Figure 5. Block Diagram of an FRF.

Since both force and motion are vector quantities (they have directions associated with them), each FRF is actually defined between an input DOF (point and direction), and an output DOF.

An FRF is a complex valued function of frequency that can be displayed in various forms, as shown in Figure 6.

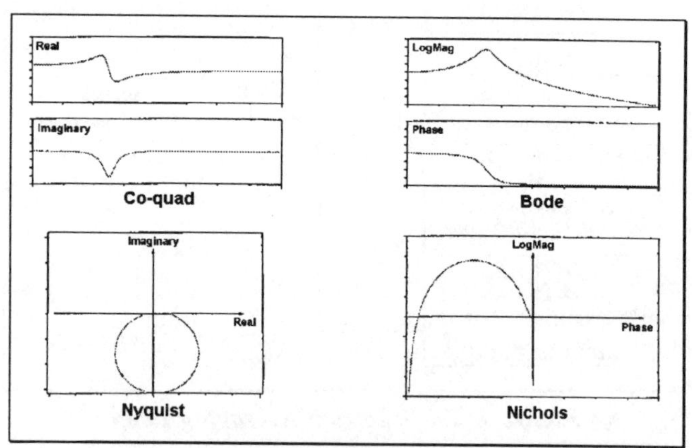

Figure 6. Alternate Forms of the FRF.

Depending on whether motion is measured as displacement, velocity, or acceleration, the FRF and its inverse have a variety of names,

- •· Compliance (displacement / force)
- •· Mobility (velocity / force)
- •· Inertance (acceleration / force)
- •· Dynamic Stiffness (1 / Compliance)
- •· Impedance (1 / Mobility)
- •· Dynamic Mass (1 / Inertance)

On a real structure, an unlimited number of FRFs can be measured between pairs of input and output DOFs, as shown in Figure 7.

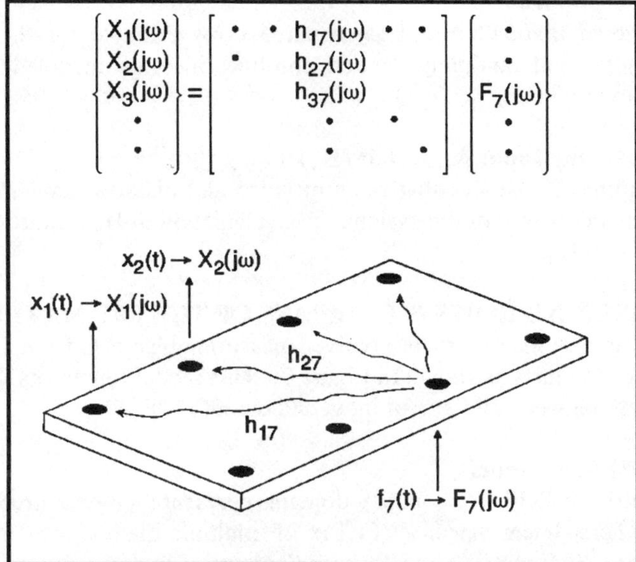

Figure 7. Measuring FRFs on a Structure

Although the FRF is defined as a ratio of Fourier transforms, is it actually computed differently using APS and XPS estimates. This is done to remove random noise and non-linearities (distortion) from the FRF, by using spectrum averaging as described earlier. There are several different ways to calculate the FRF. These are called FRF estimators.

8.1   Noise on the Output (H1)
This FRF estimator assumes that random noise and distortion are summing into the output, but not the input of the system. For this model, the FRF is calculated as,

$$H_1 = \frac{XPS}{Input\ APS}$$

It can be shown that H1 is a least squared error estimate for the FRF when extraneous noise and randomly excited non-linearities are modeled as Gaussian noise added to the output.

8.2 Noise on the Input (H2)

This FRF estimator assumes that random noise and distortion are summing into the input, but not the output of the system. For this model, the FRF is calculated as,

$$H_2 = \frac{\text{Output APS}}{\text{XPS}}$$

It can be shown that H2 is a least squared error estimate for the FRF when extraneous noise and randomly excited non-linearities are modeled as Gaussian noise added to the input.

8.3 Noise on the Input & Output (H_V)

This FRF estimator assumes that random noise and distortion are summing into both the input and output of the system. The calculation of H_V requires more steps, and is detailed in [2].

9. MEASURING ROWS & COLUMNS OF THE FRF MATRIX

Structural dynamics measurement involves measuring elements from a FRF matrix model for the structure, as shown in Figure 7. This model represents the dynamics of the structure between all pairs of input and output DOFs.

9.1 FRF Matrix Model

The FRF matrix model is a frequency domain representation of a structure's linear dynamics, where linear spectra (FFTs) of multiple inputs are multiplied by elements of the FRF matrix to yield linear spectra of multiple outputs. The FRF matrix model is written as,

$$\{X(\omega)\} = [H(\omega)]\{F(\omega)\}$$

where:

$$\{X(\omega)\} = \textbf{Linear spectra of output motions} \ldots (\textbf{n vector})$$
$$[H(\omega)] = \textbf{FRF matrix} \ldots (\textbf{n by m})$$
$$\{F(\omega)\} = \textbf{Linear spectra of input forces} \ldots (\textbf{m vector})$$
$$\textbf{m} = \textbf{number of inputs}$$
$$\textbf{n} = \textbf{number of outputs}$$
$$\omega = \textbf{frequency variable}$$

Columns of the FRF matrix correspond to inputs, and rows correspond to outputs. Each input and output corresponds to a measurement Point or DOF of the test structure.

9.2   Modal Testing

In modal testing, FRF measurements are usually made under controlled conditions, where the test structure is artificially excited with an impactor, or by one or more shakers driven with broadband signals. A multi-channel FFT analyzer is then used to make FRF measurements between input and output DOF pairs on the test structure.

9.3   Measuring FRF Matrix Rows or Columns

Modal testing requires that FRFs be measured from at least one row or column of the FRF matrix. Modal frequency & damping can be obtained from any FRF measurement. A row or column of FRF measurements is required to obtain mode shapes.

When the input is fixed and FRFs are measured for multiple outputs, this corresponds to measuring elements from a single column of the FRF matrix. This is typical of a shaker test.

On the other hand, when the output is fixed and FRFs are measured for multiple inputs, this corresponds to measuring elements from a single row of the FRF matrix. This is typical of a roving hammer impact test.

9.4   Single Reference (or SIMO) Testing

The most common type of modal testing is done with either a single fixed input or a single fixed output. A roving hammer impact test using a single fixed motion transducer is a common example of single reference testing. The single fixed output is called the reference in this case.

When a single fixed input (such as a shaker) is used, this is called SIMO (Single Input Multiple Output) testing. In this case, the single fixed input is called the reference.

9.5   Multiple Reference (or MIMO) Testing

When two or more fixed inputs are used, and FRFs are calculated between each of the inputs and multiple outputs, then FRFs from multiple columns of the FRF matrix are obtained. This is called Multiple Reference or MIMO (Multiple Input Multiple Output) testing. In this case, the inputs are the references.

Likewise, when two or more fixed outputs are used, and FRFs are calculated between each output and multiple inputs, this is also multiple reference testing, and the outputs are the references.

10.   IMPACT MEASUREMENTS

Impact testing is the most commonly used method for finding the resonances of structures and machines. A typical impact test is depicted in Figure 8. Impact testing requires a minimum of equipment,

354

1. A hammer with a load cell attached to its head to measure the impact force,
2. An accelerometer fixed to the structure to measure response motion,
3. A 2-channel FFT analyzer.

A wide variety of structures and machines can be impact tested. Of course, different sized hammers are necessary to provide the appropriate impact force to the structure. Not all structures can be impact tested, however. A structure or machine with delicate surfaces probably should not be impact tested. Typical signals from an impact test are shown in Figure 9.

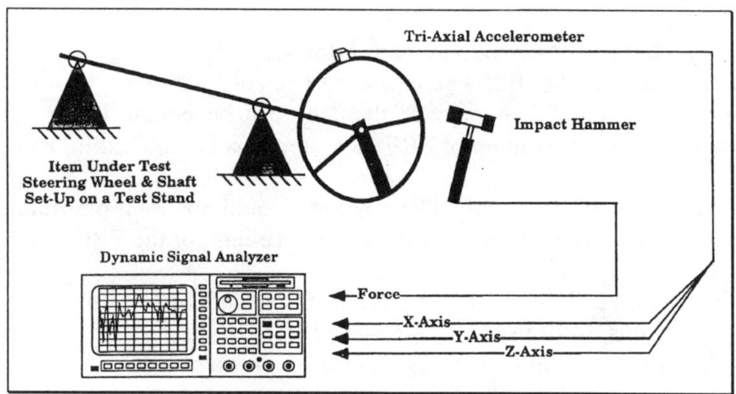

Figure 8. Impact Test Setup.

10.1   Roving Hammer Test

A roving hammer test is the most common type of impact test. In this test, the accelerometer is fixed at a single DOF (point and direction), and the structure is impacted at as many DOFs as desired to define the mode shapes of the structure.

10.2   Tri-axial Measurements

The only drawback to the roving hammer approach is that many points on a structure cannot be impacted in three directions, so tri-axial (3D) motion cannot be obtained for all points. When 3D motion is desired at each test point, a roving tri-axial accelerometer can be used, and the structure impacted at a fixed DOF. However, in order to process the tri-axial and force data together, a multi-channel FFT analyzer with at least 4 channels is required.

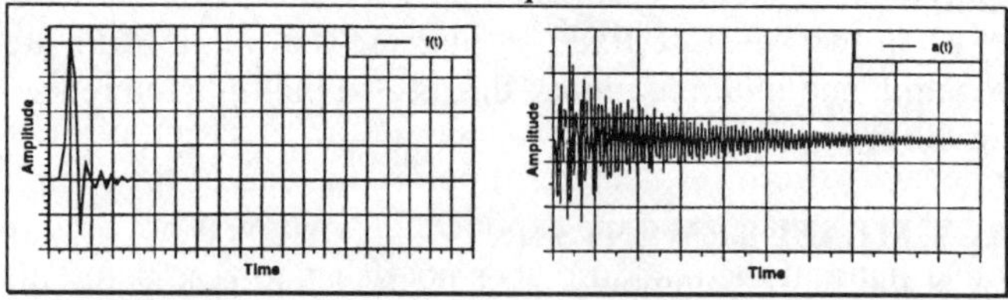

Figure 9A. Impact Force and Response Signals.

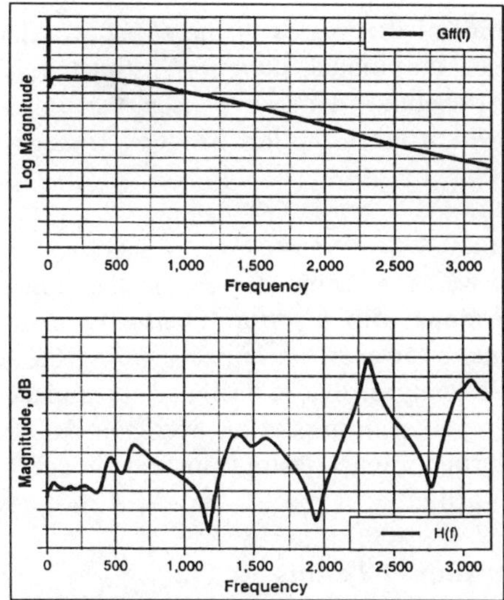

Figure 9B. Impact APS and FRF.

10.3   Pre-Trigger Delay

Because the impulse signal exists for such a short period of time, it is important to capture all of it in the sampling window. To insure that the entire signal is captured, the analyzer must be able to capture the impulse and impulse response signals prior to the occurrence of the impulse. This is called a pre-trigger delay. In other words, the analyzer must begin sampling data before the trigger point occurs. The triggering level is usually set to a small percentage of the peak value of the impulse, and this cannot be determined until the peak value has been acquired.

10.4   Force & Exponential Windows

Two common time domain windows that are used in impact testing are the force and exponential windows. These windows are applied to the signals after they are sampled, but before the FFT is applied to them.

The force window is used to remove noise from the impulse (force) signal. Ideally, an impulse signal is non-zero for a small portion of the sampling window, and zero for the remainder of the window time period. Any non-zero data following the impulse signal in the sampling window is assumed to be measurement noise. The force window preserves the samples in the vicinity of the impulse (peak), and zeros all of the other samples in the sampling window.

The exponential window is applied to the impulse response signal. The exponential window is used to reduce leakage in the spectrum of the response. If the response decays to zero (or near zero) before the end of the sampling window, then there will be no leakage, and the exponential window is not required.

On the other hand, if the response does not decay to zero before the end of the window, then the exponential window must be used to reduce the leakage effects on the response spectrum. The exponential window adds a known amount of "artificial damping" to all of the modes of the structure. This artificial damping can be subtracted from the modal damping estimates later on. But more importantly, if the exponential window causes the impulse response to be completely contained within the sampling window, leakage is removed from its spectrum.

10.5   Accept/Reject

Because impact testing relies, to some degree, on the skill of the one doing the impacting, it should be done with spectrum averaging, using 3 to 5 impacts per measurement. Since one or two of the impacts during the measurement process may be bad hits, an FFT analyzer that is designed for impact testing will have the capability to accept or reject each impact, or average. An accept/reject capability saves a lot of time during impact testing since you don't have to restart the measurement after each bad hit.

10.6   Advantages of Impact Testing

The advantages of impact testing are,

- Low equipment cost.
- Easy test set up.
- Fast measurement time.
- Signals are periodic (or near periodic) in the sampling window.

10.7   Disadvantages of Impact Testing

The disadvantages of impact testing are,

- Special analyzer capabilities are required.
- Some skill required to impact correctly.
- Low energy density in impact signal.
- Doesn't remove non-linear behavior.
- Can't be used on some structures.

11.   SHAKER MEASUREMENTS

When impact testing cannot be used, then structural dynamic measurements are made by providing excitation with one or more shakers attached to the structure. Common types of shakers are electro-dynamic and hydraulic shakers. A typical shaker test is depicted in Figure 10.

A shaker is usually attached to the structure using a stinger (long slender rod), so that the shaker will only impart force to the structure along the axis of the stinger, the axis of force measurement.

A load cell is attached between the structure and the stinger to measure the excitation force. At least a 2-channel FFT analyzer and a uni-axial accelerometer are required to make FRF measurements using a shaker. If an analyzer with 4 or more channels is used, then a tri-axial accelerometer can be used and 3D motion of the structure measured at each test point.

In a SIMO test, one shaker is used and the shaker is the (fixed) reference. In a MIMO test, multiple shakers are used, and the shakers are the multiple references.

When multiple shakers are used, care must be taken to insure that the shaker signals are not completely correlated (the same signal). Furthermore, special matrix processing software is required to calculate FRFs from the multiple input APSs and XPSs resulting from tri-spectrum averaging.

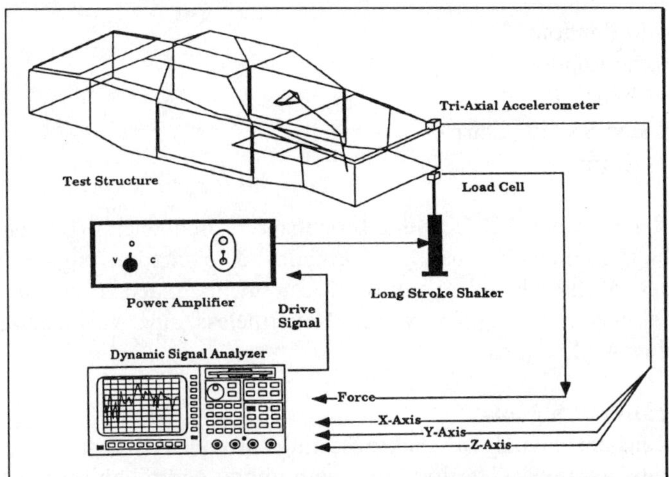

Figure 10. Shaker Test Setup.

11.1   Step Sine, Slow Swept Sine

The sine wave excitation signal has been used since the early days of structural dynamic measurement. It was the only signal that could be effectively used with traditional analog instrumentation, as described earlier.

Even though other broadband testing methods (like impact testing) have been developed for use with FFT analyzers, sine wave excitation is still useful in some applications. The primary purpose for using a sine wave excitation signal is to put energy into a structure at a specific frequency. Slowly sweeping sine wave excitation is also useful for characterizing non-linearities in structures.

11.1.1   Advantages of Sine Testing

Sine wave excitation has the following advantages,

- •· Best signal-to-noise and RMS-to-peak ratios of any signal.
- •· Controlled amplitude and bandwidth.
- •· Useful for characterizing non-linearities.
- •· Long history of use.

11.1.2   Disadvantages of Sine Testing

The disadvantages of sine wave excitation are,

- •· Distortion due to over-excitation.
- •· Extremely slow for broadband measurements.

358

11.2   Broadband Excitation Signals

A variety of new broadband excitation signals have been developed for making shaker measurements with FFT analyzers. These signals include,

- • Transient
- • True Random
- • Pseudo Random
- • Periodic Random
- • Burst Random
- • Fast Sine Sweep (Chirp)
- • Burst Chirp

Since the FFT provides a DFT over a broadband of frequencies (0 to nearly half of the sampling frequency), using a broadband excitation signal makes the measurement of broadband spectral measurements much faster than using a stepped or slowly sweeping sine wave. Nevertheless, sine wave excitation is still useful in some applications.

11.3   Transient Signals

Using a transient signal in shaker testing provides the same leakage free measurements as impact testing, but with more controllability over the test. Application of the force is more repeatable than impacting with a hand held hammer. However, this one advantage is usually outweighed by the disadvantages of using an impulsive force, when compared to the other broadband signals.

11.4   True Random

Probably the most popular excitation signal used for shaker testing with an FFT analyzer is the random signal. When used in combination with spectrum averaging, random excitation randomly excites the non-linearities in a structure, which are then removed by spectrum averaging.

A true random signal is synthesized with a random number generator, and is an unending (non-repeating) random sequence. The main disadvantage of a true random signal is that it is always non-periodic in the sampling window. Therefore, a special time domain window (a Hanning window or one like it) must always be used with true random testing to minimize leakage. Typical true random signals are shown in Figure 11.

11.4.1   Advantages of True Random Excitation

The advantages of true random excitation are,

- • Removes non-linear behavior when used with spectrum averaging.
- • Fast measurement time.
- • Leakage effects reduced with Zoom measurements.

11.4.2　Disadvantages of True Random Excitation

The disadvantages of true random excitation are,

- •· Signals are non-periodic in the sampling window. Special windowing (Hanning, etc.) is needed to reduce leakage.
- •· Many averages are typically required.

11.5　Pseudo Random

A pseudo random signal is specially synthesized within an FFT analyzer to coincide with the measurement sampling window parameters. A typical random signal starts as a uniform (or shaped) magnitude and random phase signal, synthesized over the same frequency range and samples as the intended measurement. It is then inverse FFT'd to obtain a random time domain signal, which is subsequently output through a DAC (digital-to-analog converter) as the shaker excitation signal.

During the measurement process, the measured force and response signals are sampled over the same sampling time window as the output of the excitation signal. This insures that the acquired signals are periodic in the sampling window, since the synthesized excitation signal is periodic in the window.

11.5.1　Advantages of Pseudo Random Excitation

The advantages of pseudo random excitation are,

- •· Signals are periodic in the sampling window, so measurements are leakage free.
- •· Fast measurement time.
- •· The amplitude of excitation can be shaped for impedance mismatches between the shaker and structure.

11.5.2　Disadvantages of Pseudo Random Excitation

The disadvantages of pseudo random excitation are,

- •· Doesn't remove non-linearities, because they are not excited randomly between spectrum averages.

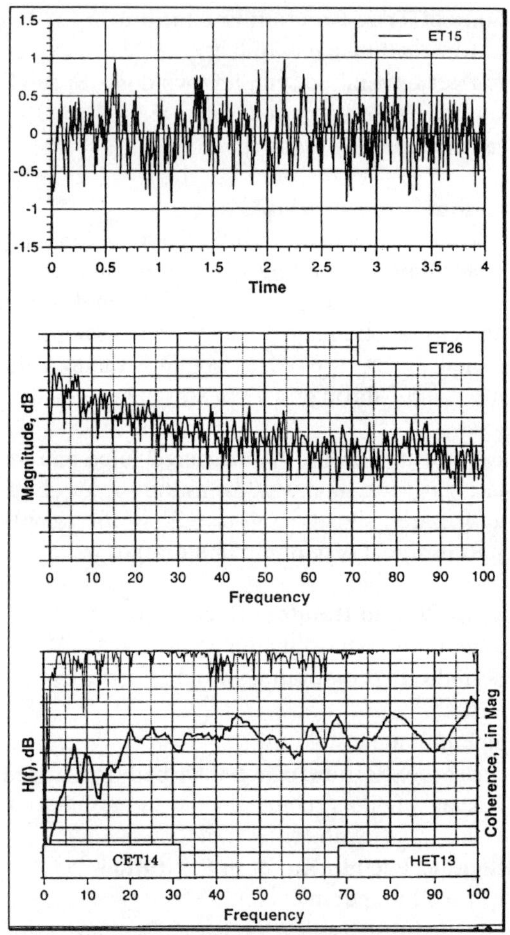

Figure 11. True Random Excitation (Time waveform, APS, FRF & Coherence).

11.6 Periodic Random

Periodic random excitation is simply a different use of a pseudo random signal, so that non-linearities can be removed from the signals with spectrum averaging. In periodic random testing, a new pseudo random sequence is generated for each new spectrum average. The advantage of this is that when multiple spectrum averages of different random signals are averaged together, randomly excited non-linearities are removed, and an effectively linear signal remains.

Although periodic random excitation overcomes the disadvantage of pseudo random excitation, it takes at least three times longer to make the same measurement. This extra time is required between spectrum averages to allow the structure to reach a new steady-state response to the new random excitation signal.

11.6.1   Advantages of Periodic Random Excitation

The advantages of periodic random excitation are,
- •· Signals are periodic in the sampling window, so measurements are leakage free.
- •· Removes non-linear behavior when used with spectrum averaging.
- •· The amplitude of excitation can be shaped for impedance mismatches between the shaker and structure.

11.6.2   Disadvantages of Periodic Random Excitation

The disadvantages of periodic random excitation are,
- •· Slower than other random test methods.
- •· Special software required for implementation.

11.7   Burst Random

Burst random excitation is similar to periodic random testing, but faster. In burst random testing, a true random signal can be used, but it is turned off prior to the end of the sampling window period. This is done in order to allow the structural response to decay within the sampling window. This insures that both the excitation and response signals are completely contained within the sampling window. Hence, they are periodic in the sampling window.

Figure 12 shows a typical burst random signal. The random generator must be turned off early enough to allow the response to decay to zero (or nearly zero) before the end of the sampling window. Of course, the length of the decay period depends on the damping in the test structure.

Burst random must therefore be set up interactively on an FFT analyzer, after observing the free decay of the structure in the sampling window. Since a pure random signal can be used with burst random testing, it does not have the disadvantages of either pseudo random or periodic random testing.

11.7.1   Advantages of Burst Random Excitation

The advantages of burst random excitation are,
- •· Signals are periodic in the sampling window, so measurements are leakage free.
- •· Removes non-linear behavior when used with spectrum averaging.
- •· Fast measurement time.

11.7.2   Disadvantages of Burst Random Excitation

The disadvantages of true random excitation are,
- •· Special software required for implementation.

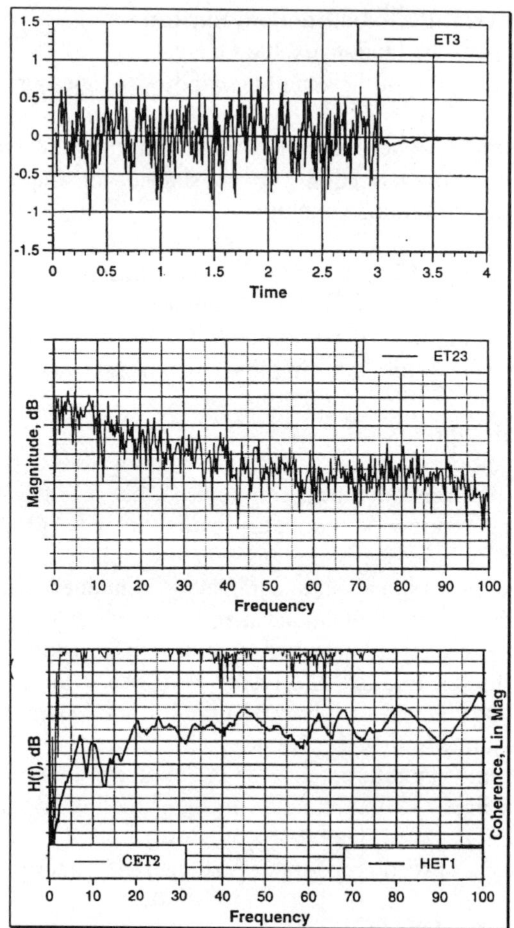

Figure 12. Burst Random Excitation (Time waveform, APS, FRF & Coherence).

11.8　Chirp & Burst Chirp

A swept sine excitation signal can also be synthesized in an FFT analyzer to coincide with the parameters of the sampling window, in a manner similar to the way a pseudo random signal is synthesized. Since the sine waves must sweep from the lowest to the highest frequency in the spectrum over the relatively short sampling window time period, this fast sine sweep often makes the test equipment sound like a bird chirping, hence the name chirp signal.

A burst chirp signal is the same as a chirp, except that it is turned off prior to the end of the sampling window, just like burst random. This is done to insure that the measured signals are periodic in the window. A typical chirp signal is shown in Figure 13.

The advantage of burst chirp over chirp is that the structure has returned to rest before the next average of data is taken. This insures that the measured response is only caused by the measured excitation, an important requirement for any multi-channel measurement such as a FRF.

11.8.1 Advantages of Burst Chirp Excitation
The advantages of burst chirp excitation are,
- High signal–to-noise and RMS-to-peak ratios.
- Signals are periodic in the sampling window, so measurements are leakage free.
- Fast measurement time.

11.8.2 Disadvantages of Burst Chirp Excitation
The disadvantages of burst chirp excitation are,
- Special software required for implementation.
- Doesn't remove non-linear behavior.

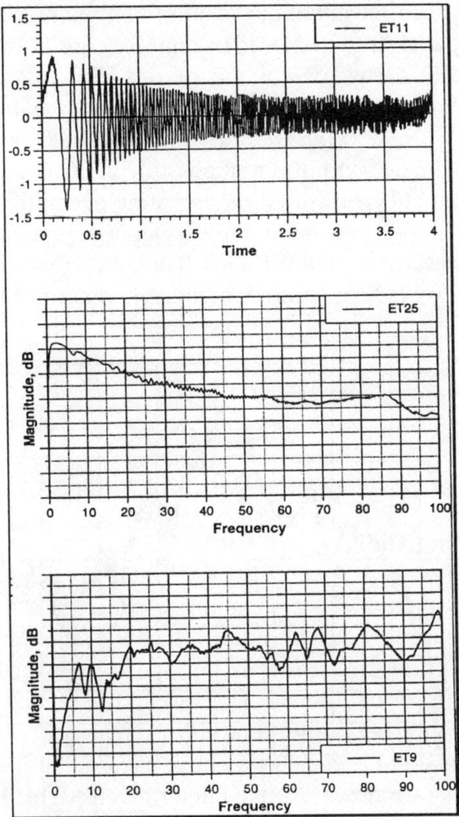

Figure 13. Chirp Excitation (Time waveform, APS, FRF & Coherence)

11.9   Comparison of Excitation Signals

Ideally, all of the shaker signals that are leakage free (periodic in the window) should yield the same result. Figure 14 shows an overlay of two FRF magnitudes, one measured with a burst random and the other with a burst chirp signal. The two FRFs match very well at low frequencies, but show some disparity at high frequencies. This could possibly be due to a small amount of non-linear behavior in the structure, which burst chirp signal processing cannot remove through averaging.

Finally, all of the previously described test methods are compared in the table shown in Figure 15. Impact testing is by far the easiest method to implement. On the other hand, when impact testing cannot be used, or when multiple shakers are needed to provide sufficient excitation, then a variety of other implementation issues must be considered.

12.   DIFFICULTY WITH FRF MEASUREMENTS

Thus far, we have talked mostly about making FRF measurements. Experimental mode shapes (the spatial amplitudes of resonances) are normally obtained from a set of FRFs, typically a row or column of the FRF matrix. Making an FRF measurement, however, carries with it the following requirement; *"All of the excitation forces causing a response must be measured simultaneously with the response"*. This can be difficult, if not impossible in many test situations.

FRFs usually cannot be measured on operating machinery or equipment where ambient forces (internally generated forces, acoustic excitation, etc.) are either unmeasured or unmeasurable. On the other hand, the vibration response caused by ambient forces can always be measured, no matter what forces are causing it.

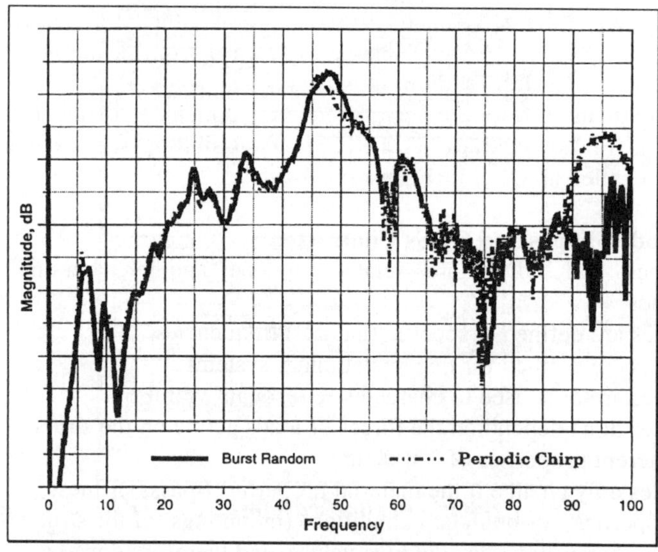

Figure 14. Burst Random Versus Burst Chirp FRF.

	Impact	Sine	Swept Sine	True Random	Pseudo Random	Periodic Random	Burst Random	Burst Chirp
Periodic	YES	NO	YES	NO	YES	YES	YES	YES
Removes Noise	YES	NO	YES	YES	YES	YES	YES	YES
Removes Non-linearities	NO	NO	NO	YES	NO	YES	YES	NO
Test Time	FAST	SLOW	FAIR	FAIR	FAST	SLOW	FAST	FAST
SNR	LOW	HIGH	HIGH	FAIR	FAIR	FAIR	FAIR	HIGH
Frequency Control	SOME	YES	YES	YES	YES	YES	YES	YES

Figure 15. Comparison of Excitation Methods

12.1   Difficulty With Operating Data Measurements

One key advantage of the FRF measurement is lost when operating data measurements are made. Without measuring the excitation forces, it is impossible to know precisely whether a peak in a response spectrum is due to a resonance or to the excitation forces. Nevertheless, valuable information can still be obtained from operating data.

13.   OPERATING DEFLECTION SHAPES

An Operating Deflection Shape (ODS) is defined as any forced motion of two or more DOFs on a structure. Specifying the motion of two or more DOFs defines a shape. Stated differently, a shape is the motion of one DOF relative to all others.

An ODS can be defined from any forced motion, either at a moment in time, or at a specific frequency. An ODS can be obtained from different types of time domain responses, be they random, impulsive, or sinusoidal. An ODS can also be obtained from many different types of frequency domain measurements, including linear spectra (FFTs), APS, XPSs, FRFs, transmissibilities, and a special type of measurement called an ODS FRF, described later.

13.1   Mode Shapes and ODSs Contrasted

Mode shapes and ODSs are related to one another, but have different characteristics,

1. Modes are defined at specific (natural) frequencies.
2. Modes are defined for linear, stationary systems.
3. Modes are only used to characterize resonant vibration.
4. Modes don't depend on the forces or loads acting on the structure. They are inherent properties of a structure.
5. Modes only change if the material properties (mass, stiffness, damping properties), or boundary conditions (mountings) of the structure change.
6. Mode shapes don't have unique values, and therefore don't have units associated with them.

7. Mode shapes are unique. They can be used to answer the question, *"What is the relative motion of one DOF to another?"*

ODSs have the following characteristics
1. ODSs can be defined for any specific frequency or moment in time.
2. ODSs can be defined for non-linear or non-stationary structures.
3. ODSs can be used to characterize structures that don't resonate.
4. ODSs depend on the forces or loads applied to a structure. They will change if the load changes.
5. ODSs also depend on the modes. ODSs will change if the modes change.
6. ODSs have unique values, and therefore have units. Typical units are displacement, velocity, or acceleration, or displacement per unit of excitation force.
7. Since ODSs have unique values, they can be used to answer the question, *"How much is the structure really moving at a particular time or frequency?"*

13.2   Modes From ODSs

Since all measurement data is forced response, whenever two of more measurements are taken spatially from two or more DOFs of a structure, this is an ODS measurement. Moreover,
- • All experimental modal parameters are obtained by post-processing ODS measurements!

14.   TRANSMISSIBILITY MEASUREMENT

We have already seen that under the assumption of either a Repeatable (more restrictive) or a Steady State (less restrictive) measurement process, spectrum averaging can be accomplished, and multi-channel measurements made. When the excitation force cannot be measured, then a reference response signal can be used instead of the force.

A transmissibility measurement is calculated in the same way as an FRF, but with a reference response signal replacing the excitation force.

14.1   Mode Shapes From Transmissibilities

A set of transmissibilities, calculated between multiple response DOFs and a single fixed reference response, can be used to find the mode shapes of structural resonances. The values of the transmissibilities at each resonant frequency is an approximation to the mode shape.

The difficulty with using a set of transmissibilities to determine mode shapes is that resonances correspond to "flat spots" instead of peaks in these measurements. Therefore, in addition to a set of transmissibilities, at least one APS is required in order to locate the resonance peaks.

15.   ODS FRF MEASUREMENT

A different cross channel measurement, called and ODS FRF, can be calculated from APS and XPS measurements [3]. An ODS FRF has two advantages over a transmissibility,

     1. It has peaks at resonant frequencies, making it easier to locate resonances and identify mode shapes.

     2. It has responses units (G's, Mils, etc.). Therefore, operating deflection shapes taken from a set of ODS FRFs have these same units.

To calculate a set of ODS FRFs between multiple response DOFs and a single fixed reference response, a tri-spectrum averaging loop is used to estimate an APS for each response, and a XPS between each response and the reference response. When tri-spectrum averaging is completed, each ODS FRF is formed by replacing the magnitude of each XPS with the APS of its corresponding response.

A set of ODS FRF measurements is useful for determining whether a structure or machine is simply undergoing excessive forced response, or whether a resonance is also being excited.

16.   NON-STEADY STATE OPERATION

All of the foregoing measurements assumed that measurement process was either repeatable or steady state. However, many types of structures and machines undergo non-steady state operation. Automobiles and machine tools are common examples.

In fact, most rotating equipment is characterized by non-steady state operation. Measurements are typically made while sweeping the speed of the machine. These are called RPM sweeps. Since the measurement process is non-steady state, the spectra cannot be averaged together. Rather, they are plotted in a waterfall plot, or spectral map.

16.1   Orders

Since the excitation forces in a rotating machine are primarily sinusoidal and usually cannot be measured, their response spectra will exhibit forced responses are peaks that vary in frequency with the speed of the machine. These peaks, called Orders, appear at frequencies that are fixed multiples of the machine speed.

Since machine speed continually changes (it is non-stationary), a portion of a rotating machine's response will also be non-stationary and exhibit peaks which "track" the cyclic forces. However, if a resonance is excited, it will always appear at its fixed (stationary) natural frequency in any spectral measurement.

17.   CONCLUSIONS

During the past 30 years, there has been a proliferation of new structural dynamics testing methods that are based upon the laboratory implementation of the FFT and related signal processing algorithms. The "parallel processing" nature of the FFT that yields the discrete frequency spectrum of a signal from one calculation, makes it a broadband tool. This created a fundamental departure from the traditional sine wave based, swept filter methods for testing structures.

For finding structural resonances, the FFT has made it convenient to excite structures using many different kinds of broadband signals. Not only are a variety of shaker signals now used, but impact testing has became very popular as a fast, convenient, and relatively low cost way of finding the mode shapes of structures.

Acknowledgements

Many of the methods and ideas reviewed here were learned over the past 25 years during on the job training at Hewlett Packard, Structural Measurement Systems, Inc. and Vibrant Technology, Inc.

Numerous co-workers, customers, and students have assisted in the development of many of these methods, and have taught me. In particular, some of this material was developed in cooperation with Jim Steadman, NAVCON Engineering Network, for use in a modal analysis training course.

References

1. Cooley, J.W. and Tukey, J.W. "An Algorithm for the Calculation of Complex Fourier Series" Mathematics and Computation, Vol. 19, 1965, pp. 297-301.
2. Rocklin, G.T, Crowley, J., and Vold, H. "A Comparison of H1, H2, and HV Frequency Response Functions", Proc. of 3rd IMAC, Orlando FL, 1985, pp. 272-278.
3. Richardson, M.H., "Is It A Mode Shape Or An Operating Deflection Shape?", Sound and Vibration Magazine, February, 1997.
4. Roth, P., "Effective Measurements Using Digital Signal Analysis", IEEE Spectrum, April 1971, pp. 62-70.
5. Richardson, M.H., "Modal Analysis using Digital Test Systems", Seminar on Understanding Digital Control and Analysis in Vibration Test Systems, Shock and Vibration Information Center publication, Naval Research Lab. , Wash. , D.C. , May 1975.
6. Ramsey, K.A., "Effective Measurements for Structural Dynamics Testing", Sound and Vibration Magazine, Part I, Nov. 1975, pp. 24-35, Part II, April 1976, pp. 18-31.
7. Potter, R. , "A General Theory of Modal Analysis for Linear Systems", Shock and Vibration Digest, Nov. 1975.
8. McKinney, W., "Band Selectable Fourier Analysis", Hewlett-Packard Journal, April 1975, pp. 20-24.

SD2000
State of the Art Review:
Damping

G R Tomlinson
Department of Mechanical Engineering
University of Sheffield, UK

1. INTRODUCTION

Damping is a physical process which dissipates energy through the conversion of work into heat. In engineering structures it is present in several forms: internal hysteresis in materials, friction via the rubbing action of surfaces or particles, viscous friction in fluids, radiation damping, electromagnetic effects.

Damping plays a very important role in structural dynamics and damping augmentation is the primary means by which resonant amplitudes are controlled, enhancing life cycle behaviour, durability, quality of life and cost reduction.

The methodologies for introducing damping into structures can include combinations of passive, semi-active, active, adaptive techniques. These make use of viscoelastic materials, ferroelectric and ferromagnetic materials, ER/MR fluids, SMAs, auxetic polymers, liquid crystals and are often employed with electric circuits (shunts) or non-linear control laws for improving structural damping.

Damping models and damping prediction/optimisation methods tend to be somewhat fragmented and limited in their ability to relate to reality. Generally linearity is assumed and the physics of the dissipation process is sometimes unclear (eg particle damping). Material models tend to be at a macro level and often employ assumptions that are clearly violated in practise. Many damping materials have extreme limitations with regard to temperature, frequency and creep behaviour and may undergo phase transformations that are excluded in the modelling process.

This contribution will attempt to address some of these issues relating to damping in structural dynamics.

2. PASSIVE DAMPING

2.1 Constrained and Free Layer Damping

Passive damping is, in general, simple to implement and cost effective, requiring no on-line control. The most well known form of passive damping is the use of a

370

viscoelastic material in either a constrained layer or a free layer configuration [1] as shown in Figure 1 [1].

Constrained layer damping (CLD) works by dissipating cyclic shear energy in a thin constrained layer of viscoelastic material of high shear modulus, thus the material should be added in regions where the surface strain (or the curvature) is a maximum.

Free layer damping (FLD) works by dissipating extensional strain energy and the modal loss factor is proportional to the loss modulus (Eη), thus a high Young's modulus is advantageous.

The important differences between FLD and CLD are related to the effective stiffness of the viscoelastic material (FLD requiring a high stiffness material) and to the fact that the loss factor for CLD is frequency dependent whereas for FLD it is independent of frequency [1].

Viscoelastic materials are frequency and temperature dependent. Their properties are usually described by Master Curves [2]. A typical Master Curve is shown in Figure 2. These properties can be employed directly with FE codes for forced harmonic response predictions or with the Relaxation curves, which are basically the inverse Fourier transforms of the frequency domain Master Curves, in time response studies.

The important parameters for optimising the damping of CLD treatments are the viscoelastic layer modulus, the thickness of the layer and the flexural stiffness of the constraining layer. It has been shown that a compromise has to be made between a high or a low viscoelastic layer modulus. For example, extensional modes require a lower constrained material modulus than flexural modes, which are dependent on the shear in the viscoelastic layer [3] .

One idea proposed to 'smooth' the damping between the two different modes of vibration is to use an anisotropic viscoelastic material which effectively reduces the transverse Young's modulus by the ratio of E(anisotropic)/E(isotropic) resulting in an improved damping performance over a wide frequency range [4]. This concept is shown schematically in Figure 3 and the effect on the modal loss factor for a model of a cantilever beam with such properties is shown in Figures 4 and 5.

Constrained layer damping techniques are, in principle, well established. However, from an applications and modelling point of view the performance of such treatments in rotating systems, where hydrodynamic loads or creep effects arising from centrifugal forces occur, is not fully understood.

Composite laminated designs in which a thin layer(s) of viscoelastic material is incorporated in the manufacturing lay-up have indicated the potential to offer significant damping without significant reduction in the stiffness characteristics of the structure. Ultra-thin finite element models of viscoelastic materials have recently been developed and have shown good correspondence with experimental results [5].

Constrained-Layer Treatment

- One of the most effective damping treatments

- Damps flexural modes with low-stiffness VEM

- Surface and integral treatment

- Weight effective

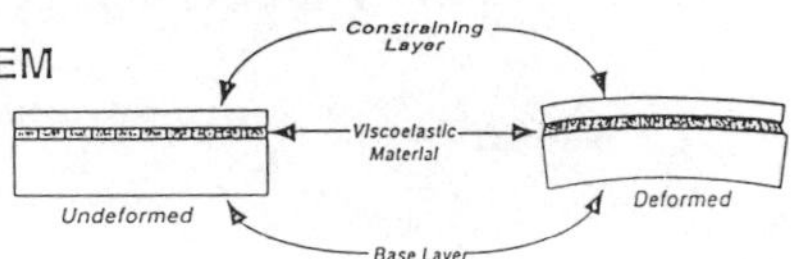

Free-Layer Treatment

- Good for extensional damping

- Requires high stiffness VEM

- May also be used to damp bending modes

- Not weight effective

Figure 1 Passive Damping Treatment Using Constrained and Free Layer Damping

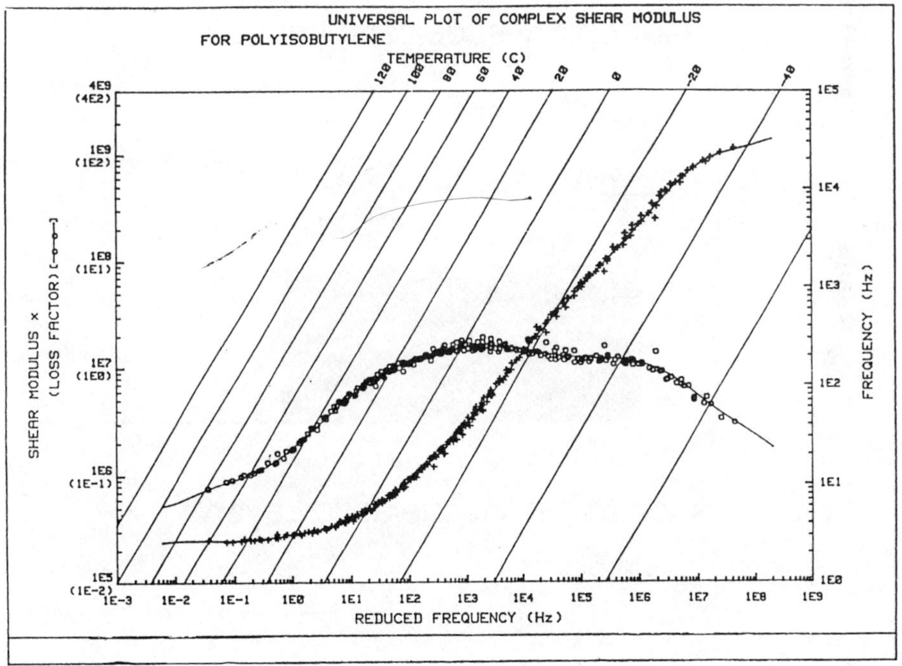

Figure 2 Typical Master Curve for a Viscoelastic material

EXTENDING "GOOD DAMPING" RANGE
Proposed alternative

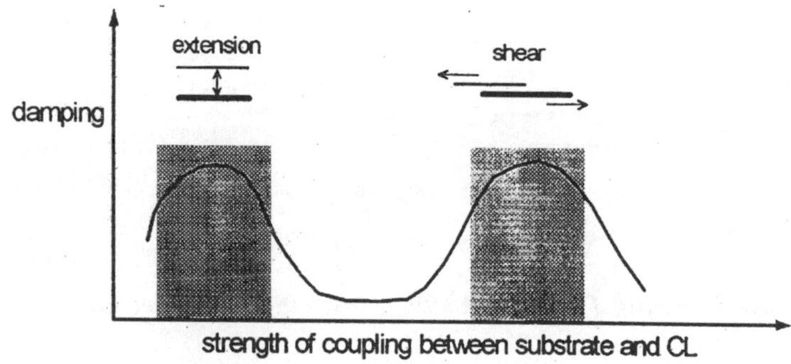

Isotropic Viscoelastic Layer

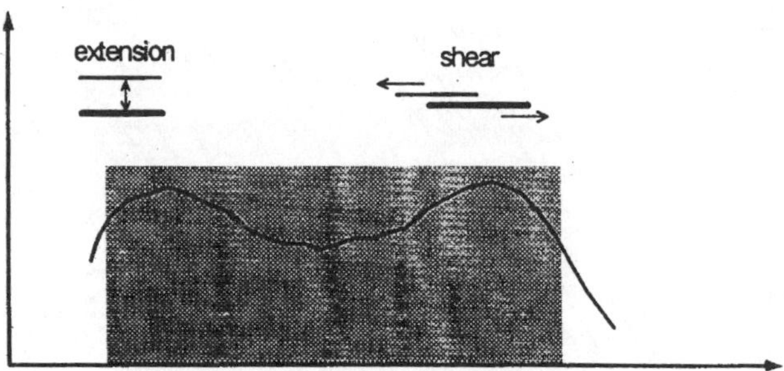

Anisotropic Viscoelastic Layer

Figure 3 The concept of using an anisotropic constrained layer material

ANISOTROPIC VISCOELASTIC LAYER
Cantilever Mode 5

Damping against VM shear modulus

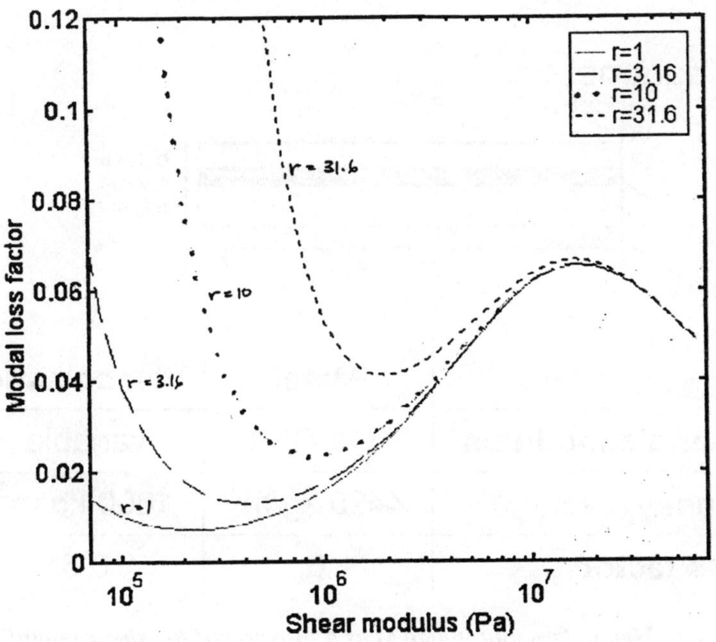

$$r = E_{anisotropic} / E_{isotropic}$$

Figure 4 Cantilever beam model properties

ANISOTROPIC VISCOELASTIC LAYER
CLD - numerical studies

- reduce transverse Young's modulus by

$$r = E_{anisotropic} / E_{isotropic}$$

- parameters

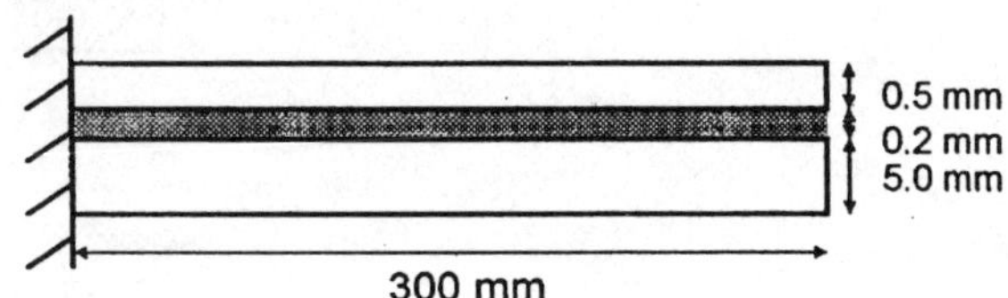

	Metal	Viscoelastic
Young's modulus	115 GPa	variable
density	4429 kg/m^3	1000 kg/m^3
loss factor	0	0.5

Figure 5 Modal damping (mode 5) as a function of the shear modulus for a simulated anisotropic material

Recent work on the use of magnetic constraining layers (referred to as electromagnetic damping) in configurations which utilise the opposing magnetic fields to induce cyclic shear in the viscoelastic constrained layer show promise and FE models have been developed to predict the performance of these systems [6].

Several important issues exist with respect to using viscoelastic materials for damping purposes. From the materials point of view, the limited temperature and frequency range tend to offer high damping in a narrow bandwidth, thus some form of optimisation is essential. The standard approach for 'optimising' the location of the CLD is to carry out modal strain energy studies and to locate the damping materials in these areas. An alternative to this using an evolutionary method has been recently reported which can automatically locate the damping material to give a vibration reduction over a range of modes of vibration [7].

2.2 Modelling

Models for predicting the performance of viscoelastic damping materials vary from the simplistic such as the Modal Strain Energy (MSE) [8] method to the more advanced models such as the Golla, Hughes, McTavish (GHM), Anelastic Displacement Fields (ADF) or Augmenting Thermodynamic Fields (ATF) models [9]. Unfortunately these advanced models tend to be research tools and are not commercially available. The simple and most commonly used approach, namely the MSE method, needs to be used with caution since the method does not currently take account of the fact that complex eigenvalues and complex modes need to be considered when significant damping is added to a structure. This tends to result in an overestimate of the damping and an underestimate in the resonant frequencies. Figures 6 and 7 give an example of these effects.

An overview of the various models and the governing equations is given below:

Current commercial FE codes use:
- viscous damping models
- Rayleigh damping
- hysteretic or structural damping via a complex modulus
- Modal Strain Energy (MSE) method

Recent models are:
- GHM   - Golla, Hughes, McTavish (1985)
- ATF   - Augmenting Thermodynamic Fields (Lesieutre et al, 1990)
- ADF   - Anelastic Displacement Fields (Lesieutre, et al, 1990)
- FD   - Fractional Derivatives (Bagley et al, 1983)

Pros and Cons

Not clear but GHM, ATF/ADF are time domain damping models using ordinary integer differential operators whereas FD method is employed in the frequency

ERRORS IN MSE ANALYSIS
Example: Ring Study

- loss factor ratio

$$\eta_R = \eta_{mode} \, / \, \eta_{viscoelastic}$$

- parameters

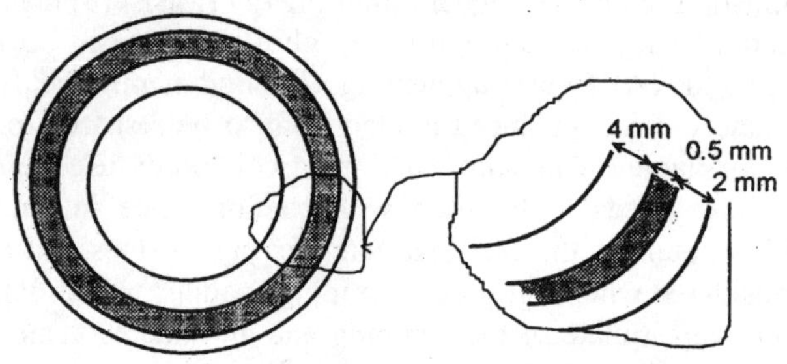

	Metal	Viscoelastic
Young's modulus	200 GPa	100 MPa
density	7900 kg/m^3	1000 kg/m^3
loss factor	0	variable

Figure 6 An example of errors in the MSE analysis

ERRORS IN MSE ANALYSIS
Results

Viscoelastic Young's modulus = 100 MPa

CE = complex eigenvalue (correct) solution
eta = loss factor of viscoelastic material

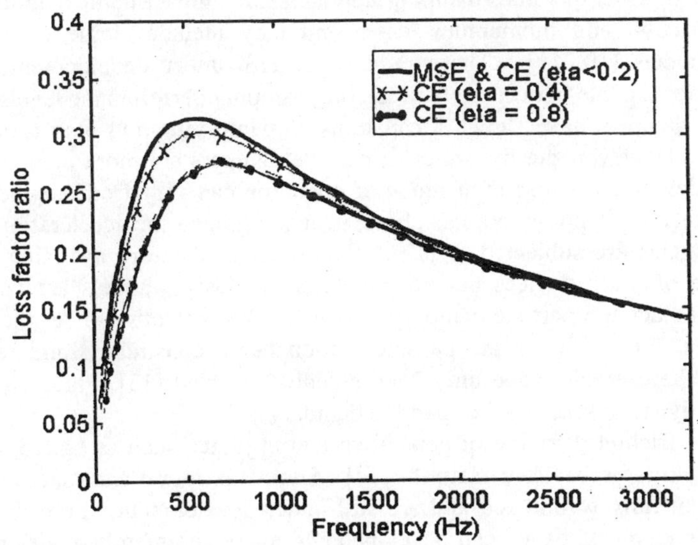

Figure 7 Errors in the MSE method for the ring study showing the discrepancy as the damping increases

domain and uses only a small number of model parameters. None of these is commercially available.

Viscous, Rayleigh and Complex Modulus damping are commercially available. The MSE method is very easy to implement.

Approaches using conformal surface coatings are also under consideration. These can be in the form of a high modulus constraining layer deposited on a thin viscoelastic coating. Numerical studies have indicated that for a 100 micron constraining layer of modulus 300GPa on a 10 micron viscoelastic layer of modulus 3GPa, loss factors in excess of 5% are possible. The use of hard coatings as a damping treatment involving plasma deposition techniques is a relatively new research area. Depending on the plasma process the microstructure of the coating can be controlled, which may give different damping mechanisms. These coatings appear to behave in a 'free-layer' configuration and test results on simple coated beam specimens have displayed good damping properties.

Research into impact and particle dampers has become topical as a result of the need for high temperature damping applications. Impact damper models are based on friction and momentum losses and may include single or multiple impacting bodies [10, 11]. These models are still under development and at present are not capable of accurately predicting damping performance levels.

Particle dampers have shown a capability to damp vibration over a range of frequencies. However, the dynamics of particle dampers are more complex than the impact dampers because their mode of operation can vary from a solid phase through a convection phase to a gas phase, depending upon the acceleration levels that the particles are subjected to [12]. Experimental data are available on the performance of these devices but no prediction or design models are currently available. Impact and particle damping devices are less effective under the action of centrifugal forces due to the particle motion being constrained and research results into these effects have only been recently reported [13]. Figure 8 shows some of the dynamic properties of particle dampers.

Interface friction damping of general structural joints, such as bolted, riveted or clamped joints, is far from maturity [14]. Stick-slip, micro and macro models have not been fully formulated and are still under development. Zero thickness friction finite elements have been developed and many commercial codes include simple contact elements. However, the prediction tools for dry friction mechanisms has still some way to go.

3. DAMPING VIA ACTIVE/ELECTRONIC METHODS

Passive CLD treatments are being re-engineered in the context of smart materials and structures by using active materials such as ferroelectrics (piezoelectric/electrostrictive materials) and magnetostrictives. In these applications the active materials may be used to enhance the energy dissipation mechanism by substituting the passive constrained layer with an active layer, referred to as Active Constrained Layer Damping (ACLD) [15]. Figure 9 shows the principles behind ACLD applied to longitudinal vibration in rods.

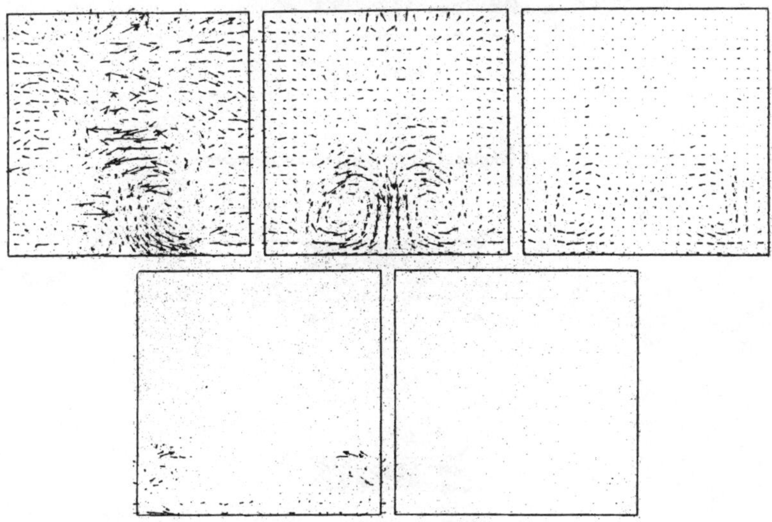

Images of the cycle-averaged motion of the system vibrated at $f = 20$ for different velocities of forcing, $A\omega$: a) 1900, b) 1250, c) 600, d) 100, e) 15. (Reference 12)

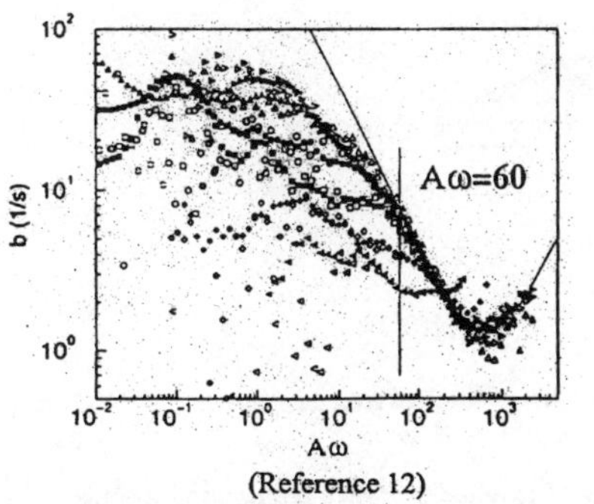

(Reference 12)

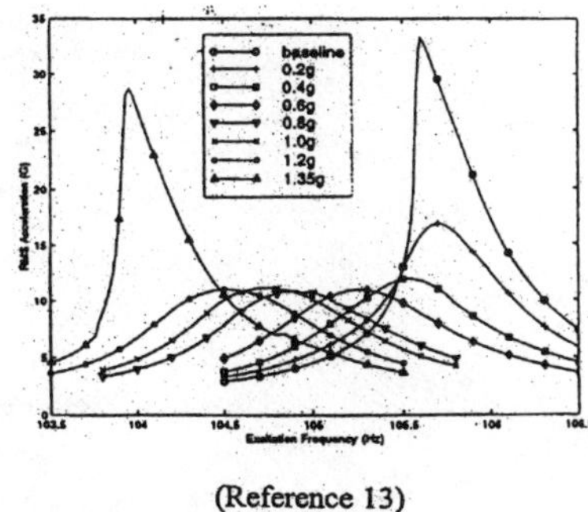

(Reference 13)

Figure 8 Particle dynamics characteristics showing convective rolls and effective damping properties

380

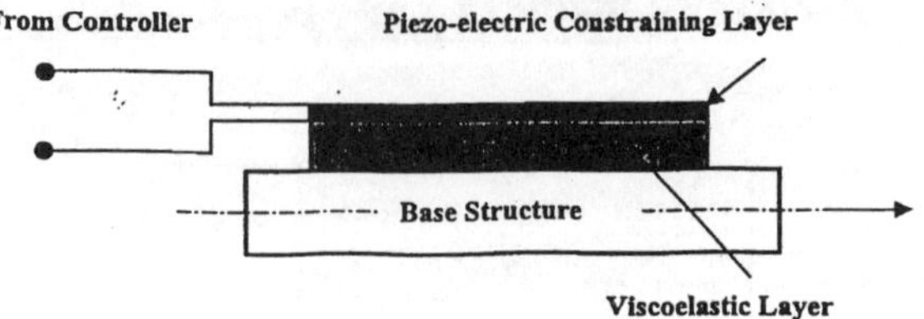

Figure (1) - Schematic drawing of the active constrained layer damping

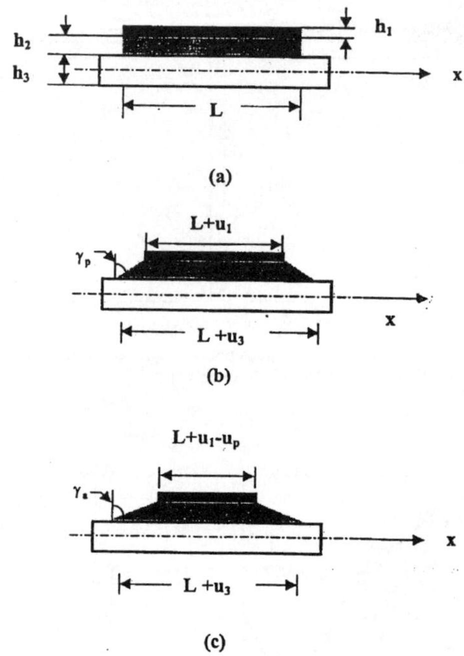

Figure (2) – Operating principle of the active constrained layer damping

Reference: A. Baz, "Spectral FE modelling of longitudinal wave propagation in rods treated with ACLD". Proc. 4th Eur Conf on Smart Materials and Structures, Harrogate, UK, 607-620, 1998.

Figure 9 Principle of ACLD

Research has focused on aspects such as the control algorithms, ACLD configurations, optimisation of the placement and sizing of the actuators, solution methods (analytical/numerical), segmented systems, noise control, boundary conditions and viscoelastic damping models. Configurations range from the active element or patch (the actuator) being attached to the host structure directly (active control) to the actuator bonded to the thin constraining layer or the actuator being bonded directly to the viscoelastic layer. Devices in which the active piezoelectric constraining layer acts as a sensor have also been reported. Numerical and experimental studies on simple beam and ring type structures have indicated good results, with the ACLD treatment outperforming the active and direct passive methods.

Developments in active intercalated embedded piezoelectric fibres [16] or embedded magnetostrictive layers/fibres [17] in composite structures have shown that good damping can be achieved without a reduction in the integrity of the structure. This is a major step forward that has potential in aerospace applications.

Electronic damping techniques in which a passive piezo patch bonded to the vibrating host structure dissipates energy via an electronic shunting circuit (usually composed of RLC components) has been shown to offer modal damping which is comparable to that obtained from passive CLD treatment. Tests on a plate with the CLD being full coverage returned similar damping properties to the electronic damper which used only a relatively small coverage [18]. Such devices act as tuned damped vibration absorbers and to damp several modes it is necessary to use an array of these.

Recent developments in single crystal materials with high internal loss factors offer promise for improved levels of damping. The ability to utilise integrated electronics to create electronic 'damping packages' with these and other materials indicate promise for electronic damping technologies [16].

4. SEMI-ACTIVE DAMPING

Semi-active damping technology combines active control to vary the properties of passive elements or materials. Semi-active friction damping via the use of an active joint composed of a piezoelectric stack replacing the washer in a bolted configuration has been reported and applied to truss structures. Figure 10 shows a schematic of the idea.

Compared to fully active control which is seen as more complex and costly, semi-active methods are being employed with materials such as electro(ER) and magneto-rheological(MR) fluids to control damping levels in a number of application areas such as suspension systems, seismic response.

ER fluids comprise a mixture of semi-conducting particles in a dielectric carrier liquid. The fluid is activated by a high electric field (of the order of 8kV per mm) using a relatively simple electrode arrangement with time constants of the order of milliseconds. On application of the electric field the particles form chain-like structures aligned nominally parallel to the applied field, producing an effective shear stress characteristic. When this is combined with an appropriate geometry, damping devices can be designed [19].

Friction Model — Active Joint

• Coulomb Friction Model

$$M_f = \mu\,N\,r\,\mathrm{sgn}(\dot\theta)$$

μ Friction coefficient
N Normal force
r Equivalent friction radius
$\dot\theta$ Relative angular velocity

• Active Joint

\# Using the normal force as control variable, one obtains a controllable friction moment.

\# Varying the normal force by means of a piezoelectric stack disc.

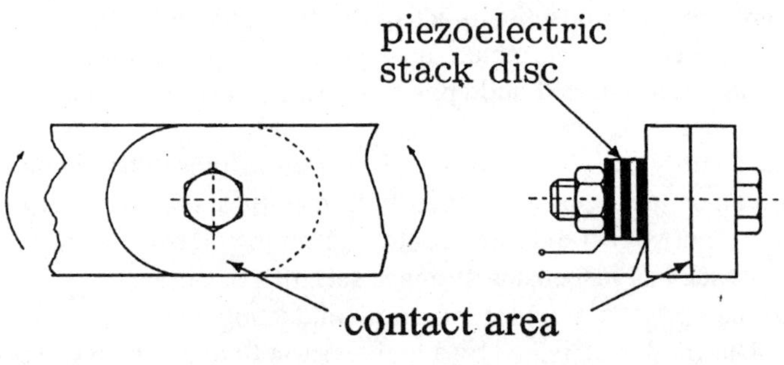

Reference 14 (Gaul et al)

Figure 10 Friction damping via an active joint

Figure 11 shows the behaviour of an ER Fluid with a low and a high electric field. The models used to represent such fluids are based on the Bingham plastic model [19]. Figure 12 shows the behaviour of this model compared to a Newtonian fluid where a finite shear stress exists at low shear rates, introducing a degree of shear stiffness in the ER fluid. Figure 13 gives an example of an ER damper.

MR fluids consist of magnetically permeable particles typically 3 to 5 microns in size suspended in a non-magnetic medium. The material is activated via a magnetic field which causes a similar effect to the ER fluids. As with the ER fluids these exhibit enhanced shear stress characteristics, which can be converted into effective damping devices [20].

MR solids have also been developed using magnetic particles in elastomer media which result in field-induced dependent moduli, offering the potential to control the dynamic properties of systems such as vibration isolation mounts.

Several important aspects relating to the use of these materials as dampers requiring further research are:

- lack of a coherent control strategy,
- improved models for the dynamic material properties,
- speed of response,
- practical issues regarding the durability and life cycle behaviour of the materials.

<u>Formation of chains in ER fluid:</u>

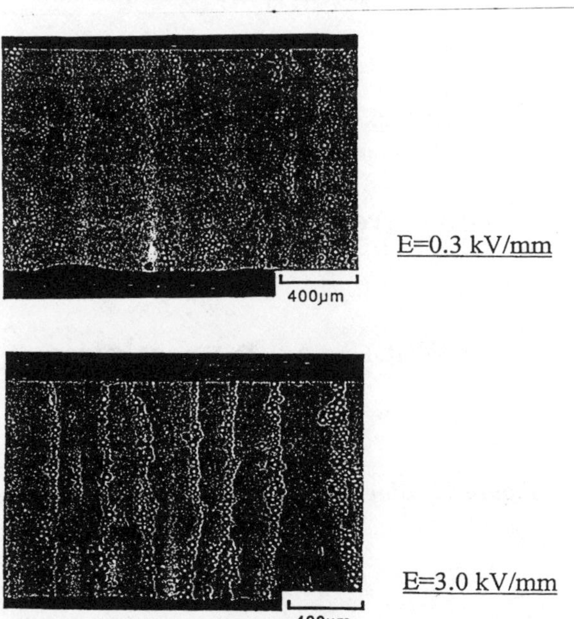

Figure 11 Formation of chain-like structures in an ER fluid

<u>A Bingham plastic model</u>

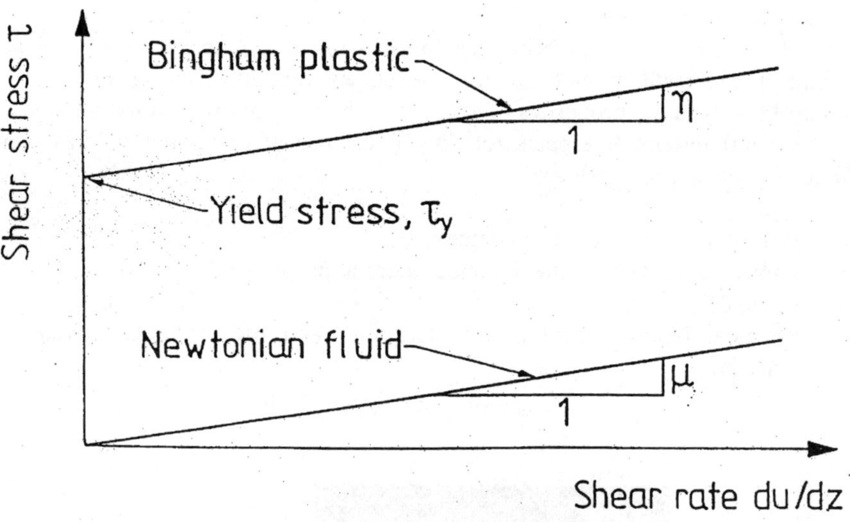

$$\text{Total stress} \quad = \quad \tau_y + \eta \frac{du}{dz}$$

$$\text{where} \quad \eta = \quad \text{plastic viscosity}$$

Figure 12 Bingham plastic model characteristics

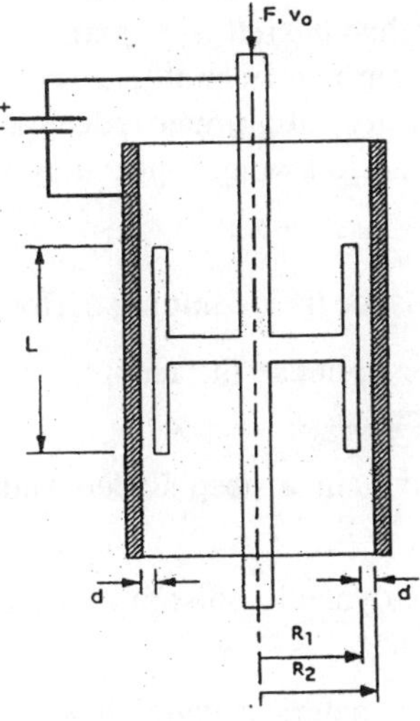

Mixed mode dashpot damper. The outer electrode is fixed and the inner electrode moves with the piston head relative to the outer electrode. Nominal design corresponds to an inner electrode or piston head length of 4 inches, piston head diameter of 2 inches, and an electrode gap of 0.1 inches.

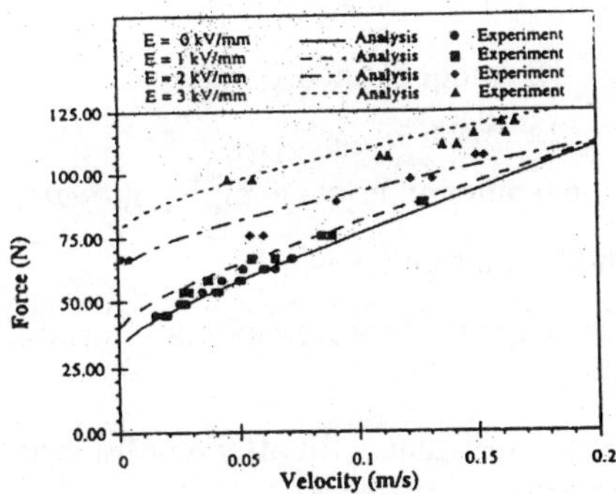

Experimental force versus velocity curves for the nominal design of the ER fluid mixed mode dashpot damper. The theoretical results are corrected to improve correlation by adjusting the field-dependent dynamic yield stress and plastic viscosity to improve curve fitting. The experimental data is shown as points and the analysis as curves.

Reference: Kamath, G.M., Hurt, M.K., Wereley, N.M., "Analysis and testing of Bingham plastic behaviour in semi-active ER fluid dampers", Smart Mater. Struct, 5 (1996), 576-590.

Figure 13 An example of an ER Damper

5. CONCLUSIONS AND RECOMMENDATIONS

Approaches to the modelling of damping would benefit from expanding the scale of understanding from the molecular, meso, micro to macro models. In addition, thermodynamic models, probabilistic and non-linear models may provide a better insight into energy dissipation capabilities, robustness and adaptability that would lead to improved design-in capabilities rather than the retro-fit approach.

In order to extend the state-of-the-art in damping technologies that would lead to the design-in of damping as opposed to retro-fitting, and reduce some of the limitations discussed in the above sections, the following topics may provide the way forward:

- introduce anisotropy into materials to enhance their damping performance,

- change the chemistry of the materials to optimise the damping performance over a wider temperature and frequency range,

- model damping at a molecular level to gain a deep understanding of the mechanisms involved,

- develop procedures for characterising the dynamic Poisson's ratio of damping materials (leading to better models),

- gain a better understanding of damping materials under high strain, large displacement conditions (non-linear behaviour),

- solve the inverse problem of defining damping requirements by designing the optimum materials/devices to achieve this,

- devise efficient and robust FE and BE models incorporating stochastic methods to accommodate variability of input data,

- quantify errors in analytical, computational and numerical damping models,

- develop end-user damping design methods and toolboxes,

- set goals for damping levels based on realistic measurements and establish damping targets for given applications,

- devise reference test methods and techniques that will offer confidence in measured data of material damping properties,

- develop 'best practise' guidelines for damping measurements,

- improve and 'keep simple', where possible, the models and methods for damping designs and predictions,

- enhance damping via active/electronic means and improve electronic damping packaging systems,

- exploit non-linearity to enhance damping properties.

ACKNOWLEDGEMENTS
The author would like to acknowledge the fact that several of the above bullet points were taken from a report arising from a workshop on Novel Damping Concepts and Materials, sponsored by the ARO and held at the VPI over the period 20-22 October 1998.

REFERENCES
1. Cremer, L., Heckl, M., and Ungar, E.E., Structure Borne Sound (Chapter III), Springer-Verlag, New York, 1973.
2. Nashif, A.D., Jones, D.I.G. and Henderson, J.P., Vibration Damping, John Wiley & Sons, 1995.
3. Rongong, J. and Tomlinson, G.R., Suppression of ring vibration modes using constrained layer damping methods, Smart Materials and Structures, 5, 672-684, 1996.
4. Rongong, J., Structures damped using viscoelastic materials - issues affecting concept designs, 1st ARO Workshop on Novel Structural Damping Concepts and Materials, VPI, Blacksburg, October 1998.
5. Remillat, C., An initial study of damped composite shells, Internal Report MDT UTC 99/02, Department of Mechanical Engineering, University of Sheffield, 1999.
6. Ebrahim, A.K. and Baz, A., Vibration control of plates using magnetic CLD, Proceedings of the SPIE Conference on Smart Structures and Materials: Passive Damping and Isolation, Vol 3327, 138-158, 1998.
7. Wardle, R. and Tomlinson, G.R., Viscoelastic damping layer optimisation using a biomimetic approach - an experimental study, Proceedings of the SPIE Conference on Smart Structures and Materials: Passive Damping and Isolation, Vol 3327, 128-136, 1998.
8. Johnson, C. D. and Kienholz, D.A., Finite element prediction of damping in structures with constrained viscoelastic layers, AIAA, (20), No 9, 1982.
9. Lesieutre, G.A. and Lee, U, A finite element for beams having segmented active constrained layers with frequency dependent viscoelastics. Smart Materials and Structures, 5, 615-627, 1996.
10. Papalou, A. and Masri, S.F., Performance of particle dampers under random excitation, Trans ASME Jnl Vibration and Acoustics, (118), 614-621, 1996.
11. Olson, S.E., Drake, M.L., Flint, E.M. and Fowler, B.L., Development of Analytical Methods for Particle Damping, Proceedings of the 3rd National Turbine Engine High Cycle Fatigue (HCF) Conference, San Antonio, TX, February 2-5, 1998.
12. Salueña, C., Esupov, S.E., Pöschel, T. and Simonian, S., Dissipative properties of granular assemblies, Proceedings of the SPIE Conference on Smart Structures and Materials: Passive Damping and Isolation, Vol 3327, 18-26, 1998.

13. Hollkamp, J.J. and Gordon, R.W., Experiments with particle damping, Proceedings of the SPIE Conference on Smart Structures and Materials: Passive Damping and Isolation, Vol 3327, 2-12, 1998.
14. Gaul, L., and Lenz, J., Nonlinear dynamics of structures assembled by bolted joints, Acta Mechanica, 125 (1-4), 169-181, 1997.
15. Baz, A. and Ro, J., The concept and performance of active constrained layer damping, Journal of Sound and Vibration, 28, 18-21, 1994.
16. Hagood, N. Reported at the 1st ARO Workshop on Novel Structural Damping Concepts and Materials, VPI, Blacksburg, October 1998.
17. Bhattacharya, B., Krishna Murty, A.V., Bhat, M.S. and Anjanappa, M., Vibration suppression in slender composite beams using magnetostrictive actuation, Journal of Aeronautical Society of India, 48 (2), 135-146, 1996.
18. Hollkamp, J.J. and Gordon, R.W., An experimental comparison of piezoelectric and constrained layer damping, Smart Materials and Structures, 5, 715-722, 1996.
19. Bullough, W.A. (ed), 5th International Conference on ER Fluids, MR Suspensions and Associated Technology, Sheffield, UK, World Scientific Publishing Co.
20. Jolly, M.R., Carlson, J.D. and Muñoz, B.C., A model of the behaviour of magnetorheological materials, Smart Materials and Structures, 5, 607-614, 1996.

Part IV

The Strategic Questions Posed at SD2000

Contributed by the Participants of the SD2000 Workshop
Edited and Compiled by

Scott W. Doebling
Engineering Analysis Group
Los Alamos National Laboratory
Los Alamos, NM, 87544

ABSTRACT

Certain attendees at the *Structural Dynamics 2000* Forum held at Los Alamos National Laboratory in April 1999 were asked to submit questions for consideration by the full body of the forum. These questions were intended to generate conversation of a strategic nature regarding the future of structural dynamics. The questions are summarized and categorized in this document for the benefit of the structural dynamics community at large.

INTRODUCTION

Throughout history, many great scientific advances have been made as a result of public debate over controversial questions. For example, historical debates over astronomy, the structure of matter, and the origins of life have sparked fierce debate over hundreds of years but have also motivated great quantities of scientific research in many different fields.

In order to promote such debate and exploration in the field of structural dynamics, certain attendees at the *Structural Dynamics 2000* Forum held at Los Alamos National Laboratory in April 1999 were asked to submit strategic questions. The original intention was to hold a structured debate based on some subset of the questions. Although such a structured debate did not fit into the forum schedule, there was still much discussion of these questions throughout the entire event. Additionally, the questions did help to structure this book.

The participants from the group designated as the "less senior" members of the community were invited to submit a small number of questions (typically three). Also, the participants who represented the interests of industry were invited to submit questions. These questions were to consist of a statement followed by a brief elaboration.

The questions are presented here for the consideration of the structural dynamics community at large. The questions are (loosely) categorized and are alphabetized by author's last name (which appears at the front of the question). The list of participants in SD2000 is presented on page 449.

CATEGORIES:

Technical Issues in Structural Dynamics
- Stochastics/Predictability/Reliability
- Nonlinearities
- Damping
- Coupled Problems
- Mid/High Frequency Problems
- Model Updating / Damage Identification
- Testing Issues
- Emerging Technologies
- Miscellaneous Technical Questions

The Culture of Structural Dynamics (Education, Image, and Role)

TECHNICAL ISSUES IN STRUCTURAL DYNAMICS

Stochastics/Predictability/Reliability

Q: (Balmes) Bias and variance in identification methods

Estimating the quality of identification results would be of significant help in assessing and improving model correlation or simply to perform robustness studies.

Typical identification algorithms estimate the poles while over-specifying model order. A major reason for doing so is that out of band modes have a strong influence on structural dynamics, so that additional computational modes are needed to account for these contributions. The representation of residual contributions by computational modes or simple asymptotic terms clearly induces bias in the estimated poles. The bias on pole estimates is usually quite small but this is not true for modeshapes (in particular for the phase of complex residues).

In this light, is it possible to determine if the (small) variations in poles estimates are representative of variance in the estimate or simply of changes in the bias effects when considering various model orders. Can something be said for the larger variations found for modeshape estimates?

Q: (Bucher) Should deterministic or statistical models of structures be used for structures having high modal density?

Large structures could have tens to even hundreds of modes (natural frequencies) in the frequency range of interest. Common inaccuracies of 1-10% in some modal parameters and even 40%-200% in damping may render such a model useless for deterministic prediction of response levels. The question arises whether dynamic prediction should be re-directed towards statistical estimation of response level based upon statistical models of the structures? Do deterministic models have some merit even in such cases that cannot be overlooked? This issue calls into question the validity of modal testing in the cases where it should be mostly required.

392

Q: **(Cogan) What criteria can be used to judge the usability of a model for a given application, that is to say, whether model-based decisions can be integrated with confidence into an engineering activity. Is it really predictive precision we are after or is it robustness of model-based design decisions to ambient uncertainties?**

Model quality has become a major concern in the structural dynamics community. The availability of attractive and highly developed software for creating mathematical models of complex behaviors (due to both local behavior laws and intricate topologies) has lured (to be provocative) the industrial community into their purchase and use. These tools have produced undeniable success stories and their usefulness is difficult to contest. However, the influence of diverse sources of uncertainties ranging from the simplifying concepts used to analyze a given problem to the various measurement errors tend to create an incompressible distance between analytical predictions and the reference behavior observed in the field. It is certainly a tribute to engineering savoir-faire (and a healthy dose of safety factors) that planes fly and nuclear reactors remain contained given the abiding inaccuracies in the mathematical models of complex technological structures.

Model-based decisions are useful only in so far as they maintain a certain fidelity with respect to real prototyping, that is to say, that they represent true pseudo-tests. The goal of modelling, measurement and updating is simply to convince ourselves that we are indeed in the presence of such a model. The question is just how this can be done in a reliable way ? Just when is a model precise enough ? Is it really 'precision' that we are looking for or is it 'robustness' of our design decisions with respect to the known (and unknown or unexpected) sources of uncertainty in both the model(s) and prototypes. A model might well provide very precise predictions with respect to a given parameterization and for a set of measurement data. These predictions however might be extremely sensitive to uncertainties in both model parameters and to the fact that a set of apparently identical prototypes present a dispersion in their behaviors. Conversely, model predictions may be relatively imprecise while the design decision in question may well be relatively robust with respect to these prediction errors . Clearly a methodology is required to allow these concerns to be addressed in a comprehensive way.

Q: **(Friswell) Has modeling structures using finite element analysis reached the limits of accuracy? Is a new paradigm required for design and analysis of structures?**

Many engineers believed that any structure may be modeled to arbitrary accuracy merely by increasing the finite element mesh density. These refined meshes are able to model the geometry of the structure more accurately, but uncertain parameters, for example geometric tolerances or joint dynamic stiffnesses, mean that modeling errors will never be resolved using this approach. Model updating can help, although for complex structures the incompleteness of the measured data makes it is impossible to identify a physically meaningful model. Thus there are definite accuracy limits for modeling a single structure, and also for modeling the variability of batches of components because of manufacturing tolerances.

Perhaps we should be looking at different ways of analyzing the dynamics of structures. Methods based on uncertainty (defined either as a statistical or convex model of uncertainty) could be used to optimize structures as part of a robust design scheme. Control engineers are more advanced in this respect and robust controller design, based on an uncertain plant model and influenced by uncertain disturbances, is a mature technology. Could any of the robust control methods be transferred to the design activity in structural dynamics? Could the design process be tailored to produce a structure where the known uncertainty in the structure has the minimum effect on its dynamic response? Is this a useful thing to do?

Q: (Hemez) Predictability of Structural Dynamics Models

Should the structural dynamics community put a strong emphasis on developing tools for assessing the predictive quality of a given numerical or experimental model? If so, which technologies need to be developed?

No practical tools are currently available to assess the predictability of a structural dynamics model. The quality of an individual finite element mesh can be assessed, the quality of test data can be characterised statistically but the structural dynamics community lacks the capability of assessing the degree of predictability of a given model with respect to a given objective. (For example, what is the accuracy associated with the response obtained from a given finite element model knowing that the objective is to minimise the distance between test data and numerical simulations?) Developing such a capability would probably require the integration of statistical analysis, multiple-model representation and inverse problem solving (for comparing the predictions of a model or family of models to test data).

394

Q: (Pickrel) Confidence Intervals

Can we estimate confidence intervals on structural dynamic parameters which affect the design and performance of our products?

In a more perfect world we would assess variability in manufacture of products, test measurements, model predictions, and the assessment of operating environments. Many products are engineered with conservatism to compensate for ignorance of these confidence limits. Improved assessment of variance and confidence intervals would lead to improved performance or reduced cost of these products or designs.

Q: (Zimmerman) Design for Reliability

As we are always reminded, we all will (or already) have Cray computing capability sitting on each and every desk of an engineer in the near future. Have we as structural dynamicists stayed in the safe "linear" world, and away from the nonlinear and stochastic real world, based on our thoughts that the computational power needed to properly perform design is not available?

Nonlinearities

Q: (Balmes) Design objectives for non-linear dynamics

Many criteria have been developed to judge the performance of linear structures. One thus minimizes the RMS response or gives a frequency domain envelope for acceptable response spectra. The assumption of linearity leads to two major simplifications. The characterization of the excitation can be limited to simple statistical representations. The relation between the excitation and the response is simply characterized by transfer functions which are easily computed using modal analysis.

With the objective of judging the dynamic performance of non-linear structures, tools seem to be missing to characterize broadband excitations with quantities that can be propagated into response characteristics in a reasonable time. Extending modal analysis notions to non-linear structures seems key to the propagation problem but this will still leave the input characterization issue open.

Q: **(Bucher) Should linear, non-linear or Floquet (periodically varying coefficient) systems be applied to rotating machines?**

Rotating structures exhibit a very rich response spectrum that is caused by the complex nature of the various phenomena co-existing in a rotating machine. A rotating structure can be excited by external forces such as fluid-structure interaction, cutting forces (in saw blades), or by speed variation and parametric excitation (e.g. modulation of the response by stiffness and mass depending upon the instantaneous angle of rotation).

It has been shown (A. Bendat) that in some cases non-linear systems possess the same input-out behavior as hypothetical linear system with periodically varying coefficients.

Inspecting the spectrum of the vibration measured on a rotating machine one can always notice numerous harmonics of the frequency of rotation as well as other frequencies. These harmonics could be attributed to the nonlinear nature of the supports (bearings) or to parametric excitation due to time varying parameters. The parametric excitation is caused an the imperfect rotating structure (not perfectly isotropic). This effect could potentially be explained or even more complicated phenomena (coupled bending torsion in shafts?).

How should this problem be approached and resolved ? How could these phenomena be separated from measurements and how should they be modeled ? Is this important or can any suitable 'black box' model that relates forces to response be used ?

Q: **(Cooper) How should non-linearities on real structures be identified?**

There has been a great deal of work devoted to the modelling and identification of non-linear systems (e.g. NARMAX, restoring force surfaces, higher order FRFs, etc.) Whereas these methods can deal with simulated systems and small purpose built laboratory experiments, their use on real structures (e.g. automotive, aerospace, civil structures) is problematic. Is the correct approach being used with improvements in the technology being made very slowly, or is a radically different methodology required to tackle the problems of detecting, quantifying and identifying non-linearities in real structures.

Q: **(Cooper) Is linear modal analysis solved?**

The process of identifying frequencies, damping ratios and mode shapes from modal test data is nowadays taken as being fairly straightforward. However, most commercial implementations of the methods rely a great deal upon engineering judgement of the user (how many modes, which modes should I select from my stability plot? How to deal with scatter in the results, etc.) The amount of research devoted to investigating new approaches has diminished and most of these (e.g. subspace methods) are viewed as research tools rather than appearing in commercially available software. Has the problem been solved totally or should further effort be devoted towards providing industry with "intelligent" modal testing that removes the need for a large amount of user interaction. (PS: When is it not OK to treat a structure as linear?)

Q: **(Davies) How can the classical single frequency excitation perturbation analysis of non-linear systems analysis become a useful tool for structural dynamicists trouble shooting problems in structures?**

When trying to understand the conditions necessary for modal interactions to occur in weakly non-linear structures, a two-stage analysis is usually performed. First, for the conditions under which the structure is operating, a reduced order modal model of the structure is derived, and secondly, the steady-state (or slowly varying) response of the modes of this model for various forms of stationary (or slowly-varying) harmonic excitation is examined. Analysis of these response versus model parameter (excitation level, natural frequency, excitation frequency, etc.) functions allow determination of regions where multiple solutions may exist, or where no stable steady-state solutions exist and energy sharing between modes may occur. Multiple solutions co-existing for the same excitation will mean that in structural testing, the response observed will be a function of initial conditions and perturbations to the system, a phenomenon not present in linear systems.

As structures become thinner and deflections become larger the role of non-linearities must be considered, and in frequency regions where many modes are present, the probability of modal interactions is high. When excitations come from rotating machinery, it is not unrealistic to consider harmonic excitations, or to assume that only a few of the structures' modes are directly excited. Therefore, this classical nonlinear dynamics analysis is suitable for practical situations. However, while people studying non-linear systems in their research use these tools, they are not widely used in the structural dynamics community. There are also some interesting issues that need to be addressed. Here are a few: (1) Is it possible to probe the structure through testing to determine with some certainty whether this modal coupling is present, and thus to justify the non-linear modal analysis? (2) Given a particular excitation, how can we systematically determine which modes to include in the model? (3) Since the response curves are highly dependant on the modal parameters, how can the parameters be estimated using excitations that will only cause the modes in the model to respond? (4) Can the analysis techniques be extended to multiple harmonic excitations? (5) Does the structural dynamicist need to be an expert in non-linear systems analysis to use these tools?

Damping

Q: **(Caesar) How can we solve the old problem of damping prediction?**

Q: **(Shaw) How can we develop effective damping models and include them into FE codes?**

Coupled Problems

Q: **(Caesar) Coupling of Control and Structure Mechanics**
How should software be designed to couple control sensor and actor design with structure behavior?

Actually the coupling is performed via standardized data package and/or model transfer from one to the other independent S/W-tool. Application example: ESA, NASA, ESO and others are working on high precision space and earth born telescopes with interferometric optics under the task "Next Generation Space Telescope" and "Very Large Telescope" for which pointing mechanisms in the delay line with accuracies in the range of 1 nm are required. Such extreme accuracy demands require new strategies.

Q: **(Hemez) Coupling Macroscopic to Microscopic Analyses**

Rather than attempting to develop "useless" mathematical models for accounting for complicated dynamics (such as energy dissipation), could the structural dynamics community focus more on the means of coupling the conventional, mechanics-based representation of structural dynamics with a localised, physics-based description of the mechanics at the microscopic (if not, nanoscopic) level?

The motivation for this question is best illustrated with a (provocative) example. Many research efforts currently focus on developing damping models starting from the equations of mechanics and a description of the material behaviour at the macroscopic level. For the past ten years or so, it seems that this research has delivered no significant improvement of our modelling capabilities. It can be argued that the reason is that most damping sources originate from the localised contact/friction between two surfaces or from the dissipation of energy in the material at the microscopic level. To model this phenomenon (and many others such as contact), it seems therefore critical to account for surface irregularity, material imperfection and anisotropy, if not molecular interactions! It seems like it would be more productive to study how macroscopic properties could be coupled to a characterisation of the material at the microscopic scale and, conversely, how microscopic state variables could be averaged to produce, for example, "global" displacement or stress components. The goal here would be to enable a fully coupled macro/micro analysis in order to improve our predictive capability by taking advantage of a physics-based representation of the dynamics, whenever necessary. Of course, such an endeavour would generate significant computational challenges.

Q: (Caesar) Regarding Coupling Macroscopic to Microscopic Analyses

Mechanisms especially high precision ones can be described by actual available mathematical models only on the basis of empirically derived data. Mostly the mathematical model is sufficiently accurate not before tuned to the hardware behavior. This causes high development risks in costs and schedule. The feasibility of a design cannot be predicted.

Mid/High Frequency Problems

Q: (Caesar) Medium and High Frequency Range Dynamics Analyses

With which approach shall the dynamics in the medium and high frequency range be treated?

The medium and high frequency range is most often important in acoustic and shock excitation. Several not fully satisfying approaches are available, one is the refinement of discretization (refinement of FE-mesh), another is the statistical energy analysis SEA. Solving this problem would help, e.g., to predict more accurate stresses and test levels for equipment on structures under acoustic or random loads, or to reduced structure radiated noise.

Q: (Cooper) How should structures in the mid-frequency range be modeled?

Conventional FE tends to be used for low frequencies whereas Statistical Energy Analysis (SEA) is applicable for very high frequencies. Little work has been undertaken to address the frequency ranges in the middle where the accuracy of FE breaks down. How should the Structural Dynamics community tackle this problem?

Q: (Pickrel) Mid-Frequency Problems

How do we approach mid-frequency and high frequency problems where modal analysis is not practical? Modal analysis provides us with a robust approach to "low frequency" problems in structural dynamics, about which much has been written. Is there a consensus for an approach to problems involving higher frequencies where the modal approach breaks down?

Model Updating / Damage Identification

Q: **(Friswell) Do uncertainties in the model, measurement noise and the changes due to environmental effects, make it impossible to robustly locate damage in structures using low frequency vibration measurements?**

The area of damage detection and location using measured low frequency vibration data has attracted considerable attention recently. The great advantage is that low frequency data are global, and thus only a limited number of sensors, located remotely from the potential damage sites, are required. However, there are considerable difficulties to be overcome before the approach is robust enough to be used in practice. The question is whether the inherent difficulties render all the proposed methods unusable in the field.

In terms of methods, the only viable approaches are based on forward identification procedures which recognise the fact that damage generally only changes a limited number of parameters. Methods that try to identify a large number of parameters simultaneously produce ill-conditioned equations, whose regularised solution destroys any capability to identify damage location.

Even forward identification methods have considerable difficulty. Is the finite element model accurate? How accurate does it need to be? How does the dynamics of the structure change with environmental conditions, for example temperature and humidity? Can these effects be modelled with sufficient accuracy so that their influence may be removed? How large are the changes in a structure's dynamics due to typical faults? Is this level of change likely to be detected reliably, given the signal to noise ratio, the changes due to environmental effects and the modelling errors? Even if damage can be detected, is the 'effective wavelength' of the low frequency modes too large to make localisation of the damage sufficient to be useful?

Given all these significant difficulties that are based on the physics of the problem not the methods used, what is the prognosis for damage detection and location using low frequency vibration?

Q: **(Imregun) Modal Updating – Is this the end of the road?**

For the last 20 years or so, significant research effort has been devoted to improving mathematical models using vibration test data. However, there are still no established and universally applicable methods that can produce an updated model that can satisfy the following criteria:

The (true) experimental and updated models must have the same modal and response properties within the measurement frequency range.
The updated model must be able to predict the effects of further changes.

It is debatable whether the corrections to the updated model should have a physical meaning. Most updating work is focussed on FE modeling, which may or may not be the way to go. Is it time to accept defeat and have a fundamental review of the modeling techniques?

Q: (Imregun) Damage detection – Given the relative lack of success with model updating methods, can we expect to do better?

In the general case where the damage location is not known, is it possible to develop detection techniques using numerical base models without having solved the model-updating problem first?

Q: (Imregun) Should we develop model-updating methods on small-size models and assume that these are equally applicable to large-size models?

It is common practice to develop updating methodologies on small-size models, say with 10-1000 DOFs. When dealing with such models, numerical accuracy, computational feasibility in terms of CPU effort and storage are taken for granted. However, there are many computational problems when working with large-size models, say 30,000-200,000 DOFs. For instance, it is not a trivial task at all to obtain the first 5,000 modes of a 100,000 DOF model. Not only will different FE codes give different results, various eigensolution extraction routes within the same code will also yield different results. Such problems will probably go unnoticed when computing the first 50 modes of a 1,000 DOF system.

Another point to remember is that the size of the model is often dictated by the amount of geometric detail that needs to be incorporated. For modelling simplicity, can we ignore such detail and hope to be able to update the model ? In other words, how can we determine the minimum amount of modelling detail that is required?

Testing Issues

Q: **(Bucher) Recommended Signal of excitation: which type excitation should be used for linear structures and which one is preferred in the slightly nonlinear ones. Sine-sweep, sine-step, random, burst, impact?**

In dynamic testing of structures many excitation methods were used and it seems that the "fashion" changes every few years and there is no agreement upon the 'best' method for each particular application.

Is it better to use an excitation that 'linearises' the identified structure (random) or an excitation that exposes the non-linear nature of the structure (sine, impulse)? Should non-linear behavior be treated in the excitation phase, signal processing phase or model-fitting stage?

Q: **(Doebling) What will be the appropriate relationship between simulation and experimentation in the early 21st century?**

Historically, in structural dynamics as in other fields, experimentation has been used to both PROVE and IMPROVE: (a) PROVE that an engineered system meets its design criteria by performing a "proof" or "qualification" test and (b) IMPROVE the mathematical models of the engineered system to better predict the response of the system under different initial and/or boundary conditions. Some engineers would claim that the IMPROVE experiments provide more useful information for engineering purposes than do the PROVE experiments, but nevertheless the PROVE experiments receive much higher priority in terms of resources (i.e. money) than do the IMPROVE experiments.

GIVEN the extreme decrease in the price-to-performance ratio of computing power over the last 30 years, is it possible that structural dynamicists can "substitute" computational simulations for many of the PROVE experiments, so that resources can be concentrated on IMPROVE experiments to actually make our modeling and analysis capabilities better?

Q: **(Hemez) Developing Accelerated Testing Procedures**

Could experimental procedures be developed that would simulate the aging of a structural system and help assess its performance after a life-span of, say, 30 years without having to wait that long to take the measurements?

For many structural dynamics applications, it is important to estimate during the design phase what the remaining performances of a system will be after it has been used for a long period of time (20 to 30 years). This is especially critical with aerospace applications for which modelling the effects of aging is almost impossible because very little data are available to validate the simulations (we are just starting to witness on a large scale the effects that aging has on aircraft systems). Another example would be civil engineering structures such as bridges that often stay operational for longer periods of time than they had been originally designed for.

Finite element simulations can be used for estimating the performance of a system after a series of load cycles. However, our current modelling capabilities do not allow us to handle complicated structural systems for which residual strength assessment still relies, to a great extent, on testing. Are experimental techniques available (or could they be developed) for reducing 30 years worth of structural dynamics in a few tests?

Emerging Technologies

Q: **(Cooper) Are there developments in the signal processing and numerical analysis fields waiting to be applied to vibration analysis?**

Great use is currently made of tools such as the Fast Fourier Transform for vibration testing and standard analysis approaches such as Least Squares. Other recent developments have included the use of the Singular Value Decomposition (SVD) and Time/Frequency methods (e.g. wavelets). Any major developments in the field from the algorithmic viewpoint will depend upon the use of new techniques. Do such tools exist, and if so, how do we apply them to vibration analysis?

Q: **(Davies) It is often possible with computer control of manufacturing to tailor the chemical content of materials during manufacture. How can we optimize the in-application macroscopic behavior of materials using these manufacturing capabilities?**

This is really a problem of linking the material and chemical level modeling with the structural dynamics models. There is a need to feed the basic material properties into the structural model, or if this is not possible to do directly, to understand how fundamental changes in material microstructure affect the terms in the macroscopic structural model. Robust system identification techniques are needed to determine the types of terms, and the parameters in those terms, that are necessary to model the macroscopic behavior of the material in the application of interest. These terms may be, for example, stiffness, viscous damping, hysteretic damping, dry friction, or viscoelastic terms. Many interesting problems arise here in system identification, not only because many of the material models will be nonlinear, but also because materials such as polyurethane foams (used in transportation seating) have both fast- and slow-time behavior and are sensitive to environmental factors such as temperature and humidity, making repeatable experimentation as well as parameter estimation challenging. The interaction between the terms in the structural model will add an additional complexity to the identification, perhaps requiring the need for excitation techniques that allow for isolation of particular types of behavior.

Development of robust system identification techniques to identify structural dynamics model parameters related to material properties, and derivation of functional relationships between these parameters and the material manufacturing parameters, is only the first stage. We then have to take this understanding and use it to design the structure, working backwards from the desired response to known excitation, to the space of effective structure dynamics model parameters, and finally to the materials manufacturing parameters required.

Miscellaneous Technical Questions

Q: **(Caesar) Transfer of CAD into Mathematical Models**

Could straight forward procedures be developed to establish mathematical models directly from the design of complex and/or assembled/integrated structures?

Actually automated procedures exist to transfer single structures into FE-meshes under certain restrictions such as simplicity of the structure design or use of tetrahedral elements. As soon as structures become more complex or there is the need of coupling several structures together, the skill of structural engineers is required to transfer the reality into the simplification of the mathematical models.

Q: **(Cooper) How should structures with non-modal looking FRFs be identified?**

When a structure with a very high modal density and relatively large damping is tested, the FRFs do not look modal and instead look very flat. Analysis using conventional curve-fitting methods does not produce very good results. Should black-box time domain approaches be used, or should some other methodology be used?

Q: (Friswell) Is state space methodology state of the art for damped structures and rotating systems?

Often damping is ignored in the analysis of dynamic models of structures, but recently there has been a resurgence of interest in the analysis of damped structures. The standard approach to analysing such structures is the transformation to a first order system (the state space formulation). The increased effort for the eigensystem computation required to progress from an undamped system with N degrees of freedom, consisting of real eigenvalues and eigenvectors, to a damped system represented by a first order system with 2N states and complex eigenvalues and eigenvectors, is significant. In terms of understanding, as soon as the state space matrices are formed, the constraint on the states (where half are displacements and half are the corresponding velocities) is lost. Indeed there is also a constraint between the real and imaginary part of the complex eigenvectors that is not used or enforced in a state space analysis. Are we missing something that could enhance our fundamental understanding of the damped system? Should we, for example, make sure these constraints for a second order system are enforced in experimental modal analysis?

What application does Geometric Algebra (for example Clifford Algebra) have in structural dynamics? Many areas of physics have benefited from the unifying force of Clifford Algebra. Most engineers are happy to use complex numbers in the analysis of dynamic systems (particularly in state space form); they afford tremendous insight and compactness of notation. Could Clifford Algebra do a similar job for second order systems, helping both understanding and the computation burden? In rotordynamics it is possible to encode multiple uses for complex numbers (for example, using complex numbers for both the eigenvalues and to encode displacements or rotations at a given node in orthogonal directions). This area is only just beginning to develop and the benefits have not as yet been fully investigated. Is this the future?

Q: (Pickrel) Dynamics and Controls

Can anything be done to increase the communication and integration between the Structural Dynamics community and the Controls community?

Increasingly, problems arise which require the feedback control of flexible structures. It seems this requires the participation of engineers from two different organizations (in a larger company), who graduated from different departments, read different journals, and speak different "language".

Q: (Shaw) Development of techniques for building reduced-order models

Several methods exist for reducing the size of computer-based, linear models of structural systems. These techniques are very useful for substantially reducing the number of degrees of freedom required for high-fidelity models over a certain frequency range. However, when nonlinearities are present, this type of model reduction becomes much more difficult. Can some basic, useful model-reduction theories be developed for this class of problems? Can these be developed into useful computer codes? Can one develop specialized methods for cases in which the nonlinearities are strictly local in nature, say at joints, or due to attached components? Or, similarly, for the case when the nonlinearities are inherently distributed, for example, due to kinematic effects?

Q: (Shaw) Structural dynamics and control at small scales

An extremely important area for the future of structural dynamics is that related to micro, and even smaller, devices. There are new exciting challenges waiting for the dynamicist here, and there are opportunities to be involved with cutting-edge technology.

THE CULTURE OF STRUCTURAL DYNAMICS (EDUCATION, IMAGE, AND ROLE)

Q: (Avitabile) Transfer of Technology - Educating Today's Engineer

Structural dynamics reaches into many areas of design today as we try to make all of our designs "lighter, quieter, more efficient, easier to manufacture, etc.". However, most graduating engineers receive little education in these applications. How do we effectively educate the working engineer to be cognizant of the approaches to be used in a realistic manner? Many of the techniques currently available (and still being developed) require "significant" background to understand the subject matter. How do we train and impart this knowledge in a usable format for the practicing engineer?

There currently exists a Shock and Vibration Handbook which is an excellent resource of information. Do we need to develop a Handbook of Structural Dynamic Modeling Techniques which addresses all the areas which should be of concern for the design of structural dynamic systems? Do we need to provide guidelines or back-of-the envelop techniques to identify if problems are expected to exist? Do we need to identify what types of tools are useful for different types of problems? Just how are we going to educate the engineer in these newly developed techniques and methodologies that are required and useful for general use in a wide variety of design scenarios?

The need for "Educating Today's Engineer" stems from some typical (uneducated) statements, questions and misconceptions that I expect that we all have heard at one time or another - some examples are:

Automotive brake engineer believes that correlation/updating of an analytical model will reduce brake squeal.

Engineers believe that model updating is a "push button" technology; salesman sold them a "bill of goods".

Operating data provides a true picture of the "actual response"; but what do I have to do to modify the design?

Performing a modal test will describe the dynamics of the system; but without a forcing function, what good is it?

And the list just goes on and on. How do we educate relatively new engineers into all the detailed technology that exists today, in a practical sense?

Q: **(Avitabile) Making Sophisticated Structural Dynamic Testing/Analysis Procedures Commonplace in Everyday Design**

About 20 to 30 years ago, finite element modeling (FEM) was reserved for sophisticated engineering applications that had extensive time and budgets. FEM analysis was reserved for the knowledgeable and capable analysis engineers working in critical applications that required its use. During that period of time, there was extensive development and new concepts/ideas in the formulation of elements and models flourished. Today, FEM is a commonplace structural design tool in almost every industry and application.

Over the past 10 to 15 years, there has been tremendous growth and development in the structural dynamics technology area from both a testing and analysis perspective. Over that period of time, more sophisticated users have been involved in the generation of very elaborate test and analysis endeavors - most of which have required extensive time and budgets.

As we move into the next century, how do we make these newly developed techniques and methodologies more readily available to the more commonplace design of everyday commercial products so that these techniques are affordable for less sophisticated industries where time and budgets are not plentiful?

Specific examples:

Computer cabinet/disk drive design must be produced rapidly with little or no time for in-depth test/analysis to be performed. Cabinets need to be designed to accommodate a wide variety of different peripherals (especially disk drives) which must perform flawlessly in their environment. Drives need to be subjected to qualification tests that are reflective of their actual installed environment (rigidly affixing the drives to a shaker for qualification may not be sufficient or proper to confirm performance). Different disk drives (with different dynamic characteristics) need to be substituted into the design configuration and perform equally well. Accuracy of the model requires that modeling techniques employed are confirmed through the use of experimentally measured data. Cabinet designs are typically constructed such that nonlinear behavior is anticipated and can have an important effect on performance. Noise and vibration (creature comfort) are issues that need to be addressed. The only time when significant money is allocated for evaluation is when problems exist and a "fire" needs to be put out.

Another example is the design of a commercial clothes dryer. Expensive tooling must be committed early in the design with the expectation that the engineering design is correct and will produce an acceptable configuration. Fairly large models are developed of the overall configuration - modal density is populated with local modes of the flimsy cabinet configuration. Test/analysis/correlation of the model is difficult due to the large number of modes. Model accuracy is important to the prediction of system performance for both noise and vibration issues.

Both of the above are excellent examples of design situations that are not much different in technical content than those of the space station design, aerospace program, or automotive designs with one big difference - there is typically very little time or budget to undertake the extremely involved techniques/methodologies required to address the designs properly. The two situations above embrace all of the aspects of structural dynamic modeling techniques required to address the problem. Development of accurate models that are validated through experimentally acquired data. Ability to address the design which is likely to include nonlinear joints, connections and component interrelations. Development of system models that are made up of a variety of different components - using a combination of analytical and experimental components which are interconnected with both translation and rotation effects. Ability to address a large number of modes (both global and local) over a wide frequency range to predict system performance due to operating conditions. Development of an accurate model that can be used for noise and vibration predictions.

Q: **(Doebling) What is the relevance of structural dynamics in the information age?**

It is generally agreed upon that the "industrial age" is largely behind us and the "information age" is fully upon us as we begin the 21st century. Structural dynamics has undoubtedly enjoyed a prime period of growth and increase in resources as the industrial age has matured. Structural dynamics is traditionally associated with the "heavy" industries such as aerospace, automotive, robotics, and civil infrastructure. But what is the relevance of such industries in the information age? An example of such relevance would be increased demand for satellite production as worldwide wireless networks are increasingly in demand. A demand for more "features" on a smaller, more lightweight satellite will certainly involve new structural dynamics issues. But aside from such "new tricks" for the "old dogs" of the industrial age, does the information age present any new opportunities for structural dynamics?

GIVEN that the majority of the economic growth in the early 21st century will most certainly come in the field of information technology, rather than industrial technology, what can we as the structural dynamics community do to "get a piece" of the information age "action"?

Q: **(Doebling) How can analysis of structural dynamics issues be used to decrease "time to market"?**

Many industries, such as those with aerospace, automotive, and civil infrastructure applications, are inherently reliant upon analysis of structural dynamics issues to ensure that the engineered systems perform their functions within specifications. However, many other industries that have high-volume, low-cost, quick-time-to-market business constraints do not have a direct requirement for structural dynamics analysis and therefore forego the expense in favor of quick turnaround time in product development and deployment. What many engineers and managers in these industries fail to consider, however, is the improvement in product quality that can be had with the use of structural dynamics in the product design and deployment phase. For example, consider the portable CD player. While undoubtedly such a product undergoes rigorous testing during the design phase, consider how much more robust the design can be through an equally rigorous application of analysis. However, the application of analysis in such instances is typically limited because it "slows down the design process".

GIVEN the potentially huge market in the consumer products (and other) industries, how can the structural dynamics community make a case for structural dynamics analysis as a tool to improve product quality while not impeding the "quick time to market"?

Q: **(Lieven) How do we improve the image of Structural Dynamics?**

All too often structural dynamics is considered in the context of failure rather than performance. If we compare this with research in say Fluids or Materials, the emphasis of these activities has been consistently related to product enhancement rather than deterioration. Historically, we have a legacy of troubleshooting, when dynamics has been used as a corrective measure "after the event" – whatever that might have been. This culture - with the notable exception of aeroelasticity – leaves dynamicists fighting a rearguard action. Inevitably this puts the issue of dynamics towards the back of the queue in the design process. Even worse the issue of dynamics may only be considered retrospectively.

The education process in dynamics often reinforces the role of dynamics as a discipline associated with structural failure. We all use the "great disasters we have known" anecdotes in our teaching – this may be doing us no favours. The challenge in teaching therefore, is to maintain the interest level imbued by eye catching events, but portrayed in a more positive light. Perhaps the dynamics should be included seamlessly in the design elements of engineering courses?

The challenge for structural dynamics research is more demanding in this respect. Industry frequently views structural dynamics as a necessary evil. The commercial case for retrospective dynamics is not strong. With the increasing representation of industrialists on government funding bodies, we need to convince industry of the benefits of dynamics research. What are these?

If we consider the apparent success Computational Fluid Dynamics in attracting funding, it is evident this has arisen through effective marketing on the basis of reduced cost and increased performance. Dynamics research strives to do the same and yet does not attract the same enthusiasm from industry. How do we address the publicity issue?

Perhaps the role of dynamics has become entrenched by its applications? When we consider dynamics we immediately think of bridges, aircraft and cars. These products are established, and do look forward to new growth technologies. We only have to consider the burgeoning NASDAQ to see where technology is expected to develop – dynamics is not in evidence. The question in this case is, how do we associate ourselves with products and technologies that will take us forward from the entrenched applications we work on today?

Q: (Pickrel) "Open Source" Research Model

Following the "open source model" (Linux software development) could a few structural dynamics problems be placed on an open web site(s), facilitating collaboration and making data and solutions widely available?

Some important problems in structural dynamics involve multiple aspects of our discipline, such as modeling, testing, and controls to name a few. Seldom does a researcher have the resources to accommodate all of these aspects of the problem. An open web site could bring together contributions from different researchers and institutions. The "publishing cycle" could be reduced from months/years down to minutes/days, accelerating the solution process. Examples of the information which might accumulate for a given problem include: drawings of structure, test plans, test data, models and predictions, modal parameter estimates, model updates, damage detection, structural modifications, solutions, discussion (questions, answers, comments and assessments).

Obviously, many issues arise, and not all research will be conducted this way any time soon. The Linux software community has found the "open source" and "release early and often" approach to be a powerful method for rapidly developing robust software. Perhaps the same would be true for the solution of multifaceted problems in structural dynamics?

Q: **(Pickrel) Dynamics in Engineering Education**

How can we increase the emphasis on dynamics in engineering education? It seems that structural dynamics has played an increasingly important roll in product design in recent years. Yet the average engineer is sorely lacking in appreciation of dynamics fundamentals. The dynamics community would benefit from a broader awareness and accessibility of dynamics fundamentals on the part of decision-makers.

Q: **(Shaw) Relevance to industry**

This is a problem that faces many areas of engineering research. Industry is typically slow to make use of the latest results from any basic and/or applied research that is done outside of their labs. In terms of structural dynamics, the aerospace industry has probably been one of the most receptive groups in this regard. However, other industries, such as the automotive companies, are slow to respond. They are often aware of advances, but are reluctant to try new ideas. How can we be more effective in making inroads into these industries?

Q: **(Zimmerman) Where does Structural Dynamics fit in?**

Recently, several traditional United States funding agencies have publicized their plans for research initiatives in the 21st century. In a recent DARPA briefing, it was stated that nearly 50% of the DARPA budget will be focussed on biotechnology (including bioengineering) within the next three years. The NSF and several other agencies (DOD, DOE, NASA, NIH and NOAA), working under the auspices of the National Science and Technology Council (NSTC) have developed a long-term research plan for Information Technology in the 21st century. At the same time, the NSF has announced that its second major focus in the 21st century will be in "Biocomplexity in the Environment". The change in the funding picture is changing rapidly ... how should the structural dynamics field react?

Q: (Zimmerman) Is there an "Economic" Based Grand Challenge Problem in Structural Dynamics?

The quick rise of Computational Fluid Dynamics research and subsequent commercial software can be traced back to President Reagan's call for the High-Speed Transport (or as he called it the Orient Express). It was estimated that after the initiation of this program, nearly 90% of all supercomputer resources in the country were associated with CFD development. Although the High-Speed transport never did materialize (in fact, NASA Langley, the NASA lead in high-speed transport is ending this program in FY00), it did serve as a catalyst to the CFD research and application community. This focussed effort was sold to the public and to Congress based on the "economic impact" and need for such transports in the "new global economy". Look at another major Grand Challenge: The Supercollider. This project folded several years ago because the physicists were unable to sell the public and congress on the economic impact of their research. Is there a Grand Challenge problem that has structural dynamics in the lead role AND has a major economic impact?

Discussions and Debates at SD2000

H. Cudney
Virginia Polytechnic Institute

1. INTRODUCTION – HOW THE DISCUSSIONS WERE CONDUCTED

In this part of the book we present a summary of the discussions which occurred during the meeting. There were several formats for these discussions. Over the course of the week the whole group met in the historic and rustic meeting house with the tables arranged in a large U-shape. This is the format for the group-wide presentations. The presentations included the current state-of-the-art practice in a broad and comprehensive set of disciplines (for example, damping), brief but thorough introductions to cutting edge research (for example, active materials), and presentations of the grand challenges.

In addition to these prearranged topics, the results of the separate group discussions were also presented in this format. The groups were created as a result of the discussion during the first session of the first day. It was determined that there were three themes the workshop could usefully and significantly address: (i) developing technologies, (ii) emerging technologies, and (iii) image. What follows in this part of the proceedings, then, is a summary of all the discussions, both those conducted as a large group, as well as the separate small-group discussions on these individual topics.

For each group-wide meeting, there was a facilitator as well as a person whose sole task it was to record the discussions. The individual group discussions had the same structure. Since these three groups met in parallel, there were three different rapporteurs, as the note-takers are called in Europe. The rapporteurs all had advanced technical degrees in structural dynamics, and hence the summaries presented here carry the exact flavor of the points being made as well as the underlying reasoning.

The discussions have been summarized below according to the three themes of the conference. Within each theme, we summarize the discussion and present several key points. After presenting the discussions, specific research projects are presented which address several of the issues identified by the workshop participants.

Probably, the single most important observation of the workshop was that *"structural dynamics is the glue that brings emerging technologies together for a broad array of applications"*. Structural dynamics is a system-wide technology and structural dynamicists are experienced system integrators. This is depicted in Figure 1.

414

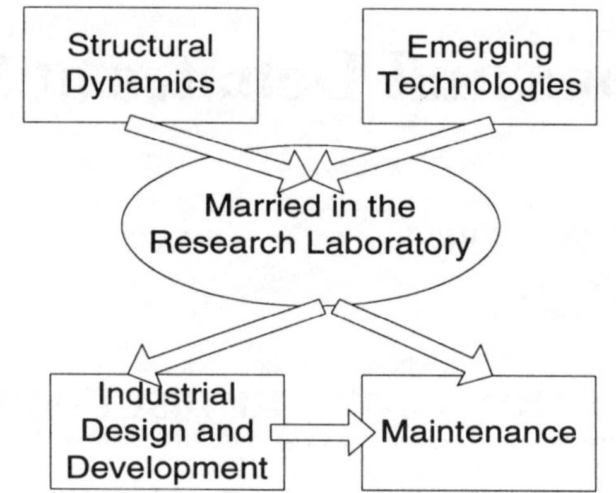

Figure 1: Role of Structural Dynamics Technology as
Integration

THE DISCUSSIONS

2.　DEVELOPING TECHNOLOGIES

Developing technologies are considered to be those technologies which already have demonstrated potential for improving industrial practice, yet still represent active areas of research.

2.1　The first of these developing technologies is Model Validation. The workshop participants developed the view that the following goals are achievable in five years:

1)　Develop guidelines for designing test and analysis model validation procedures.
2)　Add model adaptivity (discretization, element type, and geometry) within computational model updating algorithms.
3)　Develop general acceptance criteria based on probabilistic methods to assess model validity and predictability in the presence of uncertainties.
4)　Use the model updating process to identify fundamental physical principles (and not just to create finely-tuned parameters).
5)　Teach best practices for modeling to both students and practicing engineers.
6)　Identify the most common modeling and testing errors. Use "round-robin" experiments to identify these errors and provide typical solutions.
7)　Develop physics-based experiments models for joints and interfaces, etc. We need to focus on the precise physics through focused experiments rather than immediately modeling the whole integrated system.

The issues that must be considered for model validation include, first, the metrics of the problem: we must emphasize the notion of "purpose of the model" and choose metrics based on this purpose. We must address the difficulties in modeling caused by structure geometry, scale, adaptive meshing at different scales, knowledge of the operator and operator error (as well as sanity checks), uncertainty in physical parameters and boundary conditions, extrapolating results of one analysis to other similar structures, sensitivity of the response predictions in response to small changes in the modeling assumptions, and robustness, and finally, aspects of the structural interaction with its environment where the physics are not understood.

Three steps of modeling and validation include (1) Assess initial model quality, (2) Computational refinement (adaptive meshing, test and analysis correlation) model updating, and (3) Validate with respect to the final purpose (extrapolation). It was widely felt at the workshop that we needed to change our attitude toward modeling by breaking the system into components (for both testing and modeling) in an attempt to isolate individual physics first, and then validate coupled system effects.

RA1 Research Area Proposal - MODEL VALIDATION

The overall goal of this proposal is to develop a framework for validation of mathematical models. This involves developing general acceptance criteria to assess model validity and predictability for both testing and analysis in the presence of uncertainties. We also need to apply probabilistic methods to assess confidence in the model and the test data, as well as to develop experimental procedures for validation that best support the acceptance criteria. This would also involve developing methodologies to 'compare' the models and data in the presence of uncertainties (e.g., statistical tests and testing of hypothesis).

A secondary goal derives from the fact that an initial model may be inadequate due to lack of funding and lack of time. As needs arise and budgets improve, or more time becomes available, the practicing engineer will frequently seek to enhance or to improve their model. One of the identified goals of the workshop is to implement enhanced and improved ***model updating procedures***.

To accomplish this, the identified tasks include (1) improve computational inverse algorithms for model updating; (2) include model adaptivity (discretization, element type, and geometry) in the computational model updating algorithms; (3) establish indicators for identifying model physics errors; (4) develop methods to utilize results from the model updating process to identify the necessary physics; (5) develop methods for identifying/updating 'families of models', i.e., 'robust model updating'; (6) develop formal tools to identify location/type of model errors, and (7) develop a phenomenology and mechanics-based experimental approach to validating physical phenomenon in systems, (e.g., joints and interfaces). We need to specifically focus on the understanding of the precise underlying physics through focused discovery experiments rather than updating models of the complete system.

It was recognized by the workshop attendees that the most significant change in industrial practice would take place if an educational process to improve modeling and validation was developed. This includes in-service, at-distance, and other types of continuing educational programs. These programs could include: (1) developing an accreditation procedure to certify practitioners, and (2) developing a "round-robin" set of analytical and experimental cases to identify common modeling errors and testing errors in modeling and validation of structural dynamics systems.

2.2   Advanced Modeling Technologies

This was identified as the second area that is a developing technology with clear benefits to industrial practice, while still being an area of active research. The following issues were identified as key elements and challenges for advanced modeling technologies: (a) non-linearity, (b) damping, (c) mid-to-high frequency modeling, (d) modeling over the complete scale from micro-mechanics to continuum mechanics, (e) experiment design for model technology development, (f) creating evolutionary (nature based) models, and (g) using inverse methods to create models from experimental data.

2.2.1   The first sub-area to be identified as a topic where work is needed and progress would yield substantial benefits is *non-linearity*. The presence of nonlinearities presents both challenges and opportunities. Before we can address the challenges or take advantage of the opportunities we must be able to diagnose nonlinearities during testing. Nonlinearities are encountered when any of the following conditions are present during a test: (1) varying frequencies (e.g., 15% on the first mode) when test conditions vary, (2) modes that appear or disappear with various forcing levels, (3) non-linear material properties (amplitude-dependent stiffness, for example), and (4) non-Gaussian response to Gaussian excitation.

Disadvantages of nonlinearities are well known. Nonlinearities are hard to model because there is no single methodology that can effectively be used to create models of non-linear behaviour over classes of nonlinearities. Non-linear behaviour can cause unsafe bifurcation and instabilities. The presence of nonlinearities can mean there are possibly multiple states of the structure reached from the same initial conditions. Hence, it is hard to develop models, and it is difficult to validate those models. The advantages of nonlinearities are several and seldom considered. However, nonlinearities can serve (1) to limit response amplitude, (2) to redistribute or dissipate energy, and (3) to broaden frequency response peaks.

It is widely felt that engineers in the field of structural dynamics needs better tools with which to address nonlinearities. The goal identified by the workshop participants is to develop a robust methodology for detecting, identifying, and quantifying nonlinearities for the experimentalist, and to automate the techniques for diagnosis. These tools must derive from physics-based models for the nonlinear behavior. They must be valid and formulated for complex systems, not simply 1-dof or 2-dof systems. Finally, the tools need to provide the engineer with an ability to determine if the nonlinear behavior is important for the design objectives they are trying to meet.

A secondary goal developed by workshop participants is to exploit nonlinear behavior in design and control to improve performance. Specifically, nonlinear damping, dynamics localization, energy transfer and energy transfer filters, and variable material stiffness can be used to improve performance. Nonlinear elements can be introduced in linear systems as a means to validate models.

Other specific needs in the developing area of nonlinearities include (1) a non-linearity function that behaves like a coherence function, (2) a practical tool to choose when to use equivalent linear system vs. non-linear approaches (e.g., Monte Carlo), (3) a tool to determine what is the optimum linear model for a non-linear structure, (4) tools to determine how we can identify non-linear behavior (e.g. presence, type, degree), (5) how we can predict influence of nonlinearities on design targets, (6) how we can distinguish time-varying properties from non-linear properties, (7) how we can make non-linear modeling and analysis tools more useable and attractive for practical engineers, and finally, (8) what excitation we should use to develop the best model.

RA2   Research Area Proposal – Non-linearity

The workshop participants identified several specific research area proposals in the area of nonlinearity. The first project is "Detection, Identification, and Quantification of Nonlinearity". This involves, in conjunction with signal analysis and processing experts, developing robust methods for detection, identification, and quantification of nonlinearities. Specific tasks include (1) developing diagnostic methods, (2) assistance in automating testing and analysis procedures, (3) identification of physical models, (4) identification of precursors of impending instabilities, and (5) benchmark data sets to test algorithms.

The second project is exploiting nonlinearity in design. For this project, we utilize non-linear phenomena to enhance the performance of dynamic systems, and design systems to operate beyond the linear envelope. This will exploit several unique aspects of nonlinear behavior, including (1) nonlinear localization, (2) nonlinear isolators and absorbers, (3) nonlinear dissipation devices, (4) energy transfer, and (5) tailored nonlinearities in materials.

The third project is titled "Modeling and response prediction of nonlinear systems: simple to complex," and is focused on providing advanced education to practicing specialists as well as formal graduate or senior undergraduate education. The components of this education include (1) assessment of the state-of-the-art methods (analytical, computational, experimental) in nonlinear analysis, (2) developing classifications for nonlinear systems, nonlinear excitations, and nonlinear sources, (3) determining the optimal model size/equation order, (4) understanding and communicating model limitations, (5) developing linearization methods/minimization of nonlinearity, and (6) developing nonlinear multi-field problems.

418

2.2.2 The second sub-area to be identified as a topic where work is needed and progress would yield substantial benefits is **damping**. It was broadly felt by the conference attendees that the following should be the priorities:

1) Develop design procedures to include damping in applications in systematic ways.

2) Utilize micro-modeling techniques to predict macro-scale damped structural response.

3) Develop experiments to enhance our physical understanding of damping mechanisms.

4) Design damping materials from basic physical principles.

Currently, damping estimation is based on empirical data (for example, material loss factors are determined experimentally). Can we utilize advances in molecular modeling to understand and predict damping in materials, joints, and interfaces? How can we use mixed scale models (molecular to micro-scale) to enhance our ability to model damping?

The workshop attendees proposed the following classifications of damping: (a) joint, interface, i.e., hysteretic, (b) material, (c) fluid-structure, (d) viscoelastic, (e) active, (f) coatings (e.g., ceramic coatings), (g) particle, and (h) impact.

The following critical issues for improving our ability to accurately model devices and structures were (a) predicting damping in high-temperature systems, (b) efficient computation of forced response of damped structural systems given the above varieties of damping, and (c) determining the best practices for measuring damping, and educating the practicing engineer in these practices.

RA3   Research Area Proposal - Damping
There were several key project areas identified at the workshop: (1) modeling fundamental damping mechanisms (viewed as a long-term project), (2) developing molecular, micro, and macro modeling scales, (3) motivating structural dynamicists to create multidisciplinary groups leading to design of damping mechanisms and materials, and (4) understanding variance in damping performance introduced by manufacturing, and design for manufacturing.

A more developed proposal is titled "Developing design tools for damping applications". For this project, the key tasks that appear in a valid statement of work are (1) provide critical comparison of different models for passive/active damping augmentation, (2) material characterisation of properties (3) design guidelines for passive and active devices, and (4) optimisation procedures for designing damping augmentation.

Another developed proposal area is titled "Experimental validation of damping models from molecular to macroscopic scales, and from low to high damping". This project would include (1) definition of existing methods, (2) best practice/guidelines to measurement methods, (3) critical comparison of experimental characterisation methods, (4) developing damping characterizations from large-scale tests, and (5) generation of macroscopic constitutive equations from molecular/micro-scale testing.

2.2.3 Another topic addressed at the workshop is "What are the *physical models* we need to integrate into structural dynamics?". Certainly this would include (1) Micro - Quantitative models for microstructures with transitions from continuous to discrete representation, e.g., for crystals, grain-like materials molecular assemblies. This needs cooperation with physicists to determine the constraints for motions and forces (i.e., field forces, adhesion, etc.), (2) develop quantitative models for biological materials such as tissues, organs, and bones. This would be needed, for example, for connecting technical materials to life materials (which we will define as technical transplants), simulating sports actions, determining loads on the human body that cannot be measured directly, and for remotely-manipulated surgery, and (3), in a Macro level - Qualitative models to describe and simulate in a simple way, by non-experts, the motion of complex systems. This has great value to people creating virtual environment, e.g., an artist simulating the motion of a tree in the wind, a ship on the sea, and a robot picking up cloth.

RA4   Research Area Proposal – Coupled Structure Dynamics
Another research proposal area is titled "Coupling Structural Dynamics with Distributed Environments". Examples include fluids (e.g., fluid structure interaction such as aeroelasticity, with input from aerodynamics, vibro-acoustics), (2) electromagnetic (e.g. piezoelectric materials, electromagnetic bearings), (3) control (e.g., active shape control, adaptive wings), and (4) thermal (e.g., coupling thermal flux and radiation model with structural dynamics models). The goal of this project is to achieve system optimization under interdisciplinary aspects and simultaneously identify of system reliability.

The tasks involved with this proposal area are (1) theory and software development, including developing integrated multi-physics analysis tools (e.g., structure-fluid interaction models; smart materials-structure-control tool), and theory and programming of interface physics, (2) improved computing, (3) education of practicing engineers in the test and analysis methods, (4) creating toolboxes to add to standard computational packages (e.g., Matlab® or ProEngineer®), and explicitly transferring the knowledge to industrial practice through publications and newsletters.

There are several identified advantages which will accrue as a result of pursuing this project. First, we can expect new and improved products. We can also expect shorter time to market (fewer iterations in development, more integrated development, less development and risk), as well as increased reliability in these products. Finally, we will be able to develop virtual experiments in a multi-physics environment if the methodologies become sufficiently advanced and verified through experiments.

The practical applications include (1) turbine development, simulation of blade failures and resulting system consequences (overall system safety of aircraft), (2) increasing cabin comfort by reducing noise, (3) improve vehicle stability by active multi-body control, (4) improved system design to reduced loads (active structures to protect sensitive equipment), and (5) coupling structure crash models with aeroelastic and aerodynamic models in a highly transient nonlinear system).

2.3   Computing

Another developing technology area is computing. Workshop participants considered three scales of computing:

1. **Massive scale** this is generally characterized by "more," "faster," and "bigger" (rare machines). The need for large scale computing comes from the coupling of structural dynamics together with other fields to solve multidisciplinary problems. This allows tackling projects such as simultaneous domain decomposition with reduced models, as well as implementing substructuring techniques with high resolution of local events. Note that five years from now every engineer will have access to these machines. "When two technologies, each pushing computational limits, move in together, they need a bigger house."

2. **Toolboxes** (medium scale computing) on every design engineer's desk or local, many of these (100,000s per year). Analogous to Matlab toolboxes, "mechmax" runs gears, motors into Simulink; pulling information off the web from large databases which integrate structural dynamics tasks with MEMS etc. We need to create a library for FEM modeling of joints, fasteners, smart materials, and damping.

3. **Micro-computing** at the local level, such as with accelerometers with a microprocessor which self-calibrate, does on-line structural analysis for monitoring and control (perhaps millions of these), and hardware building blocks integrated into structural design.

The following tasks are associated with exploiting computing to create advanced design methods: (1) Quantify design parameters (e.g., force, bandwidth, strain, strain rate, hysteresis, duty cycle, thermal loads, time constants, internal dynamics) of advanced materials, advanced material devices, and MEMS devices so the practicing engineer begins to use them in design, (2) create a web-accessible database, containing common built up parts and complexities (mechanisms, fasteners, joints, damping, etc.) ready to be used in modeling codes, (3) develop algorithms and procedures to convert computer aided design models to finite element models automatically and efficiently for built-up complete integrated structures with all complex features (e.g., rivets) and which includes nonlinearities. (Current difficulties are interfaces between automatically-meshed substructures — the meshes don't mesh.

Currently you also have to remove complexity and details such as rivets and accompanying holes), (4) develop technology to efficiently and reliably harvest signals from large numbers of widespread sensors to allow greater integration of the sensors with the structure (currently we use wires that cause reliability, structural performance and integration problems).

3. EMERGING TECHNOLOGIES

We need to integrate all elements of emerging technologies into our research and practice, since we are responsible for the integrated system. Our first goal is to couple structural dynamics and emerging technologies through basic research projects. Our second goal is to create engineering design and analysis tools that combine structural dynamics and emerging technologies through applied research projects.

It was universally felt among workshop attendees that our community needs better marketing of the emerging technologies. Certainly, information transfer needs to be more complete and teamwork more integrated. How do we share the knowledge and experience of those working in emerging technologies that a structural dynamicist can embrace and incorporate? How can the work of recent advances in structural dynamics be better marketed and communicated to practice in industry. The answers lie partly with internet web sites (particularly educational institutes and government laboratories, since this task is explicitly part of their mission), and publications (e.g., *Sound and Vibration Magazine*).

Regarding an overall model for incorporating emerging technologies into our business of structural dynamics, there was not felt to be any good analogous models. However, we may want to consider how does the drug industry fund and decide on "blue skies" research? In a similar vein, how does the computer industry fund and decide on "blue skies" research?

So, then, the initial questions to be answered are (1) How do we integrate emerging technologies into structural dynamics? and how do we improve education to achieve this goal? (2) How do we integrate structural dynamics with emerging technologies into industrial practice? (3) What are the physical models (for example, models from the field of biomimetics) we need to integrate into structural dynamics? and (4) How do we better identify the interface between structural dynamics and emerging technologies?

Examples of Emerging Technologies include Active (also known as "smart" or "intelligent") Materials, micro-electro-mechanical systems (MEMS) and nano-technology, biomimetics, advanced computing, virtual reality, advances in related disciplines (for example, fatigue and fracture analysis, as well as micro-scale modeling), Computational fluid dynamics, and advances in both commercially-available tools and large-scale manufacturing of electronic components.

Promoting structural dynamics in these emerging technology areas means identifying liaisons and scholars who are familiar with both structural dynamics and the fundamental science and cutting-edge developments in fields such as Medicine (particularly bio-mechanics and bio-robotics), Environmental (home, work, common areas) Quality (including noise and vibration harshness), Transportation, Personalized products such as leisure, sports, entertainment equipment, Flexible Manufacturing and Adaptive Products, Information Storage systems, Precision Manufacturing (for example, using robotic manufacturing methods), and Remote Systems (alien environments, military robotics), and Packaging Sciences.

RA5   Emerging Technology Area: Robust model-based design

In mechanical design, having robust models that can reliably predict if the structure will meet response specifications greatly enhances the design engineer's productivity and shortens the time for product development. To develop a robust model, we must understand what we are going to do with our model, in order to understand what are acceptable model errors. We also need tools with which we can put bounds on model response errors.

We currently have many tools available to determine bounds on errors in the model response. These include Bayesian estimation, Monte Carlo simulation, stochastic finite element software, fuzzy analysis and other emerging design methods, and safety factor methods. However, all these tools are limited by the accuracy of our *a priori* knowledge of the physical properties of the structure and our understanding of the forces applied to the structure.

There are several challenges present in the majority of all significant design problems, and the more the current design task differs from an existing structure or product, the more pervasive and challenging these problems become. For example, consider the usual design situations: statistical data are not always extensively or readily available; it is difficult to estimate uncertainties on model form and parameters; it is difficult to estimate uncertainties from single or few tests of sample products; it is difficult to estimate uncertainties in manufactured products from prototypes, and finally, frequently it is difficult to define clearly target behavior.

What needs to be done? We need to develop guidelines to make uncertainty analysis an integrated part of the structural design and analysis process by using statistical inference and statistics in model development. We could investigate using records and past modeling experience to make improved current modeling decisions. We need to develop guidelines for quantifying uncertainties, i.e., determine what uncertainty models are appropriate given certain a priori information (e.g., probabilistic, fuzzy, and set theoretic models). We need to develop guidelines for defining appropriate design targets for components in system designs, for example, by considering system requirements, defining component level model accuracy and response requirements.

As leaders in this field, we need to provide "best practice" for the following procedure: (1) the decisions are known (for example, we may know the bounds on a car's torsional stiffness), (2) we need to find the critical components which influence the torsional stiffness, (3) next, we must find the tolerance in the critical component behaviors, and finally, (4) develop specific model refinement objectives to reduce these tolerances where necessary. This is a more productive alternative, when compared to spending time updating non-critical components. Currently the amount of computational power to be able to follow all sources of uncertainties through to the final effect on response characteristics is beyond what we have, hence, we must follow only those uncertainties which significantly affect the response.

As leaders in the community, we need to provide techniques which can develop bounds on the maximum prediction error in the response by finding the particular input from all possible inputs that leads to the greatest error. By analogy, controllers are designed to be stable in the face of parameter/input/disturbance uncertainties. So, given a maximum response in a structure, can we robustly design the structure in the face of parameter/input/disturbance uncertainties?

We need to develop guidelines for designing for better predictability. We need to develop guidelines for system-level model precision and robustness requirements. We need to determine how to quantify the diminishing returns on model refinement, and to provide guidelines for how refined a model needs to be for particular applications. It is still a problem that model updating has significant difficulties in getting a physically meaningful model. There is almost always limited measurement of some parameters. Is this reasonable? We have modeling and uncertainty parameters within a model: should we not map uncertainty in parameters to uncertainties in the response?

As a community, we need to develop education content, materials, and suggestions for modifying curricula such that graduating engineers are familiar with the concepts for robust model-based design. Are there enough statistical analysis methods incorporated into current design practice? Do we need more statistics education in our courses? Even the software rarely allows you to attach variance to the measured quantities (e.g., Young's modulus).

Let us also teach model building vs. model validation, where a model is validated if it can be used to predict the response of a slightly changed scenario. A good model can be used for prediction. A great model has known limitations to its ability to handle parameter and performance changes. We need to understand how to choose the correct parameters for the purpose of the model. We need to also prevent model "myopia," where the engineer focuses on small parts of the model while not having a significant effect on the overall usefulness of the model. We need to teach how we can tell a wrong model (for example, one based on incorrect physics) from one which simply needs better parameter estimation to give better results.

Another problem, but more systemic, is that in a particular field the collective memory "database" held by experienced designers is getting shorter because people are moved into and out of groups more quickly. Projects limited by both time and money cause this kind of problem. Industry needs a better database based on past experience, which can be used to help designers develop intuition. More thorough training for engineers implementing FEM models is needed.

Summary: it is most important that engineers predict confidence intervals in the targeted response quantities in their models. Uncertainty analysis is therefore very desirable. Currently, it is computationally expensive, time consuming, and "best methods" for modeling particular uncertainties are not yet reduced to practice. Hence, uncertainty analysis is not widely practiced.

RA6   Research Proposal: Create Engineering Design and Analysis Tools Combining Structural Dynamics and Emerging Technologies

The goal of this proposed area is to develop education goals and modified curricula to produce structural dynamicists and engineers with multidisciplinary abilities. This would require developing advanced physical modeling technologies incorporating structural dynamics plus its distributed environment. Examples include: electromagnetic fields, fluid mechanics, acoustic fields, and thermodynamics. A research objective related to this project is to quantify design parameters (e.g., force, bandwidth, strain, strain rate, hysteresis, duty cycle, thermal loads, time constants, internal dynamics) of advanced materials, advanced material devices, and MEMS devices. This is the first step towards quantifying guidelines for specifying design parameters of commercial products (e.g., advanced materials, advanced material devices, and MEMS devices) so the practicing engineer begins to use them in design.

RA7   Research Proposal – Developing Mid to High Frequency Modeling

To define this area, it is when experimental modal identification becomes impossible, because of an intractable modal density. Viewing a given structural response problem or task as "Mid- to High-Frequency" is appropriate when interest is on overall response levels and not individual resonances. Typical requirements may be bounds on response levels, and predictions of the response and the expected variance in the prediction.

The goal of the project is to seek a new descriptor which is global/small-scale and is response-based, has spatial variation and includes statistical variance – that is, develops an envelope approach to rapid amplitude variations with respect to frequency in structural response at mid- to high-frequencies. Other goals cited as part of this research effort include; (1) How do we create mid and high frequency models? (2) defining criteria for defining mid-frequency and high frequency regimes, (3) determining the influence of local vs. global information on targeted response, and (4) determining when to switch from deterministic models to statistical methods. We certainly want to define clearly if our present difficulties with mid-frequency problems are a modeling problem or something else.

The tools available include finite element models which give a deterministic approach, and statistical energy analysis (SEA) which uses a statistical approach. At present, SEA does not provide variance in the expected response levels. Determining stability margins in either modeling approach is not clear. Convergence and convergence margins are not clear. The eigen-frequencies of the problem are not apparent using SEA.

Overall, a problem which plagues solution of mid- to high-frequency structural dynamics design and analysis problems is that our modeling is greatly influenced by our assumptions about damping at joints. The response predictions are highly sensitive to these modeling assumptions.

What is missing? The workshop attendees were unanimous in citing a need for benchmark tests, which will help everyone involved with this problem to develop more general knowledge of response prediction ability. We need better descriptions of damping (this is addressed extensively as a specific topic in the section on developing technologies). We need a more complete knowledge of the effects of uncertainties.

What must be done? We should develop systematic benchmarks of predictions of modes and response, and communicate these to educators in the field. We need to explore probabilistic approaches to quantify uncertainties, and search for new approaches and/or descriptors of small order probabilities. This may result in reduced-order models in special areas. That is, we need to develop methods to model sensitive regions in detail and model insensitive regions in a global way. We need proper, useful, and accurate descriptions of damping. We need to develop means to define and model the boundary conditions and to define our interfaces (i.e., joints) better. It may be useful to develop inverse analysis procedures as well, and to develop tools linking these mid- to –high-frequency analysis tools to the acoustic response of structures for noise control and mitigation.

RA8 Research Proposal: Characterization of Devices and Interfaces

Emerging technologies incorporate different devices and phenomena that must be synthesized into a complete and combined system. Critical to this process is the efficient and comprehensive specification of the interface dynamics. For example, how is an actuator specified (maximum force, bandwidth, strain rate, time constants, *etc.*) and incorporated into a flexible wing? Since the choice of device, actuator, sensor, and perhaps MEMS device affects the structural dynamics, this is an interactive and iterative process. What dynamic characteristics are required to be specified, and how should they be specified?

For dynamic models of nonlinear components and substructures, how should these be specified to allow efficient incorporation into system models (so that the internal detail of the models does not have to be known explicitly by the system analyst)? This would allow a library of structural dynamics and other technology modules to be produced to aid the design process. Examples of these modules include joints, friction, damping, ER/MR fluids, and piezoelectric actuators and sensors. For distributed phenomena interfaces need to be specified more carefully — for example, when combining computational fluid dynamics, magnetostrictive effects, and thermal effects with structural dynamics. Can the interfaces be defined in sufficient detail for engineers to worry only about their own area of interest.

RA9   Research Proposal: Evolutionary (nature-based) Models

The general definition of nature-based models is that they utilize laws of nature, and account for adaptivity to the environment. The model methodology can produce optimal solutions with respect to a given objective. The technology embodies biomimetics.

Examples include (1) bone adaptation (internal stresses from external forces): could be applied to structures design (for example, optimal mass design or optimal stress design), (2) damping (strain energy cells): grow damping on structures where needed and optimized over a bandwidth (3) smart structures (insects, mammals, fish, etc.)--evolve the sensor, actuator, controller and composite material to enhance performance in a changing environment, and (4) cellular automata to address problems such as corrosion, cracks, and vibration.

Specific projects include creating optimal damping designs using evolutionary models (and an obvious application is space structures), developing smart sensor designs based on biological materials (which is useful for the process industry), and implementing adaptive coatings for interacting surfaces (with an application to noise reduction).

4.   IMAGE

To enhance our image with our peers (for example, academia, government funding agencies), we need to promote and develop the physics of structural dynamics. To enhance our image with practicing engineering groups we must engage in better and more prolific technology transfer and technology training activities, and develop continuing education using models not tied to existing examples. To enhance our image with young engineers, engineering students, and potential students, we must accentuate entrepreneurship as one of the core abilities of accomplished structural dynamicists. Image should include cultural shift – accenting education, excitement, legislation, and improved quality of life.

It is imperative that we embrace emerging technologies. For an illustrative example of the pitfalls of ignoring emerging technologies, we look to the emerging field of smart structures. Smart structures couldn't find a home in ASME or AIAA at first, and now the premier conference in this area is run by The International Society for Optical Engineers (SPIE). Modal analysis was rejected by ASME, and active practitioners started their own conference. Will emerging technologies find a cold shoulder or a lack of embrace in the structural dynamics community and go elsewhere? If so, will we find ourselves in less of a leading role and more of a supporting role in the future? Hence, it is key to our image and long term growth as a field to embrace emerging technologies, and be outward thinkers in these areas. If this were the case, then perhaps the more common acronym would be MDS (micro dynamical systems), rather than MEMS (micro electro-mechanical systems).

We can start this outward viewpoint by teaching and integrating broad physical aspects of biomimetics, mechatronics, optics, electronics and business into the structural dynamics curriculum. We need to influence the ABET board who decides acceptance on a classical departments by department basis, by looking to those accredited schools which don't have classical departments (UC San Diego, and Dartmouth College, for example).

An identification of the benefits and definition of structural dynamics would help promote structural dynamics for the general public. We can begin to accomplish this by accentuating how the science and engineering of structural dynamics is necessary to progress in consumer products, enhancing such characteristics as making products "smaller lighter cheap", improving noise, vibration, comfort, and quality, improving product safety and product durability, lowering maintenance costs, lowering life cycle costs, and improving insurability (e.g. in seismic areas civil engineers create loss estimates).

5.   SUMMARY

We have presented here a synopsis of the discussions at the SD2000 workshop. The three themes which emerged from the workshops were (i) Developing Technologies, (ii) Emerging Technologies, and (iii) the Image of structural dynamics. Throughout the workshop, the primary central need for the field of structural dynamics was overwhelmingly stated, in a variety of ways, as the need to develop better models.

Future activities were identified on the last day of the workshop. They include (1) development of a structural dynamics advisory group, (2) development and maintenance of a web site, (3) development and maintenance of a library of tools and software and test data, and (4) increasing the representation of structural dynamicists on industrial committees.

We conclude this part of the book with a quote from the SD2000 workshop, which reflects the real importance structural dynamics has in any mechanical engineering project:

"If it breaks, who gets the blame? The structures people, that's who!"

Hence, structural dynamicists should take the leadership role in many multidisciplinary projects.

Grand Challenges for Structural Dynamics

Charles R. Farrar
Engineering Analysis Group
Los Alamos National Laboratory

ABSTRACT

During the Structural Dynamics 2000 Forum attendees were invited to submit ideas regarding Grand Challenge problems for structural dynamics. A summary of the five ideas that were suggested is presented herein. It is hoped that this summary will motivate others to suggest additional ideas and that open discussions of Grand Challenge Problems will become an integral part of future structural dynamics conferences, forums and workshops.

1.   INTRODUCTION

For many years' physicists and researchers in large-scale computational modeling have proposed Grand Challenges for their technological field. As an example, a current Grand Challenge being proposed by engineers and scientists involved in large-scale computational modeling is the development of the technology to do global climate modeling. In the field of physics the development of Grand Challenge problems has proven to be a very effective method for publicizing technology beyond the technical physics community and generating large-scale research expenditures. Probably the most well known engineering Grand Challenge was the proposal in the early 1960s to place a man on the moon and return him safely to earth by the end of the decade. This extremely broad and ambitious Grand Challenge encompassed structural dynamics as well as many other engineering disciplines.

Attendees at the Structural Dynamics 2000 Forum were invited to submit ideas for Grand Challenges related to structural dynamics. Some features of a Grand Challenge problem were established during the solicitation process. These features include:

1.   The problem must be difficult and something that will not be readily solved in the next few years,
2.   The problem must be multi-disciplinary in nature,
3.   The problem must require development in experimental, analytical and computational methods,

4. There must be quantifiable measures indicating progress toward the problem solution,
5. The problem must be of interest to many industries, and
6. Solving the problem will have significant economical and social impact.

This contribution summarizes the five suggested Grand Challenges that were submitted at this Forum. No attempt is made to discuss the relative merits of these proposed Grand Challenges. To this end, the Grand Challenges are listed in alphabetical order of the proposing author's last name. It is hoped that this contribution will stimulate a dialog within the structural dynamics community regarding appropriate Grand Challenges and motivate others to suggest additional ideas. Also, it is hoped that proponents will use these Grand Challenges to motivate funding agencies to pursue research in these areas. Finally, it is hoped that open discussions of Grand Challenge Problems will become an integral part of future structural dynamics conferences.

2.   PLANET EARTH SEISMIC ARRAY

Proposed by Dave Brown, University of Cincinnati
Measure, monitor, image and analyze the dynamics of planet earth, its seismic fault systems and its important infrastructures using a vast array of seismic and vibration sensors which are interconnected using the Internet.

2.1   Background

This project involves measuring and modeling the dynamics of planet earth. Historically, there have been three major groups who have examined this problem. The first group is primarily basic scientists who are trying to understand the physics and dynamics of the earth. They are probing the interior of the earth with seismic waves to better understand the physics of the core, the geology of the mantel, etc. This group utilizes an array of weak wave seismic monitoring stations for monitoring seismic waves. These stations are used for basic studies and for locating seismic events and nuclear tests. The second group is primarily concerned with monitoring earthquakes utilizing strong wave monitoring stations. The third group is involved with monitoring infrastructure. There are considerable overlaps between the three groups, but in general the measurement systems are independent.

One of the main objectives of this project is to develop a common measurement, data management and computational system for addressing the needs of these different disciplines. The measurement system will be distributed worldwide and consist of massive arrays of seismic sensors, primarily concentrated around important seismic sites.

The data are collected using the Internet and distributed to computational sites located on the Internet. This project will take advantage of the rapid changes that have taken place in the last decade in the areas of measurement, signal processing, computing and networking. Unlike most past scientific technological advancements, this project will be made possible by advancements in consumer products. The personal computer revolution, digital music, wireless communications, and the Internet are all key factors in making this Grand Challenge experiment possible.

2.2 What makes measuring, monitoring, imaging and analyzing the dynamics of planet earth a Grand Challenge?

- The integration of a number of interconnected disciplines to help solve a basic scientific problem, which is to understand the dynamics of the planet.
- This project involves using a massive application of making distributed measurement and computations over the Internet. The recent advances in computers, networking, data acquisition, and consumer electronics make this possible.
- Installation of a vast array of seismic measurement nodes (100,000 or more) which can measure and image the geology of the planet.
- Utilization of low earth orbit satellite systems to synchronize and locate elements in the array (GPS) and to collect the vast amount of data and distribute it over the Internet (Teledesic Satellite System).
- Development of a series of inexpensive seismic measurement nodes (basic 4 channel node less than $1000) which includes:
- Multiple Sensors (accelerometers, strain gages, tilt sensors, etc)
- Data Acquisition Module (24 bit dynamic range)
- Re-circulating Digital Memory (128 Mbytes)
- DSP Chip
- GPS Timing and Position Module
- Internet Communication Module (Wireless – Teledesic Satellite System)
- New inexpensive multi-element seismic sensors which can measure with nano G resolution (10^{-9}g's to 10g's).
- Distributed measurements for infrastructures using local area networks interfaced with master seismic node.
- Data distributed using the Ring Buffer Network Bus (RBNB) over the Internet.
- Distributed computing using the Internet.
- Development of new beam-forming and imaging algorithms for analyzing data from large seismic arrays. These algorithms need to be developed for parallel processing using a large number of computers distributed along the Internet. These algorithms need to be optimized for network communication.
- New computational algorithms for condition monitoring of infrastructures.

2.3 Goals demonstrating that one can measure, monitor, image and analyze the dynamics of planet earth

This Challenge is a major scientific project. The final goal of this project is the development of a measurement system utilizing a vast array of seismic sensors which can be used to measure the dynamics characteristics of the planet, its seismic faults systems, and its influences on infrastructure. This vast array tremendously improves the resolution and sensitivity of the current systems. It also will allow a much larger database to be collected during the large rare events where nonlinearities and other effects are present which cannot be interpolated from a linear model based upon small events.

- Since seismic waves are the only practical type of energy which can be used to probe the core of the earth, a system which can drastically expand the capabilities of existing systems to measure and image seismic waves is being developed. The primary goal of the basic scientific experiment is imaging the core of the earth and its mantel and developing models which can explain its motion, magnetic properties, etc.

- A practical and more immediate goal concerns the imaging and monitoring of seismic fault systems and predicting the influence of these faults systems on important infrastructures. Understanding seismic fault systems has significant social and financial impact on societies that are located in areas where there are active seismic faults systems.
  - There will be significant scientific gain from this project:
  - Contributions to distributed measurement systems
  - Contributions to distributed computation systems.
  - Massive data management and distribution.
  - Measurement and signal processing which can be applied to many other types of health monitoring systems. Manufacturing, process control, and energy distribution systems are a few of the applications where there would be immediate impacts.

- The scope and size of this project rivals other scientific projects for exploring the universe.

3.   MICRO ELECTRO MECHANICAL SYSTEMS

Proposed by Izhak Bucher, Israel Institute of Technology

Develop tools and methods for studying the dynamics of Micro Electro Mechanical Systems (MEMS)

3.1   Background

The human kind often saw the greatest achievements of engineering in systems and structures having extreme size and proportions. During the 20th century huge buildings, bridges and machines were built, almost to an extent where size seems to have reached the largest reasonable proportions. New challenges arise in the other extreme namely, very small vibrating systems. Such systems although quite similar at a first glance to ordinary structures have special features attributed to their size as compared to atomic scale. Micro electro-mechanical systems (MEMS) are being used as sensors, microphones and actuating devices, all requiring a thorough understanding of their dynamical behavior. A small sensing device having a capacitive measuring device, for example, must be modeled as a coupled electro-dynamic system rather than treating the elastic and electrical parts separately. In the micron scale, the electro-static forces, the damping due to viscosity of the surrounding fluid and other effects, which could be neglected when dealing with ordinary sized structures, become very important. These tiny devices are very attractive due to their inexpensive manufacturing process (developed for micro-electronics) and the potential to include sensors and actuators in places they were never considered before.

Due to lack of knowledge, practitioners use very simple and inadequate models that postpone the appearance of MEMS in many aspects of our lives. The challenge in this field is the formulation of representative models and experimental verification of each mechanism, e.g. vibration, damping, fluid-structure interaction, influence of electrostatic forces.

3.2 What makes the development of tools and methods for studying the dynamics of MEMS a Grand Challenge?

- Micro mechanical systems (MEMS) have a great commercial potential
- The dynamic response of MEMS is one of the most important factors in their performance
- New effects that exist in small scale (electro-static), e.g. loading due to non-contacting sensing devices must be understood
- Small scale makes experimental verification very difficult
- The validity of elasticity which forms the basis for vibration theory must be validated

3.3 Goals demonstrating that adequate tools and methods have been developed for studying the dynamics of MEMS.

- Progress in this field will create a large range of products that will affect products from cars to medical instruments within 2-5 years
- In 2-4 years every car would have a MEMS 5-10 rate gyros which costs $4 and control the stability of motion, acceleration and be a navigation aid.

4. PERFORM ROBUST GLOBAL VIBRATION-BASED DAMAGE ASSESSMENT OF ENGINEERING SYSTEMS

Proposed by Charles Farrar, Los Alamos National Laboratory

4.1 Background

The interest in the ability to monitor a structure and detect damage at the earliest possible stage is pervasive throughout the aerospace, civil and mechanical engineering communities. Current damage-detection methods are either visual or localized experimental methods that require the vicinity of the damage to be known *a priori* and that the portion of the structure being inspected is readily accessible. The need for quantitative global damage detection methods that can be applied to complex structures

434

has led to research of methods that examine changes in the vibration characteristics of the structure. The basic premise of vibration-based damage detection is that the damage will significantly alter the stiffness, mass or energy dissipation properties of a system, which, in turn, will alter the measured dynamic response of that system. Although the basis for vibration-based damage detection appears intuitive, its actual application poses many significant technical challenges. The most fundamental challenge is the fact that damage is typically a local phenomenon and may not significantly influence the lower-frequency global response of structures that is typically measured during vibration tests. This challenge is supplemented by many practical issues associated with making accurate and repeatable vibration measurements at a limited number of locations on structures often operating in adverse environments. Over the last thirty years global vibration-based damage detection has been applied to numerous aerospace, civil and mechanical structures as part of research studies. However, to date, only in the rotating machinery industry has this technology made the transition from a topic of research to actual implementation as a standard diagnostic tool.

4.2 What makes performing robust, global vibration-based damage assessment a Grand Challenge?

- All portions of our technical infrastructure require damage detection.
- Successful development of this technology will have tremendous economic impact by reducing unscheduled down time of manufacturing equipment, making damage assessment after earthquakes a quantifiable process and maintaining our transportation infrastructure in operating order.
- Early detection of damage in systems such as bridges and aircraft will have positive life safety implications.
- This problem has been worked on for many years and, most likely, will not be solved in the next 2-3 yrs.
- This problem requires a multi-disciplinary approach to its solution (including vibration analysis (linear and nonlinear); vibration measurement; signal processing; sensor development; statistical analysis; and remote data acquisition, processing and transmission).
- Current measurement and data analysis technology does not allow for sufficiently precise quantification of damage-sensitive dynamic properties.

4.3 Goals demonstrating that one can perform robust global vibration-based damage assessment

- Within fifteen years the state of California mandates that every new building requiring strong motion instrumentation is also fitted with a vibration-based structural health monitoring system.
- Within ten years the micro-electronic manufacturing industry can reduce plant costs by eliminating 50% of their redundant mechanical equipment.

- Within fifteen years annual scheduled maintenance costs of commercial aircraft are reduced 10% because inspection intervals have been increased as the result of in-service structural health monitoring.

5. ACCURATE PREDICTION AND MINIMISATION OF NOISE FOR ENGINEERING STRUCTURES

Proposed by Mehmet Imergum, Imperial College

5.1 Background

The ability to minimize the noise produced by engineering structures at the earliest possible design stage is a major requirement throughout the aerospace, civil, mechanical and marine engineering communities. Current noise prediction methods are either based on experimental techniques that require trial-and-error adjustments to an existing prototype, or consist of semi-analytical techniques that can only deal with a limited number of simplified geometries. The subject area is truly multi-disciplinary as it is not always possible to distinguish between various origins of noise: structural vibration through a multitude of transmission paths, aero-acoustic effects, fluid-structure interaction, electro-magnetic effects, propagation of noise in air, water or other media, etc. Typical examples include train noise for passengers and the environment, aircraft engine noise for landing and take-off, submarine noise, noise generated by everyday tools that have rotating parts, etc.

The most difficult challenge is the formulation of accurate and representative models that can contain all required ingredients: structural vibration, unsteady aerodynamics, fluid-structure interaction, propagation of noise in compressible and incompressible flows. Currently, some of the required analytical/numerical tools are available but huge gaps exist between the various disciplines involved.

5.2 What makes accurate prediction and minimization of noise a Grand Challenge?

- All portions of our technical infrastructure require noise prediction.
- The rules regarding noise emission are becoming more and more stringent.
- For general geometries, there are no clear theoretical links between structural vibration and structure-borne sound.
- The structural (FE) models are not accurate for predicting higher modes of vibration and for dealing with damping.
- There are no established rules for ranking similar designs.
- It is not clear if statistical methods or large numerical models should be used.
- This problem has been worked on for many years and, most likely, will not be solved in the next 5-10 years.
- The amount of detail that must be incorporated into the numerical models is not known.

436

- The noise source and the required location of the prediction can be separated by large distances.

- This problem requires a multi-disciplinary approach to its solution (including linear and non-linear vibration analysis, fluid-structure interaction, unsteady aerodynamics, sound propagation in air, water, etc.)

5.3 Goals demonstrating that accurate prediction and minimization of noise has been achieved

The permissible noise levels for engineering products (aero-engines, car exhausts, lawnmowers, submarines, etc) are reviewed almost every year. A reduction of about 1 dB per year is becoming the expected norm.

6. COMPUTATION OF PROBABILISTIC STRUCTURAL DYNAMIC RESPONSE

Proposed by Tom Paez, Sandia National Laboratory
Perform accurate structural dynamic analysis of randomly excited, stochastic systems.

6.1 Background

Numerical models of complex, structural dynamic systems have been constructed since the 1950s and used to analyze system response. Such analyses have been used to predict and assess system behavior, to design and optimize systems, and for many other purposes. Normally, numerical models of structural dynamics systems use nominal, deterministic values for parameters, and predict a single deterministic response. Yet, it is widely acknowledged that actual physical systems have parameters that vary randomly because system geometries are random, as are material properties, initial conditions, boundary conditions, and other system characteristics and conditions. Moreover, most of the inputs that excite dynamic system responses are random. These physical systems, their conditions, and their excitations may be termed stochastic, and their responses must be characterized in a probabilistic framework.

In view of these things, we require analytic approaches and software implementations of these approaches to accurately predict the responses of real systems. To assure that predictions produced by the software and the mathematical models constructed within the software framework are satisfactorily accurate, predictions must be capable of being validated using formal, perhaps statistical, procedures. The software should at least have the following features:

- The computer code should be robust and user-friendly, usable by anyone with a basic understanding of structural dynamics and probability and statistics.

- It should handle relatively large, practical problems over a frequency range of interest to structural dynamicists (say 0 through 2000 Hz) where there is discrete or continuous randomness in the excitations, material properties, geometry, initial conditions, boundary conditions, component connections, etc.

- The analysis must permit random quantities to be Gaussian or non-Gaussian.

- It must permit the definition of temporal excitations as deterministic, random stationary, random non-stationary, or a combination of all three.
- It must permit the definition of random fields (spatial random processes) as homogeneous (steady state in a spatial sense) or non-homogeneous.
- It must permit the modeling of structural characteristics as linear or nonlinear.
- It must efficiently handle both shock and vibration problems.
- The computer code must have options for time and frequency domain analysis, and must permit the extraction of structural characteristics in any form.
- The mathematical models constructed within the software framework must be capable of being updated as new, experimentally measured excitation and response data become available.
- All this must be done accurately and efficiently, and the results must be capable of being validated using experimental results.

6.2 What makes performing accurate structural dynamic analysis of randomly excited, stochastic systems a Grand Challenge?

This is a Grand Challenge because:

- Even current deterministic computer codes that support the mathematical modeling of mechanical systems cannot accurately predict the detailed response of simple mechanical systems, much less the detailed response of complex systems over a practical range of frequencies.
- The probabilistic approaches required to create the software described above do not yet exist in their entirety.

6.3 Goals demonstrating that performing accurate structural dynamic analysis of randomly excited, stochastic systems exists

The challenge will be considered to have been met when:

- General techniques for solution of this problem are developed and proven via multiple comparisons to experiment.
- The solution techniques are made efficient enough for widespread application.
- Analysts, designers, and testers use code for the prediction of mechanical system response.

7.   SUMMARY

The concept of proposing "Grand Challenges" immediately generates some controversy as it reflects individuals, or groups of individuals, opinions regarding the relative difficulty, importance, and impact of various technical activities. As one might expect, lively discussions ensued after the Grand Challenges were presented at the Structural Dynamics 2000 Forum. Various technical aspects of each proposal were discussed along with issues regarding the possibility for achieving the stated goals for each activity.

The discussions that followed the presentations of the Grand Challenges addressed the issue that many of the proposed challenges have already been the focus of significant research efforts for many years. This statement is certainly true for all the challenges proposed at this Forum and for Grand Challenges proposed in other fields as well. The concept of a Grand Challenge for structural dynamics is not intended to develop some new application for structural dynamics that has not been thought of previously. Rather, the intent is to focus attention of the technical community and funding agencies on important large-scale problems in an effort to achieve a more coordinated and cost-effective means of developing the proposed technology needed to address these problems.

A general consensus arrived at in these discussions was that a certain amount of synergy exists between the various topics. As an example, the development of the planet earth seismic array will be closely coupled to developments of MEMS technology, the vibration-based damage assessment technology, and the probabilistic modeling technology. Similarly, accurate prediction and minimization of noise will benefit from developments in probabilistic modeling and MEMS.

Although the development of probabilistic modeling and MEMS were viewed as directly contributing to the other three Grand Challenges, it was suggested that these two technologies would develop somewhat independently from the other three Grand Challenges. The reason for this speculation is that the plant earth seismic array, global vibration-base damage detection and prediction and minimization of noise will make use of, and provide motivation for, the development of MEMS and probabilistic modeling technology. However, in general, these three challenges will not provide technologies that will address issues associated with the development of MEMS sensors or probabilistic modeling algorithms.

Discussions also focused on issues related to the impact of traditional and emerging structural dynamics technology on the various Grand Challenges. In general, it was recognized that the proposed topics are *challenging* because current and traditional technologies do not adequately solve these problems. As discussed previously, some of the proposed challenges themselves are related to the development of the technologies needed to address the other challenges.

Solving the proposed challenges will rely significantly on the development of a variety of emerging technologies spanning different engineering and mathematical disciplines. In turn, it is hoped that the Grand Challenge concept will then motivate further development of these various emerging technologies.

A characteristic of a Grand Challenge discussed in the introduction was that the solution to, or even progress toward the solution of, the Grand Challenge should have significant and positive economic and societal impact. To some degree, market economies provide a system that predisposes the solution of the Grand Challenge to have a positive economic and societal impact. Government agencies do not have a history of funding long-term, high-expenditure engineering projects that do not have positive economic or life-safety impact. Private industry is even less inclined to provide funding for projects that do not have such impact. It may be argued that some large scale government funded engineering projects that would fit the Grand Challenge concept, the most well known being manned-space mission to the moon, may not have started out with economic and societal issues as their primary objective. However, an after-the-fact review shows that the technology developed for such projects has had tremendous economic and societal impact. In the case of the manned mission to the moon, technology developed for this project has been used to provide satellite communications, which have a tremendous impact on global business. The space program has also had positive impact on people's day to day life through the enhanced weather tracking capabilities provided by satellites. In the case of extreme weather, this tracking capability can have significant life safety implications as recently demonstrated during a hurricane on the east coast of the United States.

With regards to economic and societal impact, the proposed Grand Challenges must be viewed in terms of enabling technologies versus end-use technologies. Micro electromechanical sensors and probabilistic modeling technology do not in themselves provide significant and positive economic or societal impact. Rather these technologies provide tools for use in applications that can have such impact. The seismic array, noise reduction and damage detection challenges all have the potential to provide significant societal benefits, such as life safety and quality of life, as well as economic benefits.

It is realized that there are many other possible Grand Challenges for structural dynamics. The question that arises is how to develop a forum where various Grand Challenge ideas can be proposed, discussed/debated, archived and disseminated. A subsequent question that must be addressed is how to best use the concept of Grand Challenges to advance the state of the art in structural dynamics. Finally, in a related matter, how does the structural dynamics community get the buy-in from funding agencies and the private sector that the proposed Grand Challenges are of interest and worth pursuing? Addressing these issues will be an ongoing activity for the Structural Dynamics 2000 participants.

The Future of Structural Dynamics

Stephen H. Crandall
Massachusetts Institute of Technology, USA

First of all I'd like to thank David and Dan and their co-conspirators, Scott and Chuck here at LANL for inviting me to this Forum.

I'd also like to thank the many speakers. I have enjoyed listening to their presentations. I have been brought up to date in several fields and I have been given visions of some exciting new avenues of advancement. I was particularly stimulated by the presentations of Dan Inman on multifunctional structures, and of Gerhart Schweitzer on Mechatronics.

The title of my remarks this evening is "The Future of Structural Dynamics". I will soon confess to you that my batting average on predicting the future of structural dynamics is pretty low. Never-the-less here is my fearless forecast.

Structural Dynamics is presently a strong and healthy niche technology.

It cannot stand still, but must advance.

I believe it will advance along two axes: it will advance in depth as present techniques are improved and extended, and it will advance in breadth as it joins with the field of control in accommodating smart structures and multifunctional structures.

One of the hurdles in front of cooperation between structural dynamics and control is the perceived cultural difference between these two fields. This is unfortunate and I think the older generation, my generation, is to blame. Let me expand on this point by telling you about my own experience.

I was initiated into the secret world of Structural Dynamics as an undergraduate Mechanical Engineer at Steven's Institute of Technology in 1940. I sat in on two graduate courses given by Professor Diemel. He was not a world famous scholar but he published a very nice practical text on rotordynamics. He taught me about Lagrange's Equations and Rayleigh's Theory on Vibrations. It was Diemel who told us that the linear theory of Vibration was essentially completed by Rayleigh, except for the treatment of damping. I think that now we have mastered the problem of non-proportional damping we can say the Linear Theory of Vibration is indeed complete but damping still remains one of the major stumbling blocks in the Application of the theory.

I graduated from Stevens a few months after Pearl Harbour. My first job was as a staff engineer in the Radiation Laboratory at MIT. This was a top secret defense Lab working on the development of Radar. Our group was led by a Havard professor of Physics. We worked on the problem of creating stable platforms for Radar antennas on the top of masts on ships as sea. We took signals from the ship's main gyrocompass and fed them up to actuators supporting the platform. I now realise that this is the classic problem of control. But in 1942, before the existence of formal courses on control, it was black magic to a green mechanical engineer. I was unable to see any connection with Lagrange or Rayleigh. I did useful work in the Lab but put my real efforts into studying classical mathematics and mechanics in the evenings. In retrospect, I realise that if I had made the effort I could have become one of the pioneers of control theory.

I think you might be interested in one little sidelight about wartime security. One Monday morning in 1943 our group leader failed to appear and we never saw him again until after the war. After some weeks we heard the rumour that he had been transferred to an even more secret project called Manhattan. We had no idea what this project was about but we all assumed it was located in New York and we were never disabused of that idea. In 1944 I was offered the opportunity to teach Naval officers calculus, so I left the Lab and got an early start into the teaching profession.

When the war was over there was a lot of publicity about the success of these two big Labs and the physicists were not shy about taking credit for it. This had the effect of giving many engineers a bad case of Physics Envy.

I joined the MIT Mechanical Engineering department in 1946 and fairly soon we were involved in revising our curriculum to make it more like physics. We called it the engineering science revolution. I led a group which transformed utilitarian Strength of Materials into scientific Mechanics of Solids and other groups upgraded the other core courses. On balance this was a healthy revolution, but, in retrospect, I think we dropped the ball in not introducing control ideas into classical dynamics.

After the Engineering Science movement ran out of steam the pendulum began to swing back. The last two decades have seen an increase of emphasis on Design and Manufacture.

In 1988, in connection with a Rayleigh Memorial Session at the Acoustical Society meeting in Seattle, I wrote a note pointing out that the first volume of the Theory of

Sound was in fact the first textbook on Vibration and that all subsequent Vibration texts have followed the grand design of Rayleigh. I noted that Rayleigh's major analysis tool was modal analysis, although he was familiar with the concept of frequency response functions. I speculated that Rayleigh might have emphasised frequency response functions more if he had had access to more sophisticated sensors and data-processing equipment. I also indicated that I expected future vibration texts to move toward frequency response functions. To put it kindly the movement has been glacial. Clouded crystal ball! One new text, David Newland's, does introduce the concept of frequency response functions at a very early stage. I still think that frequency response functions deserve a more central position in structural dynamics. From a mathematician's point of view, the theory of linear lumped parameter electrical circuits is identical to the theory of vibrations of linear lumped parameter systems. Never-the-less there is an important cultural difference in the traditional treatments of these two subjects. In electric circuits the tool of choice is the frequency response function with emphasis on the poles and zeros. Modal analysis is seldom mentioned. In vibrations the tool of choice is modal analysis with emphasis on the mode shapes and modal parameters. Frequency response functions are seldom mentioned.

As I said earlier, I am enthusiastic about the outreach of Structural Dynamics toward Control, that I heard in the presentations of Dan Inman and Gerhart Schweitzer. One of the reasons for my enthusiasm is that at MIT we are in the process of making a major revision of our undergraduate curriculum, emphasising integration of subject matter. I have volunteered to sit in on the revision of the dynamics and control sequence and I am looking forward to helping bring together the concepts of Structural Dynamics and Control in our curriculum for the next generation.

I have heard that some of you are planning a Structural Dynamics Forum for the year 3000. I don't think I'll be able to make that one, but if you have one a little sooner than that, I hope you will invite me again. Thank you.

The Outcome of SD2000

D.J. Inman
Centre for Intelligent Materials and Structures
Department of Mechanical Engineering, MC 0261
Virginia Polytechnic Institute and State University
Blacksburg, VA 24061, USA
D.J. Ewins
Professor of Vibration Engineering,
Imperial College of Science, Technology & Medicine, London, UK

INTRODUCTION

This Chapter is an attempt to summarise the significant outcomes of our week of deliberation. The hope of our group was to help point the way forward for structural dynamicists. Perhaps the greatest wisdom in this regard came from our most experienced member. He pointed out that during World War II he came in to contact with the newly formed discipline of automatic control and promptly dismissed it as an intellectual pursuit not worthy of structural dynamics. In reflecting back over his career, he now views this as a mistake. We find ourselves exactly in this same situation. New technologies are emerging at a rapid pace (computing, communications, micro-mechanical devices, smart materials, mechatronics, etc.) and determining the appropriate interaction between these new advances and structural dynamics is one of the major concerns addressed by the forum. In addition, significant effort was spent assessing the current state of the art, examining developing technologies, deciding ways to improve the image of structural dynamics and formulating "Grand Challenge" problems for structural dynamics.

SUMMARY OF ACHIEVEMENTS AND OBJECTIVES

The basic objectives of the original program were to:

- Determine future directions of Structural Dynamics
- Define/develop a strategic plan to advance the discipline
- Provide an overview of the state of the art in structural dynamics.

These broad and simply stated objectives were expanded on and developed by the group as the week unfolded. The very nature of the SD2000 Forum dictated a fluid and open definition of determining the best path forward for the structural dynamics community.

To lend further direction to the Forum, the following goals were outlined:

- Determine what Structural Dynamics is expected to do for the engineering community,
- Determine how effective Structural Dynamics is in meeting these expectations,
- Determine what Structural Dynamics should do differently.

In addressing these objectives and goals, the deliberations naturally became divided into three areas:

- Developing Technologies
- Emerging Technologies
- The Image and Culture of Structural Dynamics.

Here, "Developing Technologies" refers to existing methodologies whose current state-of-the-art is challenged by examining what can be done in the next five years in the context of the stated goals. As an example, consider the field of model updating, which proposes to develop systematic means to fine tune analytical models to bring them in line with measured data in one form or another. This field has been developing over the past 15-20 years, yet still needs several accomplishments before becoming a mature and trusted discipline. One such identified need was to provide and develop measures of model integrity by defining "what makes a good model".

The topic of "Emerging Technologies" refers to new concepts and advances that have a potentially large impact on the nature and practice of structural dynamics. Some examples of emerging technologies currently include: Smart Materials, MEMS and, in some sense, Digital Computing.

The "Image and Culture" of Structural Dynamics refers to how we are perceived by other communities, including the public sector, students, other engineers, politicians, and society at large. Although the concept of addressing the culture and image was not an original agenda item, it rapidly surfaced as an important consideration during the meeting, and one whose discussion generated several valuable ideas.

EXPECTATIONS OF STRUCTURAL DYNAMICS IN THE FUTURE

In assessing where to go next in Structural Dynamics it is important to consider what is expected of our discipline. In many circumstances, the structural dynamics community is expected to provide failure-free host designs in lighter, smaller and more durable formats. As increases in efficiency are demanded, and mass and size limits become reduced, significant increase in performance is expected from the contemporary designs of all manner of structures, vehicles and machines. This single expectation is a strong driver for many of the problems identified in the Forum. In addition, structural dynamicists are expected to provide models suitable for design that are computationally-friendly and predictive in a variety of circumstances. Furthermore, the structural dynamics community is expected to provide understandable models and to educate students and practising engineers in user-friendly ways that allow these to be communicated to and understood by people with limited technical backgrounds.

APPRAISAL OF EFFECTIVENESS

In assessing how effective the structural dynamics community has been in meeting these expectations, it is clear that current techniques of structural dynamics are very effective for low frequency, linear, deterministic, relatively low-order structures in 'nice' environments. It is equally clear that the current state of practice in structural dynamics is not very effective at treating non-linear, stochastic, mid-frequency, mixed-field problems. Our definitions of ways forward have been greatly influenced by our perception of what we have failed to be effective at. The following paragraphs present a brief summary of the important future directions in structural dynamics that emerged from the deliberations of the SD2000 group, organised according to the needs in each of the three categories: Developing Technologies, Emerging Technologies and Image and Culture.

NEEDS

The needs which will drive further advances in the Developing Technologies centre strongly on modelling issues. In particular, guidelines for procedures in analytical modelling and in validation testing are urgently needed. An acceptance criterion for assessment of model validity in the presence of uncertainties is clearly needed in order to formulate these guidelines. A particularly difficult modelling task arises when seeking to describe the behaviour of structures that involve joints and interfaces between the various components. Better physics-based models of joints and interfaces are required that are presented in ways compatible with practice. In addition, new modelling techniques and procedural guidelines should be fined verified and fine-tuned by a series of round robin modelling error exercises. Once these are in place, the community needs to address methods for learning/teaching best practice procedures.

The needs for advances in the Emerging Technologies are focused on education. Emerging Technologies will, of course, change over time. Current examples of Emerging Technologies include: MEMS, Smart Materials, Mechatronics, Fluid-Structure Interactions, Large-scale Computing etc. Integrating into these new fields requires education at a variety of levels. Clearly, there should be a focus on identifying the physical models needed, and how to interface these models into structural dynamics. In addition, it is critical to integrate these new technologies into industrial practice. The combination of structural dynamics with new technologies has all the elements of defining a new engineering discipline within the expectations placed on structural dynamics solutions.

Alongside the two technical themes, it is important to emphasise the growing needs – and opportunities - to improve the image and culture of the profession of structural dynamics. Peers, practising engineers and students need and deserve an improved image of the culture and abilities of the structural dynamics discipline to attract the best students. Our image should be expanded by interfacing with other disciplines and markets. The general public needs to be made more aware of our successes and our contributions to the safety and well-being of society in general. We need to define how to interact with the emerging technologies and inform these groups about what Structural Dynamics has to offer and why they should care about interfacing with us.

SUMMARY OF FINDINGS

In summary, the deliberations produced nine key issues to be examined in the future, and developed into more detailed strategies and plans for future research. These key issues are:

- Improved physical modelling
- Impact of improved computing capabilities
- Determining model limitations (e.g. linear vs. non-linear, stochastic characterisations)
- Improvement in non-linear techniques and, in particular, absorbing non-linear analysis into normal practice
- Mid- to high-frequency modelling
- Structural Dynamics keyed to industrial design and maintenance (a current failure)
- Finding new frontiers for Structural Dynamics
- Enhancing the image of Structural Dynamics

A strong consensus was found in support of the notion that perhaps these problems should now be addressed at a global rather than at national levels. The final chapter of the way forward is not yet written and will unfold only by continued self-examination and reflection as provided by Forums (Fora?) such as this one, and its successors.

Today, the foundations are being laid for an infrastructure to facilitate the propagation of the ideas generated at SD2000, and the next generation of activity: namely, that of developing detailed plans in the various areas identified above as those which represent the issues of highest priority for the Future Directions of the subject referred to throughout this work as "Structural Dynamics". The ensuing evolution of this next phase can be followed via the Virtual Forum SD2K which can be found at: www.sd2k.org.

Appendices

Appendix 1

SD2000 Participants

Dr P Avitabile	University of Massachusetts Lowell, USA
Dr E Balmes	Ecole Centrale de Paris, France
Professor L A Bergmann	University of Illinois at Urbana-Champaign, USA
Professor D L Brown	University of Cincinnati, USA
Dr I Bucher	Technion, Institute of Technology, Israel
Dr B Caeser	Daimler Benz Aerospace
Dr S Cogan	Universite de Franche-Comte, Besancon, France
Dr J E Cooper	University of Manchester, UK
Professor S H Crandall	MIT, USA
Dr H H Cudney	Virginia Polytechnic Institute and State Univ., USA
Dr P Davies	Purdue University, USA
Dr S W Doebling	Los Alamos National Laboratory, USA
Professor D J Ewins	Imperial College, London, UK
Dr C Farrar	Los Alamos National Laboratory, USA
Dr M Friswell	University of Wales, Swansea, UK
Professor L Gaul	Inst. of Mechanics, Univ. of Stuttgart, Germany
Professor J K Hammond	ISVR, Univ. Southampton, UK
Dr F M Hemez	Los Alamos National Laboratory, USA
Dr N Hunter	Los Alamos National Laboratory, USA
Dr M Imregun	Imperial College, London, UK
Professor D J Inman	Virginia Polytechnic Institute and State Univ., USA
Dr N Lieven	University of Bristol, UK
Professor M Link	Universität Gesamthochschule Kassel, Germany
Dr D R Martinez	Sandia National Laboratories, USA
Professor C Mote	University of Maryland, USA
Professor A Nayfeh	Virginia Polytechnic, USA
Professor R Ohayon	CNAM, France
Dr T L Paez	Sandia National Laboratory, USA
Mr C Pickrel	Boeing Commercial Airplane Group
Professor R R Randall	University of New South Wales, Australia
Dr M Richardson	Vibrant Technology, USA
Professor G Schweitzer	ETH, Zurich, Switzerland
Professor A Sestieri	University of Rome, Italy
Dr S Shaw	Michigan State University, USA
Professor G Tomlinson	University of Sheffield, UK
Dr H Vold	Vold Solutions, USA
Dr B Yang	University of Southern California, USA
Dr D Zimmerman	University of Houston, USA

BACK ROW: 1 Balmes; 2 Pickrel; 3 Cooper; 4 Sestieri; 5 Zimmerman; 6 Tomlinson; 7 Cudney; 8 Yang; 9 Friswell; 10 Cogan; 11 Gaul; 12 Doebling; 13 Hemez
MIDDLE ROW: 1 Avitabile; 2 Sohn; 3 Shaw; 4 Mote; 5 Bergman; 6 Randall; 7 Lieven; 8 Link; 9 Hammond; 10 Martinez; 11 Farrar; 12 Davies
INTERMEDIATE ROW: 1 Womack; 2 Reeves; 3 Nayfeh; 4 Imregun; 5 Caesar; 6 Ohayon
FRONT ROW: 1 Hunter; 2 Bucher; 3 Robertson; 4 Inman; 5 Brown; 6 Crandall; 7 Paez; 8 Ewins; 9 Richardson; 10 Vold

Appendix 2

Full List of Questions Submitted to SD2000

P Avitabile
University of Massachusetts Lowell, USA

Transfer of Technology - Educating Today's Engineer

Structural dynamics reaches into many areas of design today as we try to make all of our designs "lighter, quieter, more efficient, easier to manufacture, etc.". However, most graduating engineers receive little education in these applications. How do we effectively educate the working engineer to be cognizant of the approaches to be used in a realistic manner? Many of the techniques currently available (and still being developed) require "significant" background to understand the subject matter. How do we train and impart this knowledge in a usable format for the practicing engineer?

There currently exists a Shock and Vibration Handbook which is an excellent resource of information. Do we need to develop a Handbook of Structural Dynamic Modeling Techniques which addresses all the areas which should be of concern for the design of structural dynamic systems? Do we need to provide guidelines or back-of-the-envelop techniques to identify if problems are expected to exist? Do we need to identify what types of tools are useful for different types of problems? Just how are we going to educate the engineer in these newly developed techniques and methodologies that are required and useful for general use in a wide variety of design scenarios?

The need for "Educating Today's Engineer" stems from some typical (uneducated) statements, questions and misconceptions that I expect that we all have heard at one time or another - some examples are:

Automotive brake engineer believes that correlation/updating of an analytical model will reduce brake squeal.

Engineers believe that model updating is a "push button" technology; salesman sold them a "bill of goods".

Operating data provides a true picture of the "actual response"; but what do I have to do to modify the design?

Performing a modal test will describe the dynamics of the system; but without a forcing function, what good is it?

And the list just goes on and on. How do we educate relatively new engineers into all the detailed technology that exists today, in a practical sense?

Making Sophisticated Structural Dynamic Testing/Analysis Procedures Commonplace in Everyday Design

About 20 to 30 years ago, finite element modeling (FEM) was reserved for sophisticated engineering applications that had extensive time and budgets. FEM analysis was reserved for the knowledgeable and capable analysis engineers working in critical applications that required its use. During that period of time, there was extensive development and new concepts/ideas in the formulation of elements and models flourished. Today, FEM is a commonplace structural design tool in almost every industry and application.

Over the past 10 to 15 years, there has been tremendous growth and development in the structural dynamics technology area from both a testing and analysis perspective. Over that period of time, more sophisticated users have been involved in the generation of very elaborate test and analysis endeavors - most of which have required extensive time and budgets.

As we move into the next century, how do we make these newly developed techniques and methodologies more readily available to the more commonplace design of everyday commercial products so that these techniques are affordable for less sophisticated industries where time and budgets are not plentiful?

Specific examples:
Computer cabinet/disk drive design must be produced rapidly with little or no time for in-depth test/analysis to be performed. Cabinets need to be designed to accommodate a wide variety of different peripherals (especially disk drives) which must perform flawlessly in their environment. Drives need to be subjected to qualification tests that are reflective of their actual installed environment (rigidly affixing the drives to a shaker for qualification may not be sufficient or proper to confirm performance). Different disk drives (with different dynamic characteristics) need to be substituted into the design configuration and perform equally well. Accuracy of the model requires that modeling techniques employed are confirmed through the use of experimentally measured data. Cabinet designs are typically constructed such that nonlinear behavior is anticipated and can have an important effect on performance. Noise and vibration (creature comfort) are issues that need to be addressed. The only time when significant money is allocated for evaluation is when problems exist and a "fire" needs to be put out.

Another example is the design of a commercial clothes dryer. Expensive tooling must be committed early in the design with the expectation that the engineering design is correct and will produce an acceptable configuration. Fairly large models are developed of the overall configuration - modal density is populated with local modes of the flimsy cabinet configuration. Test/analysis/correlation of the model is difficult due to the large number of modes. Model accuracy is important to the prediction of system performance for both noise and vibration issues.

Both of the above are excellent examples of design situations that are not much different in technical content than those of the space station design, aerospace program, or automotive designs with one big difference - there is typically very little time or budget to undertake the extremely involved techniques/methodologies required to address the designs properly. The two situations above embrace all of the aspects of structural dynamic modeling techniques required to address the problem. Development of accurate models that are validated through experimentally acquired data. Ability to address the design which is likely to include nonlinear joints, connections and component interrelations. Development of system models that are made up of a variety of different components - using a combination of analytical and experimental components which are interconnected with both translation and rotation effects. Ability to address a large number of modes (both global and local) over a wide frequency range to predict system performance due to operating conditions. Development of an accurate model that can be used for noise and vibration predictions.

E Balmes
Ecole Centrale de Paris

Bias and variance in identification methods

Estimating the quality of identification results would be of significant help in assessing and improving model correlation or simply to perform robustness studies.

Typical identification algorithms estimate the poles while over-specifying model order. A major reason for doing so is that out of band modes have a strong influence on structural dynamics, so that additional computational modes are needed to account for these contributions. The representation of residual contributions by computational modes or simple asymptotic terms clearly induces bias in the estimated poles. The bias on pole estimates is usually quite small but this is not true for modeshapes (in particular for the phase of complex residues).

In this light, is it possible to determine if the (small) variations in poles estimates are representative of variance in the estimate or simply of changes in the bias effects when considering various model orders. Can something be said for the larger variations found for modeshape estimates.

Design objectives for non-linear dynamics

Many criteria have been developed to judge the performance of linear structures. One thus minimizes the RMS response or gives a frequency domain envelope for acceptable response spectra. The assumption of linearity leads to two major simplifications. The characterization of the excitation can be limited to simple statistical representations. The relation between the excitation and the response is simply characterized by transfer functions which are easily computed using modal analysis.

With the objective of judging the dynamic performance of non-linear structures, tools seem to be missing to characterize broadband excitations with quantities that can be propagated into response characteristics in a reasonable time. Extending modal analysis notions to non-linear structures seems key to the propagation problem but this will still leave the input characterization issue open.

I Bucher
Technion, Institute of Technology, Israel

Should deterministic or statistical models of structures be used for structures having high modal density?

Large structures could have tens to even hundreds of modes (natural frequencies) in the frequency range of interest. Common inaccuracies of 1-10% in some modal parameters and even 40%-200% in damping may render such a model useless for deterministic prediction of response levels. The question arises whether dynamic prediction should be re-directed towards statistical estimation of response level based upon statistical models of the structures? Do deterministic models have some merit even in such cases that cannot be overlooked? This issue calls into question the validity of modal testing in the cases where it should be mostly required.

Should linear, non-linear or Floquet (periodically varying coefficient) systems be applied to rotating machines?

Rotating structures exhibit a very rich response spectrum that is caused by the complex nature of the various phenomena co-existing in a rotating machine. A rotating structure can be excited by external forces such as fluid-structure interaction, cutting forces (in saw blades), or by speed variation and parametric excitation (e.g. modulation of the response by stiffness and mass depending upon the instantaneous angle of rotation).

It has been shown (A. Bendat) that in some cases non-linear systems possess the same input-out behavior as hypothetical linear system with periodically varying coefficients.

Inspecting the spectrum of the vibration measured on a rotating machine one can always notice numerous harmonics of the frequency of rotation as well as other frequencies. These harmonics could be attributed to the nonlinear nature of the supports (bearings) or to parametric excitation due to time varying parameters such as can arise in an imperfect rotating structure (not perfectly isotropic).

How should this problem be approached and resolved ? How could these phenomena be separated from measurements and how should they be modeled ? Is this important or can any suitable 'black box' model that relates forces to response be used ?

Recommended Signal of excitation: which type excitation should be used for linear structures and which one is preferred in the slightly nonlinear ones. Sine-sweep, sine-step, random, burst, impact?

In dynamic testing of structures many excitation methods were used and it seems that the "fashion" changes every few years and there is no agreement upon the 'best' method for each particular application.

Is it better to use an excitation that 'linearises' the identified structure (random) or an excitation that exposes the non-linear nature of the structure (sine, impulse)? Should non-linear behavior be treated in the excitation phase, signal processing phase or model-fitting stage?

B Caeser
Daimler Benz Aerospace, Germany

How can we solve the old problem of damping prediction?

Coupling of Control and Structure Mechanics
How should software be designed to couple control sensor and actor design with structure behavior?

Actually the coupling is performed via standardized data package and/or model transfer from one to the other independent S/W-tool. Application example: ESA, NASA, ESO and others are working on high precision space and earth born telescopes with interferometric optics under the task "Next Generation Space Telescope" and "Very Large Telescope" for which pointing mechanisms in the delay line with accuracies in the range of 1 nm are required. Such extreme accuracy demands require new strategies.

Regarding Coupling Macroscopic to Microscopic Analyses

Mechanisms especially high precision ones can be described by actual available mathematical models only on the basis of empirically derived data. Mostly the mathematical model is sufficiently accurate not before tuned to the hardware behavior. This causes high development risks in costs and schedule. The feasibility of a design cannot be predicted.

Medium and High Frequency Range Dynamics Analyses

With which approach shall the dynamics in the medium and high frequency range be treated?

The medium and high frequency range is most often important in acoustic and shock excitation. Several not fully satisfying approaches are available, one is the refinement of discretization (refinement of FE-mesh), another is the statistical energy analysis SEA. Solving this problem would help, e.g., to predict more accurate stresses and test levels for equipment on structures under acoustic or random loads, or to reduced structure radiated noise.

Transfer of CAD into Mathematical Models

Could straight forward procedures be developed to establish mathematical models directly from the design of complex and/or assembled/integrated structures?

Actually automated procedures exist to transfer single structures into FE-meshes under certain restrictions such as simplicity of the structure design or use of tetrahedral elements. As soon as structures become more complex or there is the need of coupling several structures together, the skill of structural engineers is required to transfer the reality into the simplification of the mathematical models.

S Cogan
University de Franche-Comte, Besancon, France

What criteria can be used to judge the usability of a model for a given application, that is to say, whether model-based decisions can be integrated with confidence into an engineering activity. Is it really predictive precision we are after or is it robustness of model-based design decisions to ambient uncertainties?

Model quality has become a major concern in the structural dynamics community. The availability of attractive and highly developed software for creating mathematical models of complex behaviors (due to both local behavior laws and intricate topologies) has lured (to be provocative) the industrial community into their purchase and use. These tools have produced undeniable success stories and their usefulness is difficult to contest. However, the influence of diverse sources of uncertainties ranging from the simplifying concepts used to analyze a given problem to the various measurement errors tend to create an incompressible distance between analytical predictions and the reference behavior observed in the field. It is certainly a tribute to engineering savoir-faire (and a healthy dose of safety factors) that planes fly and nuclear reactors remain contained given the abiding inaccuracies in the mathematical models of complex technological structures.

Model-based decisions are useful only in so far as they maintain a certain fidelity with respect to real prototyping, that is to say, that they represent true pseudo-tests. The goal of modelling, measurement and updating is simply to convince ourselves that we are indeed in the presence of such a model. The question is just how this can be done in a reliable way ? Just when is a model precise enough ? Is it really 'precision' that we are looking for or is it 'robustness' of our design decisions with respect to the known (and unknown or unexpected) sources of uncertainty in both the model(s) and prototypes. A model might well provide very precise predictions with respect to a given parameterization and for a set of measurement data. These predictions however might be extremely sensitive to uncertainties in both model parameters and to the fact that a set of apparently identical prototypes present a dispersion in their behaviors. Conversely, model predictions may be relatively imprecise while the design decision in question may well be relatively robust with respect to these prediction errors. Clearly a methodology is required to allow these concerns to be addressed in a comprehensive way.

J E Cooper
University of Manchester, UK

How should non-linearities on real structures be identified?

There has been a great deal of work devoted to the modelling and identification of non-linear systems (e.g. NARMAX, restoring force surfaces, higher order FRFs, etc.) Whereas these methods can deal with simulated systems and small purpose built laboratory experiments, their use on real structures (e.g. automotive, aerospace, civil structures) is problematic. Is the correct approach being used with improvements in the technology being made very slowly, or is a radically different methodology required to tackle the problems of detecting, quantifying and identifying non-linearities in real structures.

458

Is linear modal analysis solved?

The process of identifying frequencies, damping ratios and mode shapes from modal test data is nowadays taken as being fairly straightforward. However, most commercial implementations of the methods rely a great deal upon engineering judgement of the user (how many modes, which modes should I select from my stability plot? How to deal with scatter in the results, etc.) The amount of research devoted to investigating new approaches has diminished and most of these (e.g. subspace methods) are viewed as research tools rather than appearing in commercially available software. Has the problem been solved totally or should further effort be devoted towards providing industry with "intelligent" modal testing that removes the need for a large amount of user interaction. (PS: When is it not OK to treat a structure as linear?)

How should structures in the mid-frequency range be modeled?

Conventional FE tends to be used for low frequencies whereas Statistical Energy Analysis (SEA) is applicable for very high frequencies. Little work has been undertaken to address the frequency ranges in the middle where the accuracy of FE breaks down. How should the Structural Dynamics community tackle this problem?

Are there developments in the signal processing and numerical analysis fields waiting to be applied to vibration analysis?

Great use is currently made of tools such as the Fast Fourier Transform for vibration testing and standard analysis approaches such as Least Squares. Other recent developments have included the use of the Singular Value Decomposition (SVD) and Time/Frequency methods (e.g. wavelets). Any major developments in the field from the algorithmic viewpoint will depend upon the use of new techniques. Do such tools exist, and if so, how do we apply them to vibration analysis?

How should structures with non-modal looking FRFs be identified?

When a structure with a very high modal density and relatively large damping is tested, the FRFs do not look modal and instead look very flat. Analysis using conventional curve-fitting methods does not produce very good results. Should black-box time domain approaches be used, or should some other methodology be used.

P Davies
Purdue University, USA

How can the classical single frequency excitation perturbation analysis of non-linear systems analysis become a useful tool for structural dynamicists trouble shooting problems in structures?

When trying to understand the conditions necessary for modal interactions to occur in weakly non-linear structures, a two-stage analysis is usually performed. First, for the conditions under which the structure is operating, a reduced order modal model of the structure is derived, and secondly, the steady-state (or slowly varying) response of the modes of this model for various forms of stationary (or slowly-varying) harmonic excitation is examined. Analysis of these response versus model parameter (excitation level, natural frequency, excitation frequency, etc.) functions allow determination of regions where multiple solutions may exist, or where no stable steady-state solutions exist and energy sharing between modes may occur. Multiple solutions co-existing for the same excitation will mean that in structural testing, the response observed will be a function of initial conditions and perturbations to the system, a phenomenon not present in linear systems.

As structures become thinner and deflections become larger the role of non-linearities must be considered, and in frequency regions where many modes are present, the probability of modal interactions is high. When excitations come from rotating machinery, it is not unrealistic to consider harmonic excitations, or to assume that only a few of the structures' modes are directly excited. Therefore, this classical nonlinear dynamics analysis is suitable for practical situations. However, while people studying non-linear systems in their research use these tools, they are not widely used in the structural dynamics community. There are also some interesting issues that need to be addressed. Here are a few: (1) Is it possible to probe the structure through testing to determine with some certainty whether this modal coupling is present, and thus to justify the non-linear modal analysis? (2) Given a particular excitation, how can we systematically determine which modes to include in the model? (3) Since the response curves are highly dependant on the modal parameters, how can the parameters be estimated using excitations that will only cause the modes in the model to respond? (4) Can the analysis techniques be extended to multiple harmonic excitations? (5) Does the structural dynamicist need to be an expert in non-linear systems analysis to use these tools?

It is often possible with computer control of manufacturing to tailor the chemical content of materials during manufacture. How can we optimize the in-application macroscopic behavior of materials using these manufacturing capabilities?

This is really a problem of linking the material and chemical level modeling with the structural dynamics models. There is a need to feed the basic material properties into the structural model, or if this is not possible to do directly, to understand how fundamental changes in material microstructure affect the terms in the macroscopic structural model. Robust system identification techniques are needed to determine the types of terms, and the parameters in those terms, that are necessary to model the macroscopic behavior of the material in the application of interest. These terms may be, for example, stiffness, viscous damping, hysteretic damping, dry friction, or viscoelastic terms. Many interesting problems arise here in system identification, not only because many of the material models will be nonlinear, but also because materials such as polyurethane foams (used in transportation seating) have both fast- and slow-time behavior and are sensitive to environmental factors such as temperature and humidity, making repeatable experimentation as well as parameter estimation challenging. The interaction between the terms in the structural model will add an additional complexity to the identification, perhaps requiring the need for excitation techniques that allow for isolation of particular types of behavior.

Development of robust system identification techniques to identify structural dynamics model parameters related to material properties, and derivation of functional relationships between these parameters and the material manufacturing parameters, is only the first stage. We then have to take this understanding and use it to design the structure, working backwards from the desired response to known excitation, to the space of effective structure dynamics model parameters, and finally to the materials manufacturing parameters required.

S W Doebling
Los Alamos National Laboratories, USA

What will be the appropriate relationship between simulation and experimentation in the early 21st century?

Historically, in structural dynamics as in other fields, experimentation has been used to both PROVE and IMPROVE: (a) PROVE that an engineered system meets its design criteria by performing a "proof" or "qualification" test and (b) IMPROVE the mathematical models of the engineered system to better predict the response of the system under different initial and/or boundary conditions. Some engineers would claim that the IMPROVE experiments provide more useful information for engineering purposes than do the PROVE experiments, but nevertheless the PROVE experiments receive much higher priority in terms of resources (i.e. money) than do the IMPROVE experiments.

GIVEN the extreme decrease in the price-to-performance ratio of computing power over the last 30 years, is it possible that structural dynamicists can "substitute" computational simulations for many of the PROVE experiments, so that resources can be concentrated on IMPROVE experiments to actually make our modeling and analysis capabilities better?

What is the relevance of structural dynamics in the information age?

It is generally agreed upon that the "industrial age" is largely behind us and the "information age" is fully upon us as we begin the 21st century. Structural dynamics has undoubtedly enjoyed a prime period of growth and increase in resources as the industrial age has matured. Structural dynamics is traditionally associated with the "heavy" industries such as aerospace, automotive, robotics, and civil infrastructure. But what is the relevance of such industries in the information age? An example of such relevance would be increased demand for satellite production as worldwide wireless networks are increasingly in demand. A demand for more "features" on a smaller, more lightweight satellite will certainly involve new structural dynamics issues. But aside from such "new tricks" for the "old dogs" of the industrial age, does the information age present any new opportunities for structural dynamics?

GIVEN that the majority of the economic growth in the early 21st century will most certainly come in the field of information technology, rather than industrial technology, what can we as the structural dynamics community do to "get a piece" of the information age "action"?

How can analysis of structural dynamics issues be used to decrease "time to market"?

Many industries, such as those with aerospace, automotive, and civil infrastructure applications, are inherently reliant upon analysis of structural dynamics issues to ensure that the engineered systems perform their functions within specifications. However, many other industries that have high-volume, low-cost, quick-time-to-market business constraints do not have a direct requirement for structural dynamics analysis and therefore forego the expense in favor of quick turnaround time in product development and deployment. What many engineers and managers in these industries fail to consider, however, is the improvement in product quality that can be had with the use of structural dynamics in the product design and deployment phase. For example, consider the portable CD player. While undoubtedly such a product undergoes rigorous testing during the design phase, consider how much more robust the design can be through an equally rigorous application of analysis. However, the application of analysis in such instances is typically limited because it "slows down the design process".

GIVEN the potentially huge market in the consumer products (and other) industries, how can the structural dynamics community make a case for structural dynamics analysis as a tool to improve product quality while not impeding the "quick time to market"?

M Friswell
University of Wales, Swansea, UK

Has modeling structures using finite element analysis reached the limits of accuracy? Is a new paradigm required for design and analysis of structures?

Many engineers believed that any structure may be modeled to arbitrary accuracy merely by increasing the finite element mesh density. These refined meshes are able to model the geometry of the structure more accurately, but uncertain parameters, for example geometric tolerances or joint dynamic stiffnesses, mean that modeling errors will never be resolved using this approach. Model updating can help, although for complex structures the incompleteness of the measured data makes it is impossible to identify a physically meaningful model. Thus there are definite accuracy limits for modeling a single structure, and also for modeling the variability of batches of components because of manufacturing tolerances.

Perhaps we should be looking at different ways of analyzing the dynamics of structures. Methods based on uncertainty (defined either as a statistical or convex model of uncertainty) could be used to optimize structures as part of a robust design scheme. Control engineers are more advanced in this respect and robust controller design, based on an uncertain plant model and influenced by uncertain disturbances, is a mature technology. Could any of the robust control methods be transferred to the design activity in structural dynamics? Could the design process be tailored to produce a structure where the known uncertainty in the structure has the minimum effect on its dynamic response? Is this a useful thing to do?

Do uncertainties in the model, measurement noise and the changes due to environmental effects, make it impossible to robustly locate damage in structures using low frequency vibration measurements?

The area of damage detection and location using measured low frequency vibration data has attracted considerable attention recently. The great advantage is that low frequency data are global, and thus only a limited number of sensors, located remotely from the potential damage sites, are required. However, there are considerable difficulties to be overcome before the approach is robust enough to be used in practice.

The question is whether the inherent difficulties render all the proposed methods unusable in the field.

In terms of methods, the only viable approaches are based on forward identification procedures which recognise the fact that damage generally only changes a limited number of parameters. Methods that try to identify a large number of parameters simultaneously produce ill-conditioned equations, whose regularised solution destroys any capability to identify damage location.

Even forward identification methods have considerable difficulty. Is the finite element model accurate? How accurate does it need to be? How does the dynamics of the structure change with environmental conditions, for example temperature and humidity? Can these effects be modelled with sufficient accuracy so that their influence may be removed? How large are the changes in a structure's dynamics due to typical faults? Is this level of change likely to be detected reliably, given the signal to noise ratio, the changes due to environmental effects and the modelling errors? Even if damage can be detected, is the 'effective wavelength' of the low frequency modes too large to make localisation of the damage sufficient to be useful?

Given all these significant difficulties that are based on the physics of the problem not the methods used, what is the prognosis for damage detection and location using low frequency vibration?

Is state space methodology state of the art for damped structures and rotating systems?

Often damping is ignored in the analysis of dynamic models of structures, but recently there has been a resurgence of interest in the analysis of damped structures. The standard approach to analysing such structures is the transformation to a first order system (the state space formulation). The increased effort for the eigensystem computation required to progress from an undamped system with N degrees of freedom, consisting of real eigenvalues and eigenvectors, to a damped system represented by a first order system with 2N states and complex eigenvalues and eigenvectors, is significant. In terms of understanding, as soon as the state space matrices are formed, the constraint on the states (where half are displacements and half are the corresponding velocities) is lost. Indeed there is also a constraint between the real and imaginary part of the complex eigenvectors that is not used or enforced in a state space analysis. Are we missing something that could enhance our fundamental understanding of the damped system? Should we, for example, make sure these constraints for a second order system are enforced in experimental modal analysis?

What application does Geometric Algebra (for example Clifford Algebra) have in structural dynamics? Many areas of physics have benefited from the unifying force of Clifford Algebra. Most engineers are happy to use complex numbers in the analysis of dynamic systems (particularly in state space form); they afford tremendous insight and compactness of notation. Could Clifford Algebra do a similar job for second order

464

systems, helping both understanding and the computation burden? In rotordynamics it is possible to encode multiple uses for complex numbers (for example, using complex numbers for both the eigenvalues and to encode displacements or rotations at a given node in orthogonal directions). This area is only just beginning to develop and the benefits have not as yet been fully investigated. Is this the future?

F M Hemez
Los Alamos National Laboratories, USA

Predictability of Structural Dynamics Models

Should the structural dynamics community put a strong emphasis on developing tools for assessing the predictive quality of a given numerical or experimental model? If so, which technologies need to be developed?

No practical tools are currently available to assess the predictability of a structural dynamics model. The quality of an individual finite element mesh can be assessed, the quality of test data can be characterised statistically but the structural dynamics community lacks the capability of assessing the degree of predictability of a given model with respect to a given objective. (For example, what is the accuracy associated with the response obtained from a given finite element model knowing that the objective is to minimise the distance between test data and numerical simulations?) Developing such a capability would probably require the integration of statistical analysis, multiple-model representation and inverse problem solving (for comparing the predictions of a model or family of models to test data).

Coupling Macroscopic to Microscopic Analyses

Rather than attempting to develop "useless" mathematical models for accounting for complicated dynamics (such as energy dissipation), could the structural dynamics community focus more on the means of coupling the conventional, mechanics-based representation of structural dynamics with a localised, physics-based description of the mechanics at the microscopic (if not, nanoscopic) level?

The motivation for this question is best illustrated with a (provocative) example. Many research efforts currently focus on developing damping models starting from the equations of mechanics and a description of the material behaviour at the macroscopic level. For the past ten years or so, it seems that this research has delivered no significant improvement of our modelling capabilities. It can be argued that the reason is that most damping sources originate from the localised contact/friction between two surfaces or from the dissipation of energy in the material at the microscopic level. To model this phenomenon (and many others such as contact), it seems therefore critical to account for surface irregularity, material imperfection and anisotropy, if not molecular interactions!

It seems like it would be more productive to study how macroscopic properties could be coupled to a characterisation of the material at the microscopic scale and, conversely, how microscopic state variables could be averaged to produce, for example, "global" displacement or stress components. The goal here would be to enable a fully coupled macro/micro analysis in order to improve our predictive capability by taking advantage of a physics-based representation of the dynamics, whenever necessary. Of course, such an endeavour would generate significant computational challenges.

Developing Accelerated Testing Procedures

Could experimental procedures be developed that would simulate the aging of a structural system and help assess its performance after a life-span of, say, 30 years without having to wait that long to take the measurements?

For many structural dynamics applications, it is important to estimate during the design phase what the remaining performances of a system will be after it has been used for a long period of time (20 to 30 years). This is especially critical with aerospace applications for which modelling the effects of aging is almost impossible because very little data are available to validate the simulations (we are just starting to witness on a large scale the effects that aging has on aircraft systems). Another example would be civil engineering structures such as bridges that often stay operational for longer periods of time than they had been originally designed for.

Finite element simulations can be used for estimating the performance of a system after a series of load cycles. However, our current modelling capabilities do not allow us to handle complicated structural systems for which residual strength assessment still relies, to a great extent, on testing. Are experimental techniques available (or could they be developed) for reducing 30 years worth of structural dynamics in a few tests?

M Imregun
Imperial College, UK

Modal Updating – Is this the end of the road?

For the last 20 years or so, significant research effort has been devoted to improving mathematical models using vibration test data. However, there are still no established and universally applicable methods that can produce an updated model that can satisfy the following criteria:

The (true) experimental and updated models must have the same modal and response properties within the measurement frequency range.

The updated model must be able to predict the effects of further changes.

It is debatable whether the corrections to the updated model should have a physical meaning. Most updating work is focussed on FE modeling, which may or may not be the way to go. Is it time to accept defeat and have a fundamental review of the modeling techniques?

Damage detection – Given the relative lack of success with model updating methods, can we expect to do better?

In the general case where the damage location is not known, is it possible to develop detection techniques using numerical base models without having solved the model-updating problem first?

Should we develop model-updating methods on small-size models and assume that these are equally applicable to large-size models?

It is common practice to develop updating methodologies on small-size models, say with 10-1000 DOFs. When dealing with such models, numerical accuracy, computational feasibility in terms of CPU effort and storage are taken for granted. However, there are many computational problems when working with large-size models, say 30,000-200,000 DOFs. For instance, it is not a trivial task at all to obtain the first 5,000 modes of a 100,000 DOF model. Not only will different FE codes give different results, various eigensolution extraction routes within the same code will also yield different results. Such problems will probably go unnoticed when computing the first 50 modes of a 1,000 DOF system.

Another point to remember is that the size of the model is often dictated by the amount of geometric detail that needs to be incorporated. For modelling simplicity, can we ignore such detail and hope to be able to update the model ? In other words, how can we determine the minimum amount of modelling detail that is required?

N Lieven
University of Bristol, UK

How do we improve the image of Structural Dynamics?

All too often structural dynamics is considered in the context of failure rather than performance. If we compare this with research in say Fluids or Materials, the emphasis of these activities has been consistently related to product enhancement rather than deterioration. Historically, we have a legacy of troubleshooting, when dynamics has been used as a corrective measure "after the event" – whatever that might have been. This culture - with the notable exception of aeroelasticity – leaves dynamicists fighting a rearguard action. Inevitably this puts the issue of dynamics towards the back of the queue in the design process. Even worse the issue of dynamics may only be considered retrospectively.

The education process in dynamics often reinforces the role of dynamics as a discipline associated with structural failure. We all use the "great disasters we have known" anecdotes in our teaching – this may be doing us no favours. The challenge in teaching therefore, is to maintain the interest level imbued by eye catching events, but portrayed in a more positive light. Perhaps the dynamics should be included seamlessly in the design elements of engineering courses?

The challenge for structural dynamics research is more demanding in this respect. Industry frequently views structural dynamics as a necessary evil. The commercial case for retrospective dynamics is not strong. With the increasing representation of industrialists on government funding bodies, we need to convince industry of the benefits of dynamics research. What are these?

If we consider the apparent success Computational Fluid Dynamics in attracting funding, it is evident this has arisen through effective marketing on the basis of reduced cost and increased performance. Dynamics research strives to do the same and yet does not attract the same enthusiasm from industry. How do we address the publicity issue?

Perhaps the role of dynamics has become entrenched by its applications? When we consider dynamics we immediately think of bridges, aircraft and cars. These products are established, and do look forward to new growth technologies. We only have to consider the burgeoning NASDAQ to see where technology is expected to develop – dynamics is not in evidence. The question in this case is, how do we associate ourselves with products and technologies that will take us forward from the entrenched applications we work on today?

C Pickrel
Boeing Commercial Airplane Group, USA

Confidence Intervals

Can we estimate confidence intervals on structural dynamic parameters which affect the design and performance of our products?

In a more perfect world we would assess variability in manufacture of products, test measurements, model predictions, and the assessment of operating environments. Many products are engineered with conservatism to compensate for ignorance of these confidence limits. Improved assessment of variance and confidence intervals would lead to improved performance or reduced cost of these products or designs.

Mid-Frequency Problems

How do we approach mid-frequency and high frequency problems where modal analysis is not practical? Modal analysis provides us with a robust approach to "low frequency" problems in structural dynamics, about which much has been written. Is there a consensus for an approach to problems involving higher frequencies where the modal approach breaks down?

Dynamics and Controls

Can anything be done to increase the communication and integration between the Structural Dynamics community and the Controls community?

Increasingly, problems arise which require the feedback control of flexible structures. It seems this requires the participation of engineers from two different organizations (in a larger company), who graduated from different departments, read different journals, and speak different "language".

"Open Source" Research Model

Following the "open source model" (Linux software development) could a few structural dynamics problems be placed on an open web site(s), facilitating collaboration and making data and solutions widely available?

Some important problems in structural dynamics involve multiple aspects of our discipline, such as modeling, testing, and controls to name a few. Seldom does a researcher have the resources to accommodate all of these aspects of the problem. An open web site could bring together contributions from different researchers and institutions. The "publishing cycle" could be reduced from months/years down to minutes/days, accelerating the solution process. Examples of the information which might accumulate for a given problem include: drawings of structure, test plans, test data, models and predictions, modal parameter estimates, model updates, damage detection, structural modifications, solutions, discussion (questions, answers, comments and assessments).

Obviously, many issues arise, and not all research will be conducted this way any time soon. The Linux software community has found the "open source" and "release early and often" approach to be a powerful method for rapidly developing robust software. Perhaps the same would be true for the solution of multifaceted problems in structural dynamics?

Dynamics in Engineering Education

How can we increase the emphasis on dynamics in engineering education? It seems that structural dynamics has played an increasingly important roll in product design in recent years. Yet the average engineer is sorely lacking in appreciation of dynamics fundamentals. The dynamics community would benefit from a broader awareness and accessibility of dynamics fundamentals on the part of decision-makers.

S Shaw
Michigan State University, USA

Development of techniques for building reduced-order models

Several methods exist for reducing the size of computer-based, linear models of structural systems. These techniques are very useful for substantially reducing the number of degrees of freedom required for high-fidelity models over a certain frequency range. However, when nonlinearities are present, this type of model reduction becomes much more difficult. Can some basic, useful model-reduction theories be developed for this class of problems? Can these be developed into useful computer codes? Can one develop specialized methods for cases in which the nonlinearities are strictly local in nature, say at joints, or due to attached components? Or, similarly, for the case when the nonlinearities are inherently distributed, for example, due to kinematic effects?

Structural dynamics and control at small scales

An extremely important area for the future of structural dynamics is that related to micro, and even smaller, devices. There are new exciting challenges waiting for the dynamicist here, and there are opportunities to be involved with cutting-edge technology.

Relevance to industry

This is a problem that faces many areas of engineering research. Industry is typically slow to make use of the latest results from any basic and/or applied research that is done outside of their labs. In terms of structural dynamics, the aerospace industry has probably been one of the most receptive groups in this regard. However, other industries, such as the automotive companies, are slow to respond. They are often aware of advances, but are reluctant to try new ideas. How can we be more effective in making inroads into these industries?

How can we develop effective damping models and include them into FE codes

B Yang
University of Southern California

Mid-frequency analysis of multi-body flexible structures

Applications: vehicles, ships and airplanes.

Current status: low frequency range – finite element method
High frequency range – statistic energy method
In between (mid-frequency range) – analysis tools need to be developed.

Analysis of Membrane Structures

Applications: aerospace structures, military equipment, and civil structures, material processing and handling.

Characteristics:
Bending stiffness almost zero
Membrane stress distributions not known
Complex shapes and unconventional boundary conditions
Current status: wrinkling difficult to predict
Iteration-based numerical schemes may not converge, or may converge to wrong solutions
Analytical solutions difficult to construct
No commercial codes available

Nonlinear analyses in practical engineering

The importance of nonlinear analyses has been well accepted and appreciated by the academia. Although some commercial codes of nonlinear analyses are available, many practical engineers still use linear analysis tools, partially because these tools are simple to use and the results from linear analyses easy to understand.

The question is: How to make nonlinear analysis tools more accepted by and more user friendly to practical engineers?

D Zimmerman
University of Houston, Texas, USA

Where does Structural Dynamics fit in?

Recently, several traditional United States funding agencies have publicized their plans for research initiatives in the 21st century. In a recent DARPA briefing, it was stated that nearly 50% of the DARPA budget will be focussed on biotechnology (including bioengineering) within the next three years. The NSF and several other agencies (DOD, DOE, NASA, NIH and NOAA), working under the auspices of the National Science and Technology Council (NSTC) have developed a long-term research plan for Information Technology in the 21st century. At the same time, the NSF has announced that its second major focus in the 21st century will be in "Biocomplexity in the Environment". The change in the funding picture is changing rapidly ... how should the structural dynamics field react?

Is there an "Economic" Based Grand Challenge Problem in Structural Dynamics?

The quick rise of Computational Fluid Dynamics research and subsequent commercial software can be traced back to President Reagan's call for the High-Speed Transport (or as he called it the Orient Express). It was estimated that after the initiation of this program, nearly 90% of all supercomputer resources in the country were associated with CFD development. Although the High-Speed transport never did materialize (in fact, NASA Langley, the NASA lead in high-speed transport is ending this program in FY00), it did serve as a catalyst to the CFD research and application community. This focussed effort was sold to the public and to Congress based on the "economic impact" and need for such transports in the "new global economy". Look at another major Grand Challenge: The Supercollider. This project folded several years ago because the physicists were unable to sell the public and congress on the economic impact of their research. Is there a Grand Challenge problem that has structural dynamics in the lead role AND has a major economic impact?

Design for Reliability

As we are always reminded, we all will (or already) have Cray computing capability sitting on each and every desk of an engineer in the near future. Have we as structural dynamicists stayed in the safe "linear" world, and away from the nonlinear and stochastic real world, based on our thoughts that the computational power needed to properly perform design is not available?

Appendix 3

The SD2K.ORG Website

In order to take full advantage of the outcome of the SD2000 Forum, it was decided to establish a permanent website to serve the Structural Dynamics community. This website provides information on a series of issues of interest, and will include:

- News of SD2000-related workshops and other activities which seek to progress the conclusions from the SD2000 Forum.

- News of publications of interest to Structural Dynamics community.

- A comprehensive calendar of events - conferences and meetings related to Structural Dynamics - with direct links to the organising bodies.

- Details of organisations, journals and research groups of interest to the Structural Dynamics.

- The SD2K.ORG public Forum: a message board for discussion of issues relating to Structural Dynamics, including:
  - Technical issues in structural dynamics
  - Education in structural dynamics
  - Jobs available or sought in structural dynamics.

- In addition, there will be a number of private forums restricted to SD2K members for the purposes of:
  - Planning future SD2K workshops and other meetings
  - Development of SD2K publications and documentation

Everyone with an interest in the subject of Structural Dynamics is encouraged to visit the SD2K website – sd2k.org – and to contribute to its success and effectiveness by submitting additional news and materials to make it yet more comprehensive.

Index